Studies in Systems, Decision and Control

Volume 326

The series "Studies in Systems, Decision and Control" (SSDC) covers both new developments and advances, as well as the state of the art, in the various areas of broadly perceived systems, decision making and control–quickly, up to date and with a high quality. The intent is to cover the theory, applications, and perspectives on the state of the art and future developments relevant to systems, decision making, control, complex processes and related areas, as embedded in the fields of engineering, computer science, physics, economics, social and life sciences, as well as the paradigms and methodologies behind them. The series contains monographs, textbooks, lecture notes and edited volumes in systems, decision making and control spanning the areas of Cyber-Physical Systems, Autonomous Systems, Sensor Networks, Control Systems, Energy Systems, Automotive Systems, Biological Systems, Vehicular Networking and Connected Vehicles, Aerospace Systems, Automation, Manufacturing, Smart Grids, Nonlinear Systems, Power Systems, Robotics, Social Systems, Economic Systems and other. Of particular value to both the contributors and the readership are the short publication timeframe and the world-wide distribution and exposure which enable both a wide and rapid dissemination of research output.

Indexed by SCOPUS, DBLP, WTI Frankfurt eG, zbMATH, SCImago.

All books published in the series are submitted for consideration in Web of Science.

More information about this series at http://www.springer.com/series/13304

Sinan Melih Nigdeli · Gebrail Bekdaş ·
Aylin Ece Kayabekir · Melda Yucel
Editors

Advances in Structural Engineering—Optimization

Emerging Trends in Structural Optimization

Springer

Editors
Sinan Melih Nigdeli
Department of Civil Engineering
Istanbul University - Cerrahpaşa
Istanbul, Turkey

Gebrail Bekdaş
Department of Civil Engineering
Istanbul University - Cerrahpaşa
Istanbul, Turkey

Aylin Ece Kayabekir
Department of Civil Engineering
Istanbul University - Cerrahpaşa
Istanbul, Turkey

Melda Yucel
Department of Civil Engineering
Istanbul University - Cerrahpaşa
Istanbul, Turkey

ISSN 2198-4182 ISSN 2198-4190 (electronic)
Studies in Systems, Decision and Control
ISBN 978-3-030-61847-6 ISBN 978-3-030-61848-3 (eBook)
https://doi.org/10.1007/978-3-030-61848-3

This Springer imprint is published by the registered company Springer Nature Switzerland AG
The registered company address is: Gewerbestrasse 11, 6330 Cham, Switzerland

Preface

In structural engineering, the number of problems that is needed to be optimized is very high. For that reason, it is possible to find a new structural optimization problem or different cases of existing problems for different objectives, design variables, design constraints respect to demands of users and local authorities. Also, new and advanced methods are proposed for four reasons. The first reason is to solve the optimization problems quicker than the existing methods. The second factor is robustness of methods, and it is essential to find the best results for each run of the algorithm without trapping to a local optimum. The third factor is the improvement of the objective function which is to consider more than one objective function at the same time. To reach these factors, metaheuristics are improved by modifying another aspect in numerical methods or hybridizing with other algorithms to enable the best features together.

The aim of this book is to present emerging trends in the area of the metaheuristic algorithm-based structural optimization. The included 11 chapters have different subjects and different use of metaheuristics.

The book starts with a review chapter about structural engineering. In this chapter, several algorithms are described, and the major and recent studies about truss structures, reinforced concrete (RC) members, frames, bridges and structural control are mentioned.

The second chapter is related to machine learning (ML) and artificial intelligence (AI). In this chapter, the optimum parameters are predicted via the generated AI models after machine learning of optimum results found via metaheuristic algorithms. The results and problems such as tuned mass dampers (TMDs), base isolations, trusses, fiber-reinforced polymer design and structural engineering benchmark problems are given after a review of several AI and ML method in structural engineering which is presented.

Then, a multi-objective artificial bee colony algorithm is given, and the method is presented on truss, I-beam and steel welded beam design.

The fourth, fifth and sixth chapters are related to structural control. The fourth chapter presents a fuzzy controller for seismic control of nonlinear building employing tribe-charged system search. In the fifth chapter, structural control applications are reviewed for passive, active and semi-active systems. As current trends, active control of jacket platforms, soil–structure interaction (SSI) in TMD-controlled buildings, a modified harmony search optimized active tuned mass dampers, H_2 and H_∞ optimization for TMDs are given. The sixth chapter is related with metaheuristic-based optimization of active tendons for seismic structures considering SSI.

As the seventh chapter, optimum design of truss structures is presented for different algorithms and different types of structures.

An alternative structural analysis method called total potential optimization using metaheuristic (TPO/MA) is found in the eighth chapter. This method is given for plane-stress members, and a tunnel problem is presented.

Jaya algorithm was utilized for buckling analysis and stacking sequence optimization of symmetric laminated composite plates in the ninth chapter. A sustainable optimum design considering cost and CO_2 emission as multi-objective optimization is presented for RC retaining wall in the tenth chapter.

As the last chapter of the book, statistical evaluation that includes Friedman ranking, one-way ANOVA, post hoc Bonferroni test and independent sample t-test is given.

Istanbul, Turkey
September 2020

Sinan Melih Nigdeli
Gebrail Bekdaş
Aylin Ece Kayabekir
Melda Yucel

Contents

Developments on Metaheuristic-Based Optimization in Structural Engineering

Aylin Ece Kayabekir, Gebrail Bekdaş, and Sinan Melih Nigdeli

Abstract The purpose of this state-of-the-art review is to present the latest advances in design optimization and applications of metaheuristic algorithms in structural engineering. In the first part of this chapter, the importance of optimization in structural engineering and its differences with engineering problems are emphasized. Metaheuristic methods and the most appropriate techniques for various approaches are summarized and reviewed. These algorithms are effective in dealing with nonlinear design optimization with complex constraints, practical discrete design variables, and user-defined special conditions. Modifications of these algorithms have been made and applied to structural engineering applications. Finally, the results are presented with discussion about further potential improvements.

Keyword Optimization · Metaheuristic methods · Structural engineering · Non-linear constraints

1 Introduction

Civil engineering is concerned with several applications including superstructure and infrastructure construction, public transport and resources, transportation and traffic, the transport and storage of water resources, stabilization of landfills and soil improvement. In general speaking, all applications and designs that are related with

A. E. Kayabekir · G. Bekdaş · S. M. Nigdeli (✉)
Department of Civil Engineering, Istanbul University - Cerrahpaşa, 34320 Avcılar, Istanbul, Turkey
e-mail: melihnig@istanbul.edu.tr

A. E. Kayabekir
e-mail: ecekayabekir@gmail.com

G. Bekdaş
e-mail: bekdas@istanbul.edu.tr

S. M. Nigdeli et al. (eds.), *Advances in Structural Engineering—Optimization*, Studies in Systems, Decision and Control 326,
https://doi.org/10.1007/978-3-030-61848-3_1

living habitat fall within the field of civil engineering. As a subdiscipline of civil engineering, structural engineering is the most populated area in usage of optimization techniques and especially metaheuristic algorithms.

The world has various geographical regions with different resources, different risks of natural disasters and different demands due to economic reasons and peoples' choice. Due to these differences, an optimum solution in not optimum for another case. For example, different regions may have different material and labour costs. Since the most important factor in the design of civil engineering applications is security, different constraint limits may appear. The most effective factor is originated from natural hazards and soil conditions and various safety standards are defined for that reason. As a summary conclusion, optimum designs should be user-oriented in structural engineering.

For example, physics rules are a decisive factor when designing a mechanical engineering product. For these problems, excitation conditions are well known since these conditions are easily measured. For thar reason, precise production can be only possible in a factory. However, these issues may not be accepted in the construction of a building because of probabilistic quantities of applied forces due time. Also, field conditions are special. For example, the area may be in an earthquake zone where strong ground motions may occur. Due to that, additional design analysis should be considered as the effects of various factors that cannot be neglected. In addition, structures have large components, and production is usually done in construction place. Especially, the hardening of concrete and the positioning of reinforced concrete bars are two important production during the construction of reinforced concrete structures. An exact design is never possible and design variables must be assigned with discrete variables since labours of construction may not sensitive. Similarly, in the construction of steel structures, the profile dimensions in the local market are sold in standard sizes and the supply of a special section is not an economic solution. Due to these reasons, civil engineering applications and especially structural engineering ones have numbers of nonlinear constraints. For this reason, designs must be robust against uncertainties in material properties, applied forces (live loads that change in time due people and their belongings, earthquake and snow load, etc.), manufacturability and approximations done in mechanic science. Solving design optimization problems can be very difficult or mathematical approaches are not suitable to solve nonlinear constrained structural engineering problems. In that case, numerical algorithms are more practical than any mathematical approach in optimization because of the mentioned reasons.

Recently, optimization methodologies using metaheuristic algorithms have been chosen in structural engineering applications. Metaheuristic algorithms use inspirations from nature in order to solve optimization problems, and the main goal of optimization is to obtain a robust and best option so that the design variables and parameters are realistic in practice as in documented results.

This chapter is organized as 4 general sections as follows: Sect. 2 provides a short description of some metaheuristic algorithms that are previously preferred in structural optimization. Section 3 provides a detailed review of the various applications in structural engineering. As the final section, Section 4 gives conclusions briefly.

Metaheuristic applications in civil engineering are reviewed in general fields [1, 2], structures, infrastructures [3], water, geotechnical and transportation [4]. This book chapter includes an up to date review of structural engineering applications.

2 General Definition of Optimization and Metaheuristic Algorithms

In the area of using metaheuristic algorithms, optimization is a process in which one or multiple objective functions are minimized (or maximized according to need of user), and iterative search is done via randomization and features of the algorithms in this process. For example, an optimum design formulation is written as follows to minimize M number of objective functions (shown as f_i (x) for the ith function) for set of design variables (x).

$$\text{Minimize } f_i(x), x \in \mathcal{R}^n, (i = 1, 2, \ldots M) \tag{1}$$

The optimization is subjected to equality constraints:

$$h_j(\boldsymbol{x}), (j = 1, 2, \ldots j) \tag{2}$$

and (or) inequality constraints to control limits of analysis results like displacements and stresses:

$$g_k(\boldsymbol{x}) \leq 0, (k = 1, 2, \ldots k), \tag{3}$$

The vector of design variables is defined as Eq. (4) for a set of n design variables.

$$\boldsymbol{x} = (x_1, x_2, \ldots ..x_n)^T. \tag{4}$$

Design constraints are considered in the objective function as penalty function. Thus, it is possible to find the best set of design variables without any violation for design constraints. In the sub-section of this section, a few commonly used metaheuristic methods are shortly described.

2.1 Genetic Algorithm

Genetic algorithm (GA) is the mostly used metaheuristic algorithms and it is developed by John Holland in 1975 [5]. GA based on the basis theory of natural selection of Charles Darwin. The key operators taken from the Darwin's theory are the included features are crossover and recombination, mutation and selection. The main steps of process of GA are encoding of optimization objective, definition of a fitness function or criterion for selection of an individual, initialization of a population of individuals, evaluation of the fitness function for all individuals, generation of a new population by using the features such as crossover, mutation and reproduction, evolving of the population until the satisfaction of the defined stopping criterion and finally, decoding of the results.

In 1986, Goldberg and Samton [6] employed GA for the structural engineering example considering the optimum design of the 10-bar truss structure. Since then, there have been hundreds of articles in the context of structure engineering applications, and these studies are mentioned in Sect. 3.

2.2 *Simulated Annealing*

Simulated annealing (SA) is a metaheuristic algorithm that simulating the metal annealing process. In this proses, the aim is to increase the ductility and strength of the metal by rearranging the atomic structure internally. In 1983, Kirkpatrick et al. [7] developed SA and Cerny [8] was among the first user of SA in several civil engineering in 1985. Then, optimization approaches have appeared by using SA [9–11] and hybridized version of SA with GA [12, 13].

2.3 *Particle Swarm Optimization*

Particle swarm optimization (PSO) is a swarm-intelligence-based metaheuristic algorithm and it was proposed by Kennedy and Ebarhart in 1995 [14]. PSO mimics the herd's social behaviour. In the iterations, swarm particles are created as randomly generated solutions and new solutions are updated recursively. As particles explore and move around, they tend to move towards the best available solution. PSO is a population-based algorithm. The global optimum can be achieved when the design area is well searched by multiple particles over a long enough time. For example, PSO applications in structural engineering are offered by Kaveh [15], and more examples will be given in Sect. 3.

2.4 *Harmony Search*

Harmony search (HS) is a music-inspired algorithm developed by Geem et al. [16]. Design assemblies are encoded as a series of harmonies as solutions. As a musician has the option to play similar or fully different notes to gain admiration of audience, generated candidate solutions that are the set of design variables may be in a different part of the solution range. Also, optimum results may be close to existing solutions.

Considering both options can be effective if design solutions are in the neighbourhood values or these values are only local optimum values. This method has been applied to various engineering problems including civil engineering applications presented by Yoo et al. [17].

2.5 *Firefly Algorithm*

The Firefly algorithm (FA) is a swarm-intelligence based algorithm generated by Yang [18]. This algorithm uses the shooting mechanism and firefly blinking properties. Since short-range shooting is stronger than long-range shooting, the flock can be automatically split into multiple subsets, so the firefly algorithm is particularly suitable for nonlinear multi-mode optimization problems [19]. FA has been employed in the optimization of general structural engineering problems [20], tower structures [21], steel plates [22] and truss structures [23].

2.6 *Cuckoo Search*

Cuckoo birds have strange and unique behaviour on nesting their babies. This behaviour inspired by Cuckoo search (CS) is hatching parasitism of cuckoo species. The algorithm was developed by Yang and Deb [24] in 2009 by formulating the features. In CS, the solution is mentioned as eggs and nests and the best solutions are the ones with high-quality, because the bird with the highest quality egg destroys or throw away the other eggs from the nest to gain more space. During iterations, new solutions are produced by using Lévy flights to explore an efficient search area. CS was successfully applied to classical structural optimization problems including benchmark examples [25] and design problems of steel frames [26] and trusses [27].

2.7 *Bat Algorithm*

Yang developed bat algorithms by inspiring the feature of micro bats optimizing the ecological placement behaviour. This behaviour has varying frequencies, loudness and pulse emission rates. BA is using a balance exploration and exploitation to update the solution of objective function. In civil engineering, it is mostly used in structural engineering problems [28, 29] including the detailed optimization problem of steel plate curtain walls [30], skeletal structures [31], lattice beam [32], reinforced concrete beams [33] and tuned mass dampers [34].

2.8 Teaching Learning Based Optimization

Most of the metaheuristic algorithms have specific parameters according to the used features. These parameters are effective in the performance of the optimization process in the mean of minimization (or maximization) of the objective function, convergence and computational time. For that reason, adaptive, self-tuning and parameter-free version of algorithms were proposed. Teaching-learning-based-optimization (TLBO) is one of the parameter-free algorithms that do not contain these parameters by its initial and classical form. Rao et al. [35] developed TLBO and demonstrated this algorithm on constrained mechanical design optimization problems. This algorithm used the two phases of the education process. In a class, the education is done by the teacher and the teacher is the one with the best knowledge in the class. For that reason, the teacher is updated as the best existing solution in teacher phase. In teacher phase, the solutions are scanned around the best solution by considered a teaching factor value which can randomly take 1 or 2. The second phase is named as learner phase and it imitates the self-leaning process in the class. This phase is consequently applied to modify existing solutions and the solutions is searched between two randomly chosen solutions.

2.9 Flower Pollination Algorithm

Flower pollination algorithm (FPA) is a metaheuristic algorithm developed by Yang [36]. It imitates the pollination process of flowering plants. Different types of pollinators are inspired according to pollen transfer type and reproduction between the species.

The pollen transfer can be done in two ways. In biotic pollination, a living pollinator such as bees, insects and animals carry the pollens. In abiotic pollination, the pollen is transferred without pollinators and the pollen travel distance is short according to biotic pollination.

The pollination and reproduction can also be in two ways. In cross pollination, the pollen can be transferred to different flowers, while self-pollination is done in the same flower like peach.

These pollination types are considered with flower constancy and a switch probability is used to choose one of the global and local pollination. Global pollination imitates cross and biotic pollination by using a Lévy distribution. In local pollination, self and abiotic pollinations are used and a linear distribution is used.

FPA was firstly applied to benchmark structural optimization problems [37] and space trusses.

2.10 *Other Metaheuristic Algorithms Used in Civil Engineering*

It is hard to give the definition of all metaheuristic algorithms that are used in structural optimization. In this subsection, a few of them are mentioned as examples.

Ant colony optimization (ACO) developed by Dorigo in 1922 [38] is one of the oldest metaheuristic algorithms evaluated for structural engineering problems [39, 40]. Big Bang-Big Crunch (BB-BC) happening that forms the earth was formalized by Erol and Eksin [41], and the generated algorithm was employed in the optimization of trusses [42–45], steel frames [46] and RC retaining walls [47]. Charged system search (CSS) is an algorithm developed by Kaveh and Talatahari [48], and it imitates electrostatic and Newtonian mechanical law to solve optimization problems including damage detection in skeletal structures [49], castellated beams [50], structures [51, 52] and tuned mass dampers [53, 54].

A single-phase metaheuristic algorithm called Jaya algorithm (JA) was developed by Rao [55] as a parameter-free algorithm. This algorithm was hybridized to overcome local optimum trapping problem, and several structural engineering problems [56–59].

3 Applications and Optimization in Structural Engineering

In this section, the general structural optimization topics such as truss structures, frame structures, bridge and reinforced concrete members are mentioned as a state of art study. Also, structural control applications including optimum tuning of dampers are mentioned.

3.1 *Truss Structures*

The optimum design of trusses is the mostly done structural engineering problem by employing metaheuristic algorithms. Generally, a newly developed metaheuristic algorithm was firstly applied to truss optimization in structural engineering.

For truss structures, the general objective function is the minimization of weight or volume that directly related to the weight of the structure. The displacement and stress limitations are considered as design constraints, but these values can be also considered as multi objectives. The types of optimization differ according to the design variables. The cross-sectional areas of bars of trusses can be optimized in sizing optimization. Also, nodal coordinates of bars are the design variables in shape optimization. As general, topology optimization considers the link between the bars. Multiple types of design variables can be also considered together for a general optimization.

The early-studies about optimization of trussed used GA. Rajaev and Krishnamoorthy considered discrete design variable and a penalty parameter for restriction of violation of design constraints on their study employing GA [60]. Konmousis and Georgiou considered both sizing and layout of bars for roof trusses in the GA-based optimization [61]. Rajan investigated all types such as sizing, shape and topology by using GA [62]. Coello and Christianses used a multi objective approach for truss structures that have minimized deflection and maximized allowed stress according to min-max optimization approach combined with GA [63]. Erbatur et al. applied GA based structural optimization to planar and space trusses [64]. Krishnamoorthy et al. employed an object-oriented design implemented to a core library called Genetic Algorithms Library for Learning and Engineering Optimization (GALİLEO) for truss structures [65]. Hasancebi explored the application of evolutionary strategies for truss bridges [66]. Kelesoglu considered minimum weight and maximum displacement in the multi objective approach combining GA and fuzzy formulation for truss structures [67]. Sesok and Belevioius developed modified GA for truss structures using the repair of genotype instead of using constraints [68]. By using GA, self-adaptive member grouping and an initial population strategy was used for optimization of truss structures [69]. Richardson et al. [70] used kinematic stability repair on GA based approach for truss-like structures. Li proposed an improved species-conserving GA for truss structures [71].

PSO [72, 73] and different variants of PSO including methodology based on the passive congregated particle swarm optimizer and HS [74] was used in truss optimization. Several swarm intelligence methods such as particle swarm optimizer with passive community, ACO and HS was combined in discrete optimization of truss structures [75].

ACO [76], Artificial Bee Colony [77], TLBO [78–80], Firefly algorithm [23], CS [27], BA [32], BB-BC [42–45], mine blast algorithm [81] and FPA [82], the contrast based fruit fly optimization [83], symbiotic organisms search [84], vibrating particle system algorithm [85], heat transfer search [86, 87] and JA [88] are some of the employed metaheuristic algorithms for truss structures.

Chaos theory was combined with metaheuristics to optimize truss structures [89, 90]. Automatic member grouping and multiple frequency constraints were considered by modifying the velocity expression of PSO by an operator called craziness velocity to avoid premature convergence of truss optimization problem [91].

Grey wolf algorithm was integrated into an adaptive differential evolution and fully stressed design methodology to optimize truss structures including topology, shape and sizing design variables [92].

For the analysis of trusses, metaheuristics combined with the total potential energy minimization optimization using metaheuristic algorithms (TPO/MA) was used to trusses [93, 94], cables [95] and plane-stress members [96, 97].

3.2 Reinforced Concrete (RC) Members

The dimensions of RC members such as beams and columns are determined according to experience of engineers. Despite that the design of RC members are done according to provide ductile behavior since concrete is a brittle element that can fracture suddenly. In that case, the ductility is provided by the use of steel as a ductile member. Due to that, the yielding of steel must start before the fracture of concrete to prevent sudden collapse of the structure. Additionally, this situation can only be provided for members under bending. For the member under high axial loadings like columns, additional measures like confinement of bars and limitation of internal forces (axial and shear forces) are needed to provide long lasting load bearing capacity. These limitations and load balance formulations about stresses and strain are considered as nonlinear design constraints for the optimization.

The second factor in the optimization of RC structures and member is the need of discrete design variables because of practical production in construction yard.

Due to the mentioned reason, the optimum design of RC members is a popular recent research area. GA is one of the widely used algorithm as the other topics in structural optimization for RC beams [98, 99]. For continuous beam, GA was combined with simulated annealing for cost optimization [100].

HS is also a highly used metaheuristic in optimization RC members. Methodologies include optimization of continuous beams [101], T-shaped beams [102], columns [103, 104] including the detailed optimization method considering biaxial loaded columns [105].

The design constraint about stresses can be also considered as a secondary objective in optimum design of RC member. Afshari et al. investigated the constrained multi objective optimization algorithms and evaluated the method on an RC beam example [106]. Sanchez-Olivares and Tomas developed a modified FA for RC rectangular sections under compression and biaxial bending [107]. Ulusoy et al. investigated RC beams by comparing HS, BA and TLBO [108].

For the structures interacting with soil, additional limit states related to geotechnical measures are needed to be controlled in the optimum design. These types of structures can be listed as footings, retaining walls and piles. Nigdeli et al. employed HS, TLBO and FPA for detailed optimum design of RC footings [109]. RC retaining walls are the most optimized structures that are currently the most investigated ones in structural optimization. One of the pioneer approaches of RC retaining walls is the simulated annealing based minimum cost design of Ceranic et al. [110]. Yepes et al. included the design variables about reinforcement in the methodology employing simulated annealing [111]. Similarly, all dimension and reinforcement variables are considered in the PSO based method of Ahmadi-Nedushan and Varaee [112]. In addition to cost optimization, the embedded carbon dioxide (CO_2) emission minimization is also taken as a secondary objective function for RC structures. Generally, RC retaining walls have been investigated as a multi objective problem including both cost and CO_2 emission by using a hybrid optimization method by Yepes et al. [113]. Kayabekir et al. generated a HS based multi objective optimization approach for multiple cases of optimization of Eco-Friendly RC retaining walls [114]. Other recent example for optimum design of cantilever retaining walls employed BB-BC [47], charged system search [115], collided bodies optimization [116], Biogeography-based optimization [117], black hole algorithm [118], FPA [119], TLBO and JA [120]. Also, buttressed earth-retaining walls were optimized by employing a hybrid HS [121, 122].

Cantilever solider pile retaining walls are another infrastructure that was optimized by HS based methods [123, 124].

RC frames and bridges are mentioned in separate sections. Detailed methodologies and formulations about metaheuristic-based optimization can be found in Kayabekir et al. [125] including codes and different types of RC members and structures.

3.3 *Frame Structures*

Frames are structures that are the combination of beam and column elements. For optimization, both RC and steel frames have been investigated by proposing several methodologies.

In 2000, Pezeashk et al. employed GA for geometric design of nonlinear 2D steel frames [126]. Then, GA was also used in the optimum design of semi-rigid connected nonlinear steel frames [127]. ACO used in the discrete optimization of steel frames by Camp et al. [128]. The stochastic search techniques were reviewed by Saka [129] for steel frames with discrete variables to show the advantages of the method comparing to mathematical approaches. Kociecki and Adeli optimized free-form steel space frame roof structures consisting of rectangular hollow sections for sizing and topology by using a two-phase GA [130]. The recent examples for optimization of steel frames used metaheuristic such as TLBO [131], eagle strategy with differential evolution [132], artificial bee colony with implementation of Lévy distribution to the scout bee search [133], enhanced firefly algorithm [134], school based optimization [135, 136], chaotic firefly algorithm [137], colliding body optimization [138] and hybrid taboo search [139].

For RC frames, the studies about the bridge frames are mentioned in a separate section. Various metaheuristic algorithms have been proposed to RC frame structure for cost optimization [140–149] and minimization of CO_2 emission [150, 151].

3.4 *Bridges*

Bridges are another type of structure that can be generated by steel or RC members. RC frames forming bridges for road construction were optimized by Perea et al. employing combination of two heuristic (random walk and decent local search) and two metaheuristics (threshold acceptance and simulate annealing) [152]. RC bridge piers that have hollow rectangular sections were optimized by using ACO compared with GA and threshold acceptance algorithm [153, 154]. In addition to cost minimization, reinforcing steel congestion and embedded CO_2 emission was considered in GA and simulated annealing-based method by Martinez-Martin et al. [155]. By using GA, the reliability assurance in long span suspension bridges was investigated by Sgambi et al. [156]. Post-tensioned concrete box-girder road bridges were optimized via HS to consider measures such as cost, CO_2 emission and safety [157]. Abd Elrehim et al. employed GA for concrete arch bridges [158]. Atmaca et al. developed a JA-based optimization approach to find the best design for cable sizes and pre-stressing force of a single pylon cable-stayed bridges [159].

3.5 Structural Control

For the structures subjected to time-varying excitation, undesired vibrations may occur. These vibrations may prevent the use of structure with comfort or damage non-structural components. In that case, structural control techniques have been developed to use in important and valuable structures.

The structural control techniques have passive and active types in common. Passive control uses the additional mechanical components to change the behavior of structure positively on responding to vibrating excitations. The main idea is the perfect tune of properties of additional control system. For that reason, optimization is a must in finding the best suitable values. Also, several factors such as assumptions in structural design, random and unknown index of excitations, and complex behavior of problem including the consideration of damping prevent the use of mathematical optimization.

Differently in active control, an external energy source in action. For that reason, these systems may have high cost, and economical ways may have been also proposed by developing semi-active and hybrid systems. Similarly, like passive control system, active control parameters must be tuned, and also the employed control algorithms used in generation of additional control force need tuning for the best performance. For the structures subjected to earthquake, wind and traffic excitation, several control systems have been developed. In this study, metaheuristic-based approaches for several structural control systems are given. The mentioned studies are limited with tuned mass dampers (TMDs) for passive control systems and several active control approaches using active tendons and active tuned mass dampers (ATMDs) are mentioned.

As basic form of TMDs, the system consists of mass, stiffness element and a damper. The tuned parameters are the stiffness and damping coefficient of TMD. Generally, mass of TMDs is at the upper limit if the additional weight of TMD can be securely carried by the structure. The metaheuristic are alternative to basic expressions derived for TMD [160–162] that are found by several assumptions and single degree of freedom systems. These methods are not effective to consider multiple vibration modes, damping and limited stroke capacity of TMDs. The employed methodologies based on GA [163–166], PSO [167, 168], HS [169–173], BA [34,

173], ACO [174], Artificial Bee Optimization [175], JA [59], FPA [176, 177], CSS [53], TLBO [178, 179] and Coral Reefs Optimization algorithm [180].

In addition to optimization of TMDs, the optimum parameters of TMD found via metaheuristic methods were used in machine learning to generate an artificial neural network (ANN) model predicting optimum values. Also, the ANN model was used to find expressions of TMD parameters such as linear, polynomial and exponential [181].

For active structural control, the controller parameters are also needed to be optimized. Chapters "Optimal Parameter Identification of Fuzzy Controllers in Nonlinear Buildings Based on Seismic Hazard Analysis Using Tribe-Charged System Search", "Current Trends in the Optimization Approaches for Optimal Structural Control" and "The Effect of SSI and Impulsive Motions on Optimum Active Controlled MDOF Structure" of this book contains major applications of active structural control. As a summary, metaheuristic have been employed in GA based optimizer for ATMDs and active bracing system [182], ATMDs using fuzzy logic and GA [183], semi-active TMDs optimized via Fuzzy controller using CSS [54], linear quadratic regulator (LQR) controller using several metaheuristic algorithms for active tendon control [184], Proportional Integral Derivative (PID) type controller implemented ATMDs using modified HS [185], CS optimized fractional order PID controller [186], Gases Brownian motion optimization based PID controller active control system [187] and PID controlled active tendons via FPA, TLBO and JA [188–190].

4 Conclusions

In this chapter, recent developments in metaheuristic-based optimization of structural engineering problems are reviewed. Nowadays, the number of metaheuristic algorithms is in great increase and all problems and applications in structural engineering need optimization because of existing design constraints and demands of future needs of people. For that reason, only a few methods and topics are covered and mentioned in this chapter.

The area of structural engineering is open to new problems which are not optimized until now or not widely investigated. Due to that, it is possible to see an increasing trend on structural optimization by using advance, modified, user-friendly and hybrid metaheuristic algorithms.

References

1. Yang, X.S., Bekdaş, G., Nigdeli, S.M. (eds.) Metaheuristics and Optimization in Civil Engineering. Springer International Publishing
2. Bekdaş, G., Nigdeli, S.M., Kayabekir, A.E., Yang, X.S.: Optimization in civil engineering and metaheuristic algorithms: a review of state-of-the-art developments. In: Computational

Intelligence, Optimization and Inverse Problems with Applications in Engineering, pp. 111–137. Springer, Cham (2019)
3. Gandomi, A.H., Yang, X.S., Talatahari, S., Alavi, A.H.: Metaheuristic applications in structures and infrastructures. Elsevier Publishing (2013)
4. Yang, X.S., Gandomi, A.H., Talatahari, S., Alavi, A.H.: Metaheuristics in water, geotechnical and transport engineering. Elsevier (2012)
5. Holland, J.H.: Adaptation in Natural and Artificial Systems. University of Michigan, Ann Arbor (1975)
6. Goldberg, D.E., Samtani, M.P.: Engineering optimization via genetic algorithm. In: Proceedings of Ninth Conference on Electronic Computation, New York, NY, pp. 471–482 (1986)
7. Kirkpatrick, S., Gelatt Jr., C.D., Vecchi, M.P.: Optimization by simulated annealing. Science **220**(4598), 671–680 (1983)
8. Černý, V.: Thermodynamical approach to the traveling salesman problem: an efficient simulation algorithm. J. Optim. Theory Appl. **45**(1), 41–51 (1985)
9. Costa, A.L., Cunha, M.D.C., Coelho, P.A., Einstein, H.H.: Solving high-speed rail planning with the simulated annealing algorithm. J. Transp. Eng. **139**(6), 635–642 (2013). https://doi.org/10.1061/(ASCE)TE.1943-5436.0000542
10. Tong, K.H., Bakhary, N., Kueh, A.B.H., Yassin, A.Y.: Optimal sensor placement for mode shapes using improved simulated annealing. Smart Struct. Syst. **13**(3), 389–406 (2014). https://doi.org/10.12989/sss.2014.13.3.389
11. Karovic, O., Mays, L.W.: Sewer system design using simulated annealing in excel. Water Resour. Manag. **28**(13), 4551–4565 (2014). https://doi.org/10.1007/s11269-014-0750-8
12. Liu, W., Ye, J.: Collapse optimization for domes under earthquake using a genetic simulated annealing algorithm. J. Constr. Steel Res. **97**, 59–68 (2014). https://doi.org/10.1016/j.jcsr.2014.01.015
13. Junghans, L., Darde, N.: Hybrid single objective genetic algorithm coupled with the simulated annealing optimization method for building optimization. Energy Build. **86**, 651–662 (2015). https://doi.org/10.1016/j.enbuild.2014.10.039
14. Kennedy, J., Eberhart, R.C.: Particle swarm optimization. In: Proceedings of IEEE International Conference on Neural Networks, no. IV, 27 November–1 December, pp. 1942–1948, Perth Australia (1995)
15. Kaveh, A.: Advances in metaheuristic algorithms for optimal design of structures. Springer (2014)
16. Geem, Z.W., Kim, J.H., Loganathan, G.V.: A new heuristic optimization algorithm: harmony search. Simulation **76**, 60–68 (2001). https://doi.org/10.1177/003754970107600201
17. Yoo, D.G., Kim, J.H., Geem, Z.W.: Overview of Harmony search algorithm and its applications in civil engineering. Evol. Intell. **7**(1), 3–16 (2014). https://doi.org/10.1007/s12065-013-0100-4
18. Yang, X.S.: Nature-Inspired Metaheuristic Algorithms. Luniver Press (2008)
19. Yang, X.S.: Firefly algorithms for multimodal optimization. In Stochastic algorithms: foundations and applications, pp. 169–178. Springer, Heidelberg (2009)
20. Gandomi, A.H., Yang, X.S., Alavi, A.H.: Mixed variable structural optimization using firefly algorithm. Comput. Struct. **89**(23), 2325–2336 (2011). https://doi.org/10.1016/j.compstruc.2011.08.002
21. Talatahari, S., Gandomi, A.H., Yun, G.J.: Optimum design of tower structures using firefly algorithm. Struct. Des. Tall Spec. Build. **23**(5), 350–361 (2014). https://doi.org/10.1002/tal.1043
22. Mauder, T., Sandera, C., Stetina, J., Seda, M.: Optimization of the quality of continuously cast steel slabs using the firefly algorithm. Mater. Technol. **45**(4), 347–350 (2011)
23. Miguel, L.F.F., Lopez, R.H., Miguel, L.F.F.: Multimodal size, shape, and topology optimisation of truss structures using the Firefly algorithm. Adv. Eng. Softw. **56**, 23–37 (2013). https://doi.org/10.1016/j.advengsoft.2012.11.006

24. Yang, X.S., Deb, S.: Cuckoo search via Lévy flights. In: World Congress on Nature & Biologically Inspired Computing, NaBIC 2009, pp. 210–214 (2009)
25. Gandomi, A.H., Yang, X.S., Alavi, A.H.: Cuckoo search algorithm: a metaheuristic approach to solve structural optimization problems. Eng. Comput. **29**(1), 17–35 (2013). https://doi.org/10.1007/s00366-011-0241-y
26. Kaveh, A., Bakhshpoori, T.: Optimum design of steel frames using cuckoo search algorithm with Lévy flights. Struct. Des. Tall Spec. Build. **22**(13), 1023–1036 (2013). https://doi.org/10.1002/tal.754
27. Gandomi, A.H., Talatahari, S., Yang, X.S., Deb, S.: Design optimization of truss structures using cuckoo search algorithm. Struct. Des. Tall Spec. Build. **22**(17), 1330–1349 (2013). https://doi.org/10.1002/tal.1033
28. Yang, X.S., Hossein Gandomi, A.: Bat algorithm: a novel approach for global engineering optimization. Eng. Comput. **29**(5), 464–483 (2012). https://doi.org/10.1108/02644401211235834
29. Gandomi, A.H., Yang, X.S., Alavi, A.H., Talatahari, S.: Bat algorithm for constrained optimization tasks. Neural Comput. Appl. **22**(6), 1239–1255 (2013). https://doi.org/10.1007/s00521-012-1028-9
30. Gholizadeh, S., Shahrezaei, A.M.: Optimal placement of steel plate shear walls for steel frames by bat algorithm. Struct. Des. Tall Spec. Build. **24**(1), 1–18 (2015). https://doi.org/10.1002/tal.1151
31. Kaveh, A., Zakian, P.: Enhanced bat algorithm for optimal design of skeletal structures. Asian J. Civil Eng. **15**(2), 179–212 (2014)
32. Talatahari, S., Kaveh, A.: Improved bat algorithm for optimum design of large-scale truss structures. Int. J. Optim. Civil Eng. **5**(2), 241–254 (2015)
33. Bekdas, G., Nigdeli, S.M., Yang, X.S.: Metaheuristic optimization for the design of reinforced concrete beams under flexure moments. In: Proceedings of the 5th European Conference of Civil Engineering (ECCIE 2014), Florence, Italy (2014)
34. Bekdaş, G., Nigdeli, S.M., Yang, X.S.: A novel bat algorithm based optimum tuning of mass dampers for improving the seismic safety of structures. Eng. Struct. **159**, 89–98 (2018)
35. Rao, R.V., Savsani, V.J., Vakharia, D.P.: Teaching–learning-based optimization: a novel method for constrained mechanical design optimization problems. Comput.-Aided Des. **43**(3), 303–315 (2011)
36. Yang, X.S.: Flower pollination algorithm for global optimization. In: Unconventional Computation and natural Computation 2012, Lecture Notes in Computer Science, vol. 7445, pp. 240–249 (2012)
37. Nigdeli, S.M., Bekdaş, G., Yang, X.S.: Application of the flower pollination algorithm in structural engineering. In: Metaheuristics and Optimization in Civil Engineering. Springer (2016)
38. Dorigo, M.: Optimization, learning and natural algorithms, Ph.D. thesis, Politecnico di Milano, Italy (1992)
39. Sahab, M.G., Toropov, V.V., Gandomi, A.H.: A review on traditional and modern structural optimization: problems and techniques. Metaheuristic Appl. Struct. Infrastruct. 24–47 (2013)
40. Angelo, J.S., Bernardino, H.S., Barbosa, H.J.: Ant colony approaches for multiobjective structural optimization problems with a cardinality constraint. Adv. Eng. Softw. **80**, 101–115 (2015). https://doi.org/10.1016/j.advengsoft.2014.09.015
41. Erol, O.K., Eksin, I.: A new optimization method: big bang big crunch. Adv. Eng. Softw. **37**, 106–111 (2006). https://doi.org/10.1016/j.advengsoft.2005.04.005
42. Camp, C.V.: Design of space trusses using big bang-big crunch optimization. J. Struct. Eng. **133**(7), 999–1008 (2007). https://doi.org/10.1061/(ASCE)0733-9445(2007)133:7(999)
43. Kaveh, A., Talatahari, S.: Size optimization of space trusses using big bang-big crunch algorithm. Comput. Struct. **87**(17), 1129–1140 (2009). https://doi.org/10.1016/j.compstruc.2009.04.011
44. Kaveh, A., Talatahari, S.: A discrete big bang-big crunch algorithm for optimal design of skeletal structures. Asian J. Civil Eng. **11**(1), 103–122 (2010)

45. Hasançebi, O., Kazemzadeh Azad, S.: Discrete size optimization of steel trusses using a refined big bang-big crunch algorithm. Eng. Optim. **46**(1), 61–83 (2014). https://doi.org/10.1080/0305215X.2012.748047
46. Hasançebi, O., Azad, S.K.: An exponential big bang-big crunch algorithm for discrete design optimization of steel frames. Comput. Struct. **110**, 167–179 (2012). https://doi.org/10.1016/j.compstruc.2012.07.014
47. Camp, C.V., Akin, A.: Design of retaining walls using big bang-big crunch optimization. J. Struct. Eng. **138**(3), 438–448 (2012). https://doi.org/10.1061/(ASCE)ST.1943-541X.0000461
48. Kaveh, A., Talatahari, A.: A novel heuristic optimization method: charged system search. Acta Mech. **213**, 267–289 (2010). https://doi.org/10.1007/s00707-009-0270-4
49. Kaveh, A., Maniat, M.: Damage detection in skeletal structures based on charged system search optimization using incomplete modal data. Int. J. Civil Eng. **12**(2A), 292–299 (2014)
50. Kaveh, A., Shokohi, F.: Cost optimization of castellated beams using charged system search algorithm. Iranian J. Sci. Technol. Trans. Civil Eng. **38**, 235–249 (2014)
51. Kaveh, A., Nasrollahi, A.: Charged system search and particle swarm optimization hybridized for optimal design of engineering structures. Scientia Iranica Trans. A Civil Eng. **21**(2), 295–305 (2014)
52. Kaveh, A., Massoudi, M.S.: Multi-objective optimization of structures using charged system search. Scientia Iranica Trans. A, Civil Eng. **21**(6), 1845–1860 (2014)
53. Kaveh, A., Mohammadi, S., Hosseini, O.K., Keyhani, A., Kalatjari, V.R.: Optimum Parameters of tuned mass dampers for seismic applications using charged system search. Iranian J. Sci. Technol. Trans. Civil Eng. **39**, 21–40 (2015)
54. Kaveh, A., Pirgholizadeh, S., Hosseini, O.K.: Semi-active tuned mass damper performance with optimized fuzzy controller using CSS algorithm. Asian J. Civil Eng. (BHRC) **16**(5), 587–606 (2015)
55. Rao, R.: Jaya: a simple and new optimization algorithm for solving constrained and unconstrained optimization problems. Int. J. Ind. Eng. Comput. **7**(1), 19–34 (2016)
56. Bekdaş, G., Kayabekir, A.E., Nigdeli, S.M., Toklu, Y.C.: Advanced energy-based analyses of trusses employing hybrid metaheuristics. Struct. Des. Tall Spec. Build. **28**(9), e1609 (2019)
57. Toklu, Y.C., Kayabekir, A.E., Bekdaş, G., Nigdeli, S.M., Yücel, M.: Analysis of plane-stress systems via total potential optimization method considering nonlinear behavior. J. Struct. Eng. **146**(11), 1–13 (2020). https://doi.org/10.1061/(ASCE)ST.1943-541X.0002808
58. Bekdaş, G., Kayabekir, A.E., Nigdeli, S.M., Toklu, Y.C.: Tranfer function amplitude minimization for structures with tuned mass dampers considering soil-structure interaction. Soil Dyn. Earthq. Eng. **116**, 552–562 (2019)
59. Bekdaş, G.: Optimum design of post-tensioned axially symmetric cylindrical walls using novel hybrid metaheuristic methods. Struct. Des. Tall Spec. Build. **28**(1), e1550 (2019)
60. Rajeev, S., Krishnamoorthy, C.S.: Discrete optimization of structures using genetic algorithms. J. Struct. Eng. **118**(5), 1233–1250 (1992). https://doi.org/10.1061/(ASCE)0733-9445(1992)118:5(1233)
61. Koumousis, V.K., Georgiou, P.G.: Genetic algorithms in discrete optimization of steel truss roofs. J. Comput. Civil Eng. **8**(3), 309–325 (1994). https://doi.org/10.1061/(ASCE)0887-3801(1994)8:3(309)
62. Rajan, S.D.: Sizing, shape, and topology design optimization of trusses using genetic algorithm. J. Struct. Eng. **121**(10), 1480–1487 (1995). https://doi.org/10.1061/(ASCE)0733-9445(1995)121:10(1480)
63. Coello, C.A., Christiansen, A.D.: Multiobjective optimization of trusses using genetic algorithms. Comput. Struct. **75**(6), 647–660 (2000). https://doi.org/10.1016/S0045-7949(99)00110-8
64. Erbatur, F., Hasançebi, O., Tütüncü, I., Kılıç, H.: Optimal design of planar and space structures with genetic algorithms. Comput. Struct. **75**(2), 209–224 (2000). https://doi.org/10.1016/S0045-7949(99)00084-X

65. Krishnamoorthy, C.S., Prasanna Venkatesh, P., Sudarshan, R.: Object-oriented framework for genetic algorithms with application to space truss optimization. J. Comput. Civil Eng. **16**(1), 66–75 (2002). https://doi.org/10.1061/(ASCE)0887-3801(2002)16:1(66)
66. Hasançebi, O.: Optimization of truss bridges within a specified design domain using evolution strategies. Eng. Optim. **39**(6), 737–756 (2007). https://doi.org/10.1080/03052150701335071
67. Kelesoglu, O.: Fuzzy multiobjective optimization of truss-structures using genetic algorithm. Adv. Eng. Softw. **38**(10), 717–721 (2007). https://doi.org/10.1016/j.advengsoft.2007.03.003
68. Šešok, D., Belevičius, R.: Global optimization of trusses with a modified genetic algorithm. J. Civil Eng. Manag. **4**(3), 147–154 (2008). https://doi.org/10.3846/1392-3730.2008.14.10
69. Toğan, V., Daloğlu, A.T.: An improved genetic algorithm with initial population strategy and self-adaptive member grouping. Comput. Struct. **86**(11–12), 1204–1218 (2008)
70. Richardson, J.N., Adriaenssens, S., Bouillard, P., Coelho, R.F.: Multiobjective topology optimization of truss structures with kinematic stability repair. Struct. Multidisc. Optim. **46**(4), 513–532 (2012). https://doi.org/10.1007/s00158-012-0777-5
71. Li, J.P.: Truss topology optimization using an improved species-conserving genetic algorithm. Eng. Optim. **47**(1), 107–128 (2015). https://doi.org/10.1080/0305215X.2013.875165
72. Schutte, J.F., Groenwold, A.A.: Sizing design of truss structures using particle swarms. Struct. Multidisc. Optim. **25**(4), 261–269 (2003). https://doi.org/10.1007/s00158-003-0316-5
73. Perez, R.E., Behdinan, K.: Particle swarm approach for structural design optimization. Comput. Struct. **85**(19), 1579–1588 (2007). https://doi.org/10.1016/j.compstruc.2006.10.013
74. Li, L.J., Huang, Z.B., Liu, F., Wu, Q.H.: A heuristic particle swarm optimizer for optimization of pin connected structures. Comput. Struct. **85**(7), 340–349 (2007). https://doi.org/10.1016/j.compstruc.2006.11.020
75. Kaveh, A., Talatahari, S.: Particle swarm optimizer, ant colony strategy and harmony search scheme hybridized for optimization of truss structures. Comput. Struct. **87**(5), 267–283 (2009). https://doi.org/10.1016/j.compstruc.2009.01.003
76. Camp, C.V., Bichon, B.J.: Design of space trusses using ant colony optimization. J. Struct. Eng. **130**(5), 741–751 (2004). https://doi.org/10.1061/(ASCE)0733-9445(2004)130:5(741)
77. Sonmez, M.: Artificial Bee Colony algorithm for optimization of truss structures. Appl. Soft Comput. **11**(2), 2406–2418 (2011). https://doi.org/10.1016/j.asoc.2010.09.003
78. Degertekin, S.O., Hayalioglu, M.S.: Sizing truss structures using teaching-learning-based optimization. Comput. Struct. **119**, 177–188 (2013). https://doi.org/10.1016/j.compstruc.2012.12.011
79. Camp, C.V., Farshchin, M.: Design of space trusses using modified teaching–learning based optimization. Eng. Struct. **62–63**, 87–97 (2014). https://doi.org/10.1016/j.engstruct.2014.01.020
80. Dede, T., Ayvaz, Y.: Combined size and shape optimization of structures with a new meta-heuristic algorithm Appl. Soft Comput. 28, 250–258 (2015). https://doi.org/10.1016/j.asoc.2014.12.007
81. Sadollah, A., Bahreininejad, A., Eskandar, H., Hamdi, M.: Mine blast algorithm for optimization of truss structures with discrete variables. Comput. Struct. **102**, 49–63 (2012). https://doi.org/10.1016/j.compstruc.2012.03.013
82. Bekdaş, G., Nigdeli, S.M., Yang, X.S.: Sizing optimization of truss structures using flower pollination algorithm. Appl. Soft Comput. **37**, 322–331 (2015). https://doi.org/10.1016/j.asoc.2015.08.037
83. Kanarachos, S., Griffin, J., Fitzpatrick, M.E.: Efficient truss optimization using the contrast-based fruit fly optimization algorithm. Comput. Struct. **182**, 137–148 (2017)
84. Tejani, G.G., Savsani, V.J., Patel, V.K., Mirjalili, S.: Truss optimization with natural frequency bounds using improved symbiotic organisms search. Knowl.-Based Syst. **143**, 162–178 (2018)
85. Kaveh, A., Ghazaan, M.I.: Vibrating particles system algorithm for truss optimization with multiple natural frequency constraints. Acta Mech. **228**(1), 307–322 (2017)
86. Tejani, G.G., Kumar, S., Gandomi, A.H.: Multi-objective heat transfer search algorithm for truss optimization. Eng. Comput. 1–22 (2019)

87. Kumar, S., Tejani, G.G., Pholdee, N., Bureerat, S.: Multi-objective modified heat transfer search for truss optimization. Eng. Comput. 1–16 (2020)
88. Degertekin, S.O., Lamberti, L., Ugur, I.B.: Sizing, layout and topology design optimization of truss structures using the Jaya algorithm. Appl. Soft Comput. **70**, 903–928 (2018)
89. Kaveh, A., Sheikholeslami, R., Talatahari, S., Keshvari-Ilkhichi, M.: Chaotic swarming of particles: a new method for size optimization of truss structures. Adv. Eng. Softw. **67**, 136–147 (2014). https://doi.org/10.1016/j.advengsoft.2013.09.006
90. Kaveh, A., Dadras, A., Montazeran, A.H.: Chaotic enhanced colliding bodies algorithms for size optimization of truss structures. Acta Mech. **229**(7), 2883–2907 (2018)
91. Carvalho, J.P., Lemonge, A.C., Carvalho, É.C., Hallak, P.H., Bernardino, H.S.: Truss optimization with multiple frequency constraints and automatic member grouping. Struct. Multidisc. Optim. **57**(2), 547–577 (2018)
92. Panagant, N., Bureerat, S.: Truss topology, shape and sizing optimization by fully stressed design based on hybrid grey wolf optimization and adaptive differential evolution. Eng. Optim. **50**(10), 1645–1661 (2018)
93. Toklu, Y.C.: Nonlinear analysis of trusses through energy minimization. Comput. Struct. **82**(20–21), 1581–1589 (2004)
94. Nigdeli, S.M., Bekdaş, G., Toklu, Y.C.: Total potential energy minimization using metaheuristic algorithms for spatial cable systems with increasing second order effects. In: 12th International Congress on Mechanics (HSTAM2019), 22–25 September 2019, Thessaloniki, Greece (2019)
95. Kayabekir, A.E., Nigdeli, S.M., Bekdaş, G., Toklu, Y.C.: The analyses of spatial cable structures employing new generation metaheuristic methods. In: 9th GRACM 2018 International Congress on Computational Mechanics, 4–6 June 2018, Crete, Greece (2018)
96. Kayabekir, A.E., Toklu, Y.C., Bekdaş, G., Nigdeli, S.M., Yücel, M., Geem, Z.W.: A novel hybrid harmony search approach for the analysis of plane stress systems via total potential optimization. Appl. Sci. **10**(7), 2301 (2020)
97. Toklu, Y.C., Bekdaş, G., Kayabekir, A.E., Nigdeli, S.M., Yucel, M.: Total potential optimization using metaheuristics: analysis of cantilever beam via plane-stress members. In: 6th International Conference on Harmony Search, Soft Computing and Applications (ICHSA 2020), 16–17 July 2020, Istanbul, Turkey (2020)
98. Govindaraj, V., Ramasamy, J.V.: Optimum detailed design of reinforced concrete continuous beams using genetic algorithms. Comput. Struct. **84**, 34–48 (2005). https://doi.org/10.1016/j.compstruc.2005.09.001
99. Fedghouche, F., Tiliouine, B.: Minimum cost design of reinforced concrete T-beams at ultimate loads using Eurocode2. Eng. Struct. **42**, 43–50 (2012). https://doi.org/10.1016/j.engstruct.2012.04.008
100. Leps, M., Sejnoha, M.: New approach to optimization of reinforced concrete beams. Comput. Struct. **81**, 1957–1966 (2003). https://doi.org/10.1016/S0045-7949(03)00215-3
101. Akin, A., Saka, M.P.: Optimum detailed design of reinforced concrete continuous beams using the harmony search algorithm. In: The Tenth International Conference on Computational Structures Technology, Stirlingshire, UK, Paper 131 (2010)
102. Bekdaş, G., Nigdeli, S.M.: Cost optimization of T-shaped reinforced concrete beams under flexural effect according to ACI 318. In: 3rd European Conference of Civil Engineering, Paris, France (2012)
103. Bekdaş, G., Nigdeli, S.M.: Optimization of slender reinforced concrete columns. In: 85th Annual Meeting of the International Association of Applied Mathematics and Mechanics, Erlangen, Germany (2014)
104. Nigdeli, S.M., Bekdaş, G.: Optimum design of RC columns according to effective length factor in buckling. In: The Twelfth International Conference on Computational Structures Technology, 2–5 September 2014, Naples, Italy (2014)
105. Nigdeli, S.M., Bekdas, G., Kim, S., Geem, Z.W.: A novel harmony search based optimization of reinforced concrete biaxially loaded columns. Struct. Eng. Mech. **54**(6), 1097–1109 (2015). https://doi.org/10.12989/sem.2015.54.6.1097

106. Afshari, H., Hare, W., Tesfamariam, S.: Constrained multi-objective optimization algorithms: review and comparison with application in reinforced concrete structures. Appl. Soft Comput. **83**, 105631 (2019)
107. Sánchez-Olivares, G., Tomás, A.: Improvements in meta-heuristic algorithms for minimum cost design of reinforced concrete rectangular sections under compression and biaxial bending. Eng. Struct. **130**, 162–179 (2017)
108. Ulusoy, S., Kayabekir, A.E., Bekdaş, G., Niğdeli, S.M.: Metaheuristic algorithms in optimum design of reinforced concrete beam by investigating strength of concrete. Chall. J. Concr. Res. Lett. **11**, 33–37 (2020)
109. Nigdeli, S.M., Bekdaş, G., Yang, X.S.: Metaheuristic optimization of reinforced concrete footings. KSCE J. Civil Eng. **22**(11), 4555–4563 (2018)
110. Ceranic, B., Freyer, C., Baines, R.W.: An application of simulated annealing to the optimum design reinforced concrete retaining structure. Comput. Struct. **79**, 1569–1581 (2001). https://doi.org/10.1016/S0045-7949(01)00037-2
111. Yepes, V., Alcala, J., Perea, C., Gonzalez-Vidosa, F.: A parametric study of optimum earth-retaining walls by simulated annealing. Eng. Struct. **30**, 821–830 (2008). https://doi.org/10.1016/j.engstruct.2007.05.023
112. Ahmadi-Nedushan, B., Varaee, H.: Optimal design of reinforced concrete retaining walls using a swarm intelligence technique. In: The First International Conference on Soft Computing, UK (2009)
113. Yepes, V., Gonzalez-Vidosa, F., Alcala, J., Villalba, P.: CO2-optimization design of reinforced concrete retaining walls based on a VNS-threshold acceptance strategy. J. Comput. Civil Eng. **26**(3), 378–386 (2011). https://doi.org/10.1061/(ASCE)CP.1943-5487.0000140
114. Kayabekir, A.E., Arama, Z.A., Bekdaş, G., Nigdeli, S.M., Geem, Z.W.: Eco-friendly design of reinforced concrete retaining walls: multi-objective optimization with harmony search applications. Sustainability **12**(15), 6087 (2020)
115. Talatahari, S., Sheikholeslami, R., Shadfaran, M., Pourbaba, M.: Optimum design of gravity retaining walls using charged system search algorithm, Mathematical Problems in Engineering, Article ID 301628, 1–10 (2012). http://doi.org/10.1155/2012/301628
116. Kaveh, A., Soleimani, N.: CBO and DPSO for optimum design of reinforced concrete cantilever retaining walls. Asian J. Civil Eng. **16**(6), 751–774 (2015)
117. Aydogdu, I.: Cost optimization of reinforced concrete cantilever retaining walls under seismic loading using a biogeography-based optimization algorithm with Levy flights. Eng. Optim. **49**(3), 381–400 (2017)
118. Yepes, V., Martí, J.V., García, J.: Black hole algorithm for sustainable design of counterfort retaining walls. Sustainability **12**(7), 2767 (2020)
119. Mergos, P.E., Mantoglou, F.: Optimum design of reinforced concrete retaining walls with the flower pollination algorithm. Struct. Multidisc. Optim. **61**(2), 575–585 (2020)
120. Öztürk, H.T., Dede, T., Türker, E.: Optimum design of reinforced concrete counterfort retaining walls using TLBO, Jaya algorithm. Structures **25**, 285–296 (2020)
121. Molina-Moreno, F., García-Segura, T., Martí, J.V., Yepes, V.: Optimization of buttressed earth-retaining walls using hybrid harmony search algorithms. Eng. Struct. **134**, 205–216 (2017)
122. Molina-Moreno, F., Martí, J.V., Yepes, V.: Carbon embodied optimization for buttressed earth-retaining walls: implications for low-carbon conceptual designs. J. Cleaner Prod. **164**, 872–884 (2017)
123. Bekdaş, G., Arama, Z.A., Kayabekir, A.E., Geem, Z.W.: Optimal design of cantilever soldier pile retaining walls embedded in frictional soils with harmony search algorithm. Appl. Sci. **10**(9), 3232 (2020)
124. Arama, Z.A., Kayabekir, A.E., Bekdaş, G., Geem, Z.W.: CO2 and cost optimization of reinforced concrete cantilever soldier piles: a parametric study with harmony search algorithm. Sustainability **12**(15), 5906 (2020)
125. Kayabekir, A.E., Bekdaş, G., Nigdeli, S.M. (eds.): Metaheuristic Approaches for Optimum Design of Reinforced Concrete Structures: Emerging Research and Opportunities: Emerging Research and Opportunities. IGI Global (2020)

126. Pezeshk, S., Camp, C.V., Chen, D.: Design of nonlinear framed structures using genetic optimization. J. Struct. Eng. **126**(3), 382–388 (2000). https://doi.org/10.1061/(ASCE)0733-9445(2000)126:3(382)
127. Kameshki, E.S., Saka, M.P.: Optimum design of nonlinear steel frames with semi-rigid connections using a genetic algorithm. Comput. Struct. **79**(17), 1593–1604 (2001). https://doi.org/10.1016/S0045-7949(01)00035-9
128. Camp, C.V., Bichon, B.J., Stovall, S.P.: Design of steel frames using ant colony optimization. J. Struct. Eng. **131**(3), 369–379 (2005). https://doi.org/10.1061/(ASCE)0733-9445(2005)131:3(369)
129. Saka, M.P.: Optimum design of steel frames using stochastic search techniques based on natural phenomena: a review, In: Civil Engineering Computations: Tools and Techniques, Saxe-Coburgh Publications (2007)
130. Kociecki, M., Adeli, H.: Two-phase genetic algorithm for topology optimization of free-form steel space-frame roof structures with complex curvatures. Eng. Appl. Artif. Intell. **32**, 218–227 (2014). https://doi.org/10.1016/j.engappai.2014.01.010
131. Toğan, V.: Design of planar steel frames using teaching–learning based optimization. Eng. Struct. **34**, 225–232 (2012). https://doi.org/10.1016/j.engstruct.2011.08.035
132. Talatahari, S., Gandomi, A.H., Yang, X.S., Deb, S.: Optimum design of frame structures using the eagle strategy with differential evolution. Eng. Struct. **91**, 16–25 (2015). https://doi.org/10.1016/j.engstruct.2015.02.026
133. Aydoğdu, İ., Akın, A., Saka, M.P.: Design optimization of real world steel space frames using artificial bee colony algorithm with Levy flight distribution. Adv. Eng. Softw. **92**, 1–14 (2016). https://doi.org/10.1016/j.advengsoft.2015.10.013
134. Carbas, S.: Design optimization of steel frames using an enhanced firefly algorithm. Eng. Optim. **48**(12), 2007–2025 (2016)
135. Farshchin, M., Maniat, M., Camp, C.V., Pezeshk, S.: School based optimization algorithm for design of steel frames. Eng. Struct. **171**, 326–335 (2018)
136. Degertekin, S.O., Tutar, H., Lamberti, L.: School-based optimization for performance-based optimum seismic design of steel frames. Eng. Comput. 1–15 (2020)
137. Gholizadeh, S., Baghchevan, A.: Multi-objective seismic design optimization of steel frames by a chaotic meta-heuristic algorithm. Eng. Comput. **33**(4), 1045–1060 (2017)
138. Kaveh, A., BolandGerami, A.: Optimal design of large-scale space steel frames using cascade enhanced colliding body optimization. Struct. Multidisc. Optim. **55**(1), 237–256 (2017)
139. Tayfur, B., Yilmaz, H., Daloğlu, A.T.: Hybrid tabu search algorithm for weight optimization of planar steel frames. Eng. Optim. 1–15 (2020)
140. Rajeev, S., Krishnamoorthy, C.S.: Genetic algorithm-based methodology for design optimization of reinforced concrete frames. Comput.-Aided Civ. Inf. **13**, 63–74 (1998). https://doi.org/10.1111/0885-9507.00086
141. Camp, C.V., Pezeshk, S., Hansson, H.: Flexural design of reinforced concrete frames using a genetic algorithm. J Struct. Eng.-ASCE **129**, 105–110 (2003). https://doi.org/10.1061/(asce)0733-9445(2003)129:1(105)
142. Lee, C., Ahn, J.: Flexural design of reinforced concrete frames by genetic algorithm. J. Struct. Eng.-ASCE 129(6), 762–774 (2003). https://doi.org/10.1061/(ASCE)0733-9445(2003)129:6(762)
143. Govindaraj, V., Ramasamy, J.V.: Optimum detailed design of reinforced concrete frames using genetic algorithms. Eng. Optim. **39**(4), 471–494 (2007). https://doi.org/10.1080/03052150601180767
144. Paya, I., Yepes, V., Gonzalez-Vidosa, F., Hospitaler, A.: Multiobjective optimization of concrete frames by simulated annealing. Comput.-Aided Civ. Inf. **23**, 596–610 (2008). https://doi.org/10.1111/j.1467-8667.2008.00561.x
145. Akin, A., Saka, M.P.: Harmony search algorithm based optimum detailed design of reinforced concrete plane frames subject to ACI 318-05 provisions. Comput. Struct. **147**, 79–95 (2015). https://doi.org/10.1016/j.compstruc.2014.10.003

146. Kaveh, A., Sabzi, O.: A comparative study of two meta-heuristic algorithms for optimum design of reinforced concrete frames. Int. J. Civil Eng. **9**(3), 193–206 (2011)
147. Ulusoy, S., Kayabekir, A.E., Bekdaş, G., Nigdeli, S.M.: Optimum design of reinforced concrete multi-story multi-span frame structures under static loads. Int. J. Eng. Technol. **10**(5), 403–407 (2018)
148. Gharehbaghi, S., Moustafa, A., Salajegheh, E.: Optimum seismic design of reinforced concrete frame structures. Comput. Concr. **17**(6), 761–786 (2016)
149. Boscardin, J.T., Yepes, V., Kripka, M.: Optimization of reinforced concrete building frames with automated grouping of columns. Autom. Constr. **104**, 331–340 (2019)
150. Paya-Zaforteza, I., Yepes, V., Hospitaler, A., Gonzalez-Vidosa, F.: CO2-optimization of reinforced concrete frames by simulated annealing. Eng. Struct. **31**, 1501–1508 (2009). https://doi.org/10.1016/j.engstruct.2009.02.034
151. Camp, C.V., Huq, F.: CO2 and cost optimization of reinforced concrete frames using a big bang-big crunch algorithm. Eng. Struct. **48**, 363–372 (2013). https://doi.org/10.1016/j.engstruct.2012.09.004
152. Perea, C., Alcala, J., Yepes, V., Gonzalez-Vidosa, F., Hospitaler, A.: Design of reinforced concrete bridge frames by heuristic optimization. Adv. Eng. Softw. **39**(8), 676–688 (2008). https://doi.org/10.1016/j.advengsoft.2007.07.007
153. Martínez, F.J., González-Vidosa, F., Hospitaler, A., Yepes, V.: Heuristic optimization of RC bridge piers with rectangular hollow sections. Comput. Struct. **88**(5), 375–386 (2010). https://doi.org/10.1016/j.compstruc.2009.11.009
154. Martínez, F.J., González-Vidosa, F., Hospitaler, A., Alcalá, J.: Design of tall bridge piers by ant colony optimization. Eng. Struct. **33**(8), 2320–2329 (2011). https://doi.org/10.1016/j.engstruct.2011.04.005
155. Martinez-Martin, F.J., Gonzalez-Vidosa, F., Hospitaler, A., Yepes, V.: Multi-objective optimization design of bridge piers with hybrid heuristic algorithms. J. Zhejiang Univ. Sci. A **13**(6), 420–432 (2012). https://doi.org/10.1631/jzus.A1100304
156. Sgambi, L., Gkoumas, K., Bontempi, F.: Genetic algorithms for the dependability assurance in the design of a long-span suspension bridge. Comput.-Aided Civil Infrastruct. Eng. **27**(9), 655–675 (2012). https://doi.org/10.1111/j.1467-8667.2012.00780.x
157. García-Segura, T., Yepes, V.: Multiobjective optimization of post-tensioned concrete box-girder road bridges considering cost, CO2 emissions, and safety. Eng. Struct. **125**, 325–336 (2016)
158. Abd Elrehim, M.Z., Eid, M.A., Sayed, M.G.: Structural optimization of concrete arch bridges using genetic algorithms. Ain Shams Eng. J. **10**(3), 507–516 (2019)
159. Atmaca, B., Dede, T., Grzywinski, M.: Optimization of cables size and prestressing force for a single pylon cable-stayed bridge with Jaya algorithm. Steel Compos. Struct. **34**(6), 853–862 (2020)
160. Den Hartog, J.P. (ed.): Mechanical Vibrations. McGraw-Hill, New York (1956)
161. Warburton, G.B.: Optimum absorber parameters for various combinations of response and excitation parameters. Earthq. Eng. Struct. Dyna. **10**(3), 381–401 (1982). https://doi.org/10.1002/eqe.4290100304
162. Sadek, F., Mohraz, B., Taylor, A.W., Chung, R.M.: A method of estimating the parameters of tuned mass dampers for seismic applications. Earthq. Eng. Struct. Dyn. **26**(6), 617–636 (1997). https://doi.org/10.1002/(SICI)1096-9845(199706)26:6
163. Hadi, M.N., Arfiadi, Y.: Optimum design of absorber for MDOF structures. J. Struct. Eng. **124**(11), 1272–1280 (1998). https://doi.org/10.1061/(ASCE)0733-9445(1998)124:11(1272)
164. Marano, G.C., Greco, R., Chiaia, B.: A comparison between different optimization criteria for tuned mass dampers design. J. Sound Vibr. **329**(23), 4880–4890 (2010). https://doi.org/10.1016/j.jsv.2010.05.015
165. Singh, M.P., Singh, S., Moreschi, L.M.: Tuned mass dampers for response control of torsional buildings. Earthq. Eng. Struct. Dyn. **31**(4), 749–769 (2002). https://doi.org/10.1002/eqe.119
166. Desu, N.B., Deb, S.K., Dutta, A.: Coupled tuned mass dampers for control of coupled vibrations in asymmetric buildings. Struct. Control Health Monit. **13**(5), 897–916 (2006). https://doi.org/10.1002/stc.64

167. Leung, A.Y.T., Zhang, H.: Particle swarm optimization of tuned mass dampers. Eng. Struct. **31**(3), 715–728 (2009). https://doi.org/10.1016/j.engstruct.2008.11.017
168. Leung, A.Y., Zhang, H., Cheng, C.C., Lee, Y.Y.: Particle swarm optimization of TMD by non-stationary base excitation during earthquake. Earthq. Eng. Struct. Dyn. **37**(9), 1223–1246 (2008). https://doi.org/10.1002/eqe.811
169. Bekdaş, G., Nigdeli, S.M.: Estimating optimum parameters of tuned mass dampers using harmony search. Eng. Struct. **33**(9), 2716–2723 (2011). https://doi.org/10.1016/j.engstruct.2011.05.024
170. Bekdaş, G., Nigdeli, S.M.: Mass ratio factor for optimum tuned mass damper strategies. Int. J. Mech. Sci. **71**, 68–84 (2013). https://doi.org/10.1016/j.ijmecsci.2013.03.014
171. Nigdeli, S.M., Bekdaş, G.: Optimum tuned mass damper design in frequency domain for structures. KSCE J. Civil Eng. **21**(3), 912–922 (2017)
172. Nigdeli, S.M., Bekdaş, G.: Optimum tuned mass damper design for preventing brittle fracture of RC buildings. Smart Struct. Syst. **12**(2), 137–155 (2013)
173. Bekdaş, G., Nigdeli, S.M.: Metaheuristic based optimization of tuned mass dampers under earthquake excitation by considering soil-structure interaction. Soil Dyn. Earthq. Eng. **92**, 443–461 (2017)
174. Farshidianfar, A., Soheili, S.: Ant colony optimization of tuned mass dampers for earthquake oscillations of high-rise structures including soil–structure interaction. Soil Dyn. Earthq. Eng. **51**, 14–22 (2013). https://doi.org/10.1016/j.soildyn.2013.04.002
175. Farshidianfar, A.: ABC optimization of TMD parameters for tall buildings with soil structure interaction. Interact. Multiscale Mech. **6**, 339–356 (2013). https://doi.org/10.12989/imm.2013.6.4.339
176. Nigdeli, S.M., Bekdas, G.: Optimum tuning of mass dampers employing flower pollination algorithm in frequency domain. In: 4rd International Conference on Earthquake Engineering and Seismology (4ICEES), 11–13 October 2017, Eskişehir, Turkey (2017)
177. Nigdeli, S.M., Bekdaş, G., Yang, X.S.: Optimum tuning of mass dampers by using a hybrid method using harmony search and flower pollination algorithm. In: International Conference on Harmony Search Algorithm, pp. 222–231. Springer, Singapore, February 2017
178. Niğdeli, S.M., Bekdaş, G.: Teaching-learning based optimization for parameter estimation of double tuned mass dampers. Challenge **3**(2), 90–95 (2017)
179. Alizadeh, H., Lavasani, H.H.: TMD parameters optimization in different-length suspension bridges using OTLBO algorithm under near and far-field ground motions. Earthq. Struct. **18**(5), 625–635 (2020)
180. Salcedo-Sanz, S., Camacho-Gómez, C., Magdaleno, A., Pereira, E., Lorenzana, A.: Structures vibration control via tuned mass dampers using a co-evolution coral reefs optimization algorithm. J. Sound Vibr. **393**, 62–75 (2017)
181. Yucel, M., Bekdaş, G., Nigdeli, S.M., Sevgen, S.: Estimation of optimum tuned mass damper parameters via machine learning. J. Build. Eng. **26**, 100847 (2019)
182. Arfiadi, Y., Hadi, M.N.: Passive and active control of three-dimensional buildings. Earthq. Eng. Struct. Dyn. **29**(3), 377–396 (2000)
183. Pourzeynali, S., Lavasani, H.H., Modarayi, A.H.: Active control of high rise building structures using fuzzy logic and genetic algorithms. Eng. Struct. **29**(3), 346–357 (2007)
184. Katebi, J., Shoaei-parchin, M., Shariati, M., Trung, N.., Khorami, M.: Developed comparative analysis of metaheuristic optimization algorithms for optimal active control of structures. Eng. Comput. 1–20 (2019)
185. Kayabekir, A.E., Bekdaş, G., Nigdeli, S.M., Geem, Z.W.: Optimum design of PID controlled active tuned mass damper via modified harmony search. Appl. Sci. **10**(8), 2976 (2020)
186. Zamani, A.A., Tavakoli, S., Etedali, S.: Fractional order PID control design for semi-active control of smart base-isolated structures: a multi-objective cuckoo search approach. ISA Trans. **67**, 222–232 (2017)
187. Etedali, S., Zamani, A.A., Tavakoli, S.: A GBMO-based PIλDμ controller for vibration mitigation of seismic-excited structures. Autom. Constr. **87**, 1–12 (2018)

188. Ulusoy, S., Niğdeli, S.M., Bekdaş, G.: Optimization of PID controller parameters for active control of single degree of freedom structures. Challenge **5**(4), 130–140 (2019)
189. Ulusoy, S., Bekdas, G., Nigdeli, S.M.: Active structural control via metaheuristic algorithms considering soil-structure interaction. Struct. Eng. Mech. **75**(2), 175–191 (2020)
190. Ulusoy, S., Nigdeli, S.M., Bekdaş, G.: Novel metaheuristic-based tuning of PID controllers for seismic structures and verification of robustness. J. Build. Eng. **33**, 101647 (2020)

Artificial Intelligence and Machine Learning with Reflection for Structural Engineering: A Review

Melda Yücel, Sinan Melih Nigdeli, and Gebrail Bekdaş

Abstract In recent years, artificial intelligence (AI) and one of the sub-fields for it, which is machine learning, became significant in numerous study disciplines. During the determination of some results or finding of required parameters/information, these technologies provides many advantages and easiness for the usage of people such as researchers, designers, engineers, inventors and each kind of person dealt with them in terms of ease of use, saving of time, besides cost and efficiency for effort. In this chapter, a comprehensive research was presented intended for AI and machine learning technology together with their historical evaluation and applications. Also, a review of frequently-used machine learning techniques were imparted. As addition to these, many applications respect to some fields and also several studies of the structural engineering area were explained detailly. Furthermore, prediction studies carried out in structural engineering were demonstrated with all stages. In this regard, all of studies performed with the usage of different machine learning techniques were given in six sub-headings. As it can be understood from many developments, AI and machine learning technologies and application of them are pretty significant in terms of providing of beneficial situations and the usage of them will be more substantial and remarkable day by day.

Keywords Machine learning · Artificial intelligence · Review · Prediction applications · Structural engineering

M. Yücel · S. M. Nigdeli (✉) · G. Bekdaş
Department of Civil Engineering, Istanbul University-Cerrahpaşa, Avcılar, 34320 Istanbul, Turkey
e-mail: melihnig@istanbul.edu.tr

M. Yücel
e-mail: melda.yucel@yahoo.com.tr

G. Bekdaş
e-mail: bekdas@istanbul.edu.tr

S. M. Nigdeli et al. (eds.), *Advances in Structural Engineering—Optimization*, Studies in Systems, Decision and Control 326,
https://doi.org/10.1007/978-3-030-61848-3_2

1 Introduction

Human, who is an alive, which thinks, can learn and by this means may experience, can find a way out by bringing different approaches to encounter problems thanks to these experiences, and at the same time can update own knowledge by obtaining new inferences with experiences. There exactly, a research area, which is by the name of artificial intelligence (AI), which is banded out by benefitting from this magnificent evolution feature belonging human, reached by separating to numerous branches until nowadays.

According to the expression specified of McCarthy, for AI, it is remarked that the feature, which is an engineering and science branch intended for generation of smart machines, especially smart computer programs [1]. In this regard, as it is understood from its name, it appears that actually this intelligence, which is artificial, can be thought as a virtual function created based on the usage of human intelligence, functions and working principles by a machine, software or tool.

However, machine learning, which is another area that it is frequently compared with AI, realizes in line with the usage of the mentioned intelligence. Namely, in this regard, the topic of learning of machines, can be considered as to ably to carry out some functions, which are required knowledge and experience without any supervisor from outside, such as determination of the route correctly by a robot, detection of areas, which should be cleaned, or winnow out of spam mails autonomously by a software, by using of intelligence, which is artificial.

The mentioned learning means to gradually sense of any ensured knowledge via existing intelligence, to understand, and to answer by forming reaction (deduction of information) and gain an experience. On the other hand, for learning, Grebow mentioned from that expression: "Real learning is an ability providing of internalization/embracing of things that they are known by us and be able to make, and adapt these factors to continuously changing conditions" [2]. Within this context, it is understood that it has important to realize learning by any member equipped with artificial intelligence, has importance in terms of process by sensing the gained information correctly and turn into experience, internalize and gaining the ability of answer-producing by adapting to different statements. But, features and realization principles of learning actions belonging to biological lives in real nature, especially humans, must be understood correctly in order to be provided by learning by the mentioned factor successfully. Indeed, it is also consulted to knowledge of some fields such as engineering, besides this information, in terms of performing of learning [3]. Besides of these, especially in nowadays, an area of deep learning, which is a more complex sub-field of the mentioned machine learning, is frequently used, too. In the direction of given definitions and made explanations, Fig. 1 may be shown for the connections and relationships between AI, machine and deep learning, clearly.

Fig. 1 Differences and Relations Between Artificial Intelligence, Machine and Deep Learning [4]

2 Evolution of Artificial Intelligence and Transition Process to Machine Learning

According to the definition made by McCarthy, AI is: "The usage of computers intended for understanding of humans, and in this context, science and engineering of generating of smart computer programs and machines" [1]. Thereafter, as we touched on previously, machine learning is a sub-field of AI field that it is focused on independent computers and machines by benefiting from many disciplines such as math, engineering, psychology, biology and genetic science, linguistics, and can be accepted as inversion to a humanlike state by transferring of a virtual intelligence to these kinds of formations.

Also, in the present days, it is seen that different techniques are advised on operations of smart devices, which will be improved through development of technology notably together with get diversified of work principles, besides that these techniques enhance day by day. In initial times more generally, with learning activities and reactions of a human, later on, in this process, which is based on benefiting from structure and collaboration mechanism of brain and neural system, in nowadays, it is remarked to be possible of solutions for more complex, big and variety problems such as that a system predicts by processing of a very complex image like a scanner, determination of who belonging of listened voices, detection which language has letters in a scanned text. However, even so, realizing and be applicable in real life of such as these developments are becoming possible as from the middle of the nineteenth century.

With this purpose, until from past to present, it can be seen in Table 1 that a chronologic schema, which is presented intended for whole activities such as books, articles, which are published with the aim of providing of contributions to areas of AI and machine learning in the meaning of scientific literature, and representations, conferences besides robots and machines, which are rendered to intelligent and smart, robots, which think smartly, devices improved to carry out of activities, which can be done only by any human.

3 Review for Machine Learning Techniques

Human, who has a specific-to-own intelligence and understanding capacity, possess very different alternatives for any topic, information or occupation wanted to learn. As reason of this, it can be shown that human can be trained thanks to having this intelligence and also different sub-formations belonging to it. In this direction, human experiences by himself/herself and can increase of knowledge, by using own intelligence during learning, or can operate the learning mechanism by depending on other factors, namely intelligence of other humans, and under the guidance of them. Besides of this, in initial, while human needs another intelligence and training, in the process of time, can sustain of learning without need to these due to gain skills and experiences by oneself. Thereunder, learning issue of smart machines, robots or tools, which include a virtual intelligence, realizes with a similar way to learning ways and process of human, who is inspired from intelligence and functions of it. In this regard, machines also can put forward an idea by processing the ensured knowledge of oneself, and can set the connection between information, and at the same time, can need a teacher or supervisor like human to learn about this information when may gain experience.

Various learning approaches are developed for different learning options explained and intended for machine learning concept mentioned in here. These are supervised learning, unsupervised learning and reinforcement learning. Supervised learning is a kind whether can be checked by answer given in response to any information processed in the process of learning. For this respect, in this learning, an observer exists, which can give feedback as true or false for the generated expression for target output. As to in unsupervised learning, this situation is exactly reversed, and there is no observer, which will give consultation about the condition of answer. In this regard, for response, the learner element makes a forecasting with the help of connections, which are determined between processed data by oneself. Also, in reinforcement learning, the case occurs that learner element receives an award or penalty according to true or false decision taken from environment in response to each performed action, and thus makes more correct decisions by gain experience from faults, and the learning is provided [9, 10]. Within this scope, Fig. 2, where is expressed of the mentioned machine learning types and some applications that they can perform with these, can be seen as follows:

Table 1 Important points of artificial intelligence applications during the historical process, and chronological display

Name of invention/development	Property of study	Owners of study	Year	References
Robotics	The term is proposed	Isaac Asimov	1945	[5]
–	A book was published with name as Cybernetics	Weiner	1948	[6]
–	A rule was proved for changing of the weights between neurons on realization of learning	Donald Hebb	1949	[6]
The Turing Test	The mentioned structure, which is used for measurement of smart behaviors, was published in a book named as Computing Machinery and Intelligence	Alan Turing	1950	[1]
SNARC	The first big neural network computer was established	Marvin Minsky	1951	[7]
–	The first gaming program was written for checkers	Arthur Samuel	1952–1962	[8]
Neural Nets and the Brain Model Problem	The mentioned study is a thesis, which is the first publication that it collected whole current results and theorems about neural networks	Marvin Minsky	1954	[9]
Logic Theorist	The first written AI program for a computer	Allen Newell J.C. Shaw Herbert Simon	1956	[10]
Artificial Intelligence	This is demonstrated as a term	John McCarthy	1956	[5]
–	The first operated AI program	John McCarthy	1956	[5]
LISP	It was presented as the first AI programming language in article named as “Recursive Functions of Symbolic Expressions and Their Computation by Machine, Part I”	John McCarthy	1956	[7]

(continued)

Table 1 (continued)

Name of invention/development	Property of study	Owners of study	Year	References
Perceptron	A simple neural network structure	Frank Rosenblatt	1958	[11]
Perception learning rule	It was discovered that a rule managing of how update of connection weights of neural networks	Frank Rosenblatt	1958	[12]
Dendral	A system, which can determine of mass spectrometry structure of organic molecules	Edward Feigenbaum Bruce G. Buchanan Joshua Lederberg Carl Djerassi	1965	[13]
Eliza	A program has a property that it can answer to questions of a subject just as a psychologist	Joseph Weizenbaum	1966	[7]
Shakey	The first mobile robot was improved that it includes AI algorithms	The Stanford Research Institute	1966-1972	[10]
MacHack	A good chess program was developed that it was designed depending on knowledge, at the same time as can take high rank in a tournament as	Richard Greenblatt	1967	[6]
Macsyma	An algebra system, which was benefited for solving of integration problems	Carl Engelman William Martin Joel Moses	1967	[6, 10]
Lunar	The first natural language program, which lets to ask English questions about rock/stone samples by geologists and is used by other people as practically	William Woods	1973	[8]

(continued)

Table 1 (continued)

Name of invention/development	Property of study	Owners of study	Year	References
Mycin	It presents treatment advices for bacterial infections in blood by diagnosing with the usage of expert medical information	Ted Shortliffe	1974	[10, 14]
Backpropagation	Backpropagation method, which is based on approach that network error can spread towards back from middle layer, was discovered	Paul Werbos	1975	[9]
Stanford Cart	The first autonomous vehicle was developed as computer-controlled	–	1979	[5]
HEARSAY-II	It was developed as a speech and speech-understanding system.	Lee Erman Rick Hayes-Roth Victor Lesser Raj Reddy	1980	[6]
TD Gammon	The program for backgammon	Gerald Tersauro	In early of 1990' s	[7]
–	A system was improved to can decide to assign of what a kind of technician by company, when a customer reported the phone problem	Bell Atlantic	1991	[15]
Robocup	An activity organized for playing football by autonomous robots in all around the world	–	1993	[7]
–	Usage of neural networks was started in pattern recognition technology	Bishop Ripley	1995 1996	[15]
Kismet	It is a robot, which has the feature to can simulate of behavior, feeling and appearance of a human via its facial expressions and vocalizing	Cynthia Breazeal	1997	[10]

(continued)

Table 1 (continued)

Name of invention/development	Property of study	Owners of study	Year	References
The Deep Blue	The mentioned program defeated of Garry Kasparov, who was the World champion in own time	IBM	1997	[10]
Nomad	It is a robot, which was improved in order to search of meteorite samples in Antarctica	Carneige Mellon	2005	[6]
–	The first self-driving car moved on roads	Google	2009	[7]
AtomNet	It has a property, which is the first deep neural network developed for designing of medicines	Atomwise	2015	[16]

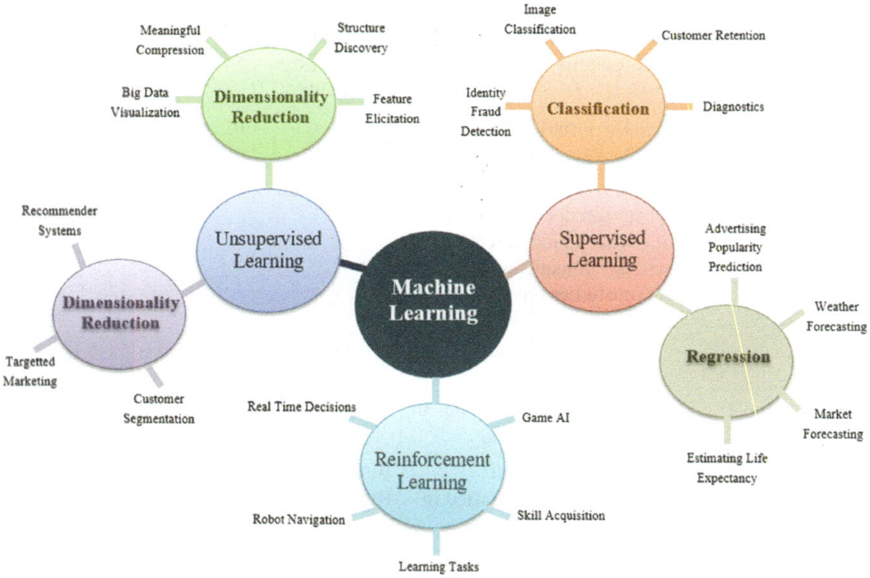

Fig. 2 Learning Schemes of Machines and Applications of Machine Learning [17]

3.1 *Frequently-Used Machine Learning Techniques*

As encountered in applied examples in Sect. 5, some machine learning techniques, which are utilized with the purpose of prediction (such as numerical regression), are consist of mentioned supervised techniques and similar generally. For this reason, in this sub-heading, an outline is presented in terms of a review of major or commonly used techniques within the mentioned AI and machine learning area.

Artificial Neural Networks (ANNs)

Artificial neural networks (ANNs) are a kind of machine learning method that are pretty mostly-used in the topics as classification, clustering, attribute selection/feature extraction, recognition of pattern, image and speech, and deep learning besides generally prediction and forecasting. Also, central neural system of human, working style of it with brain coordinately, and response formation mechanism according to experience are inspired for this method, which is widely consulted in the matter of learning of any information by machines with the help of various and numerous samples, in this context, there are many kinds of developed ANNs.

If we will say briefly from architecture and general working principles of the mentioned method, ANNs contain two layers as input and output in the basic meaning, yet multilayer ones have a layer, where hidden nodes exist, additionally. In these layers, there are nodes where input and output data are represented, and these nodes are corresponded to each parameter.

Any signal, namely information, given from input layer is processed by transfering to hidden layer (if available) firstly, and then output layer. In this duration, it is provided that network learns through the use of different techniques. On the other hand, major techniques, which are benefited to learn information and minimize error amount, are backpropagation algorithm, perceptron learning rule, method of least-mean-squares learning, k-means clustering algorithm, adaptive resonance theory, kohonen self-organizing maps, vector quantization, hebbian learning and hopfield auto-associator etc. [10, 15].

Support Vector Machines (SVM)

During 1990s, support vector machines were developed by Vapnik that this method was actually proposed for the binary classification problems constituted from two different labels/classes. These vector machines split of data space into two parts for each class by using a plane/surface. This surface is found by SVM with the usage of vectors, which are the main/basic recorded training groups, and margins, namely distances (total) between splitting plane and vectors. In here, splitting surface is determined as optimally by providing of maximum margins. By following this operation, by considering of classified data, it is determined that unclassified ones belong to which classes. This case can be seen in Fig. 3 [11, 18–20].

On the other hand, this methodology also can be used for regression problems. In this respect, data are not labelled with a class in a similar way classification application, but their numerical values are predicted [19].

k-Nearest Neighbor Classifier

As understood from the name, k-nearest neighbor (kNN) classifier is one of the basic methods that it is used in classification problems, at the same time, can be also utilized for regression that it has a similar principle with numerical estimation.

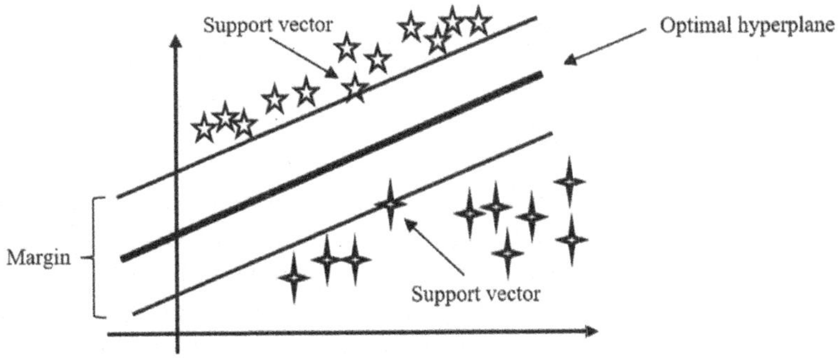

Fig. 3 The Structure and Modeling of SVMs [20]

kNN is a method that it was suggested by T.M. Cover and P.E. Hart for the first time and provides to realize of the case, which is determination/description of sample or prediction of its value through classification of the other samples, which are the closest to itself [21, 22]. In other words, this method classifies data by estimating of the closest to itself from a dataset (values are known) for any data, which have values that are not known. In here, k value, which is predefined by any user, indicates the number of neighbors as k [10, 12].

In this method, firstly, the distance between the new sample with other ones in the searching space is calculated. The closest sample, namely nearest neighbor, is class, which will be considered for placing of new sample. Calculating this distance between two sample/pattern can be generated with different formulations such as Euclidean, Chebyshev, Manhattan, and Minkowski distances [15, 22].

Decision Trees

Decision trees are the technique, which have the most widely used among the classification models due to reasons such as that set up is cheap, interpreted is easy, easily accommodate with database systems, reliability of them is high, and visually-understood is easy, too. This technique constitutes of leaves and branches, besides likes to a tree. As examples of this technique: ***TDIDT*** (Top-Down Induction of Decision Trees) algorithm is a decision tree model, which has been recognized since middle of 1960s and comprises a basis to different classification-based methods such as ID3, C4.5. As said, for this method that is used for classification, it is possible that it uses by transform of continuous attributes to categorical [21]. Moreover, another method is ***CART*** algorithm proposed by L. Breiman, J. Friedman, R. Olshen, and C. Stone in the year of 1984 and it is given a name from words as "Classification and Regression Trees" and represents these two techniques as classification and regression. This algorithm generates only binary trees, namely a tree structure separating in two directions from one node. For this reason, CART cannot process groups more than two within an attribute [11, 23, 24]. On the other side, in 1986, J. Ross Quinlan developed an algorithm known as ***ID3*** (Iterative Dichotomiser 3). This algorithm is remarkable respect to representation of learned information, potential/capacity for processing of noisy data and prediction of candidate concepts. For any attribute, it can divide of training samples to subsets, which have a common value in terms of this attribute. To do this, first, it selects the attribute, which will be tested, and then uses this test to divide of sample set. Following, the algorithm generates a tree for each part again and again [11, 14]. Likewise, in 1993, ***C4.5*** algorithm developed by Ross Quinlan, is an advanced version of ID.3 and takes place among classification methods. While comparing with CART, it is not limited with binary splittings and also can perform multiple ones. Therefore, separate branches are created for each attribute, too. Decision tree structure belonging to ***C5*** algorithm, which is the newer version of C4.5, is actually likes to the tree used by C4.5 algorithm, but tests show some differences [7, 15, 23]. Also, ***SPRINT*** (Scalable, Parallelizable Induction of Decision Trees) algorithm was developed by John Shafer, Rakeeh Agrawal and Manish Mehta in 1996. This algorithm is a decision-tree based classification algorithm that it is quick, scalable and remove all memory constraints. At the same time, the algorithm was designed in the way giving a chance to working together for generating of only one consistent model by many processors and that can parallelize easily. The other property of it is to provide that decision trees generated from very big datasets. Also, it can constitute of tree structures by using the properties have both categorical and continuous values [11, 25]. ***QUEST*** is a decision tree algorithm developed by Wei-Yin Loh and Yu-Shan Shih in 1997 that it was taken the name from initials of words of "Quick, Unbiased, and Efficient Statistical Trees". The algorithm can be used as intended for classification problems, besides that it has binary tree structure. As to the strongest sides of this algorithm, it can make calculation quickly, and variables' selection is unbiased [26].

Bagging

Bagging is an ensemble model that it uses more than one classification or prediction techniques and together evaluate of the results ensured via all of them.

Bagging (bootstrap aggregating) method was investigated by Breiman in the year of 1996. In this method, this process performs in this way that each classifier/predictor see the sample, which's value is wanted to found, and determines a prediction for this sample as parallelizing, namely simultaneously. And following, a main classifier/predictor detects the final result, which has the highest vote/rate, by dealing with of all predictions provided from each sub-techniques. Moreover, this tackled sample can be the same corresponding to any classifier due to that the sample can be occur with re-election (reloading of this sample to the main dataset) from the main dataset. As a summation, purpose in this method is to generate of an effective and successfull major predictor by evaluating with collect of the results provided from all sub-classifiers/predictors respect to improve of accuracy for output [12, 27].

Boosting

In 1990, an ensemble-based approach, which is called as boosting algorithm, was improved by Schapir [12]. Although, boosting also makes a training of sample data with the help of many different trees, like bagging, it performs this training step by step. Namely, each tree performs the training by using of information ensured from training of previous tree, and more weights are given to samples have faulty results [18]. And then, in this approach, samples are selected according to the proportion with them weights [27]. If it is required to explain about this situation more clearly, working mechanism of this method is related to selection of the major/basic method as classifier or predictor. For example, it depends on a principle such as that diagnosing for any disease that is made by more than one doctor, not only one, and the ultimate result is ensured by evaluated of each diagnose in the result of a weight determined according to accuracy of previous diagnoses made by doctors. Accordingly, in the model, it is taken care to groups classified as wrong, besides that a weight is given to each training group [11].

Random Forest

Random forest algorithm is an ensemble-based method that it was developed by Breiman (2001). This method is actually constituted from an application executed on the basis of bagging approach to decision trees [12, 27]. In this regard, random forest approach is a model, which is enhanced by combining of numerous random trees. As it is recognized from the name of this methodology, has a structure similar to a forest due to many trees exist inside, and information obtained from each classifier tree is also considered by evaluating of them together, when output prediction is tried to find. This demonstration is expressed with Fig. 4.

Also, as we can understand from Fig. 4, the ultimate prediction result is provided as an average of whole predictions ensured from many sub-classifiers, which go through a training process.

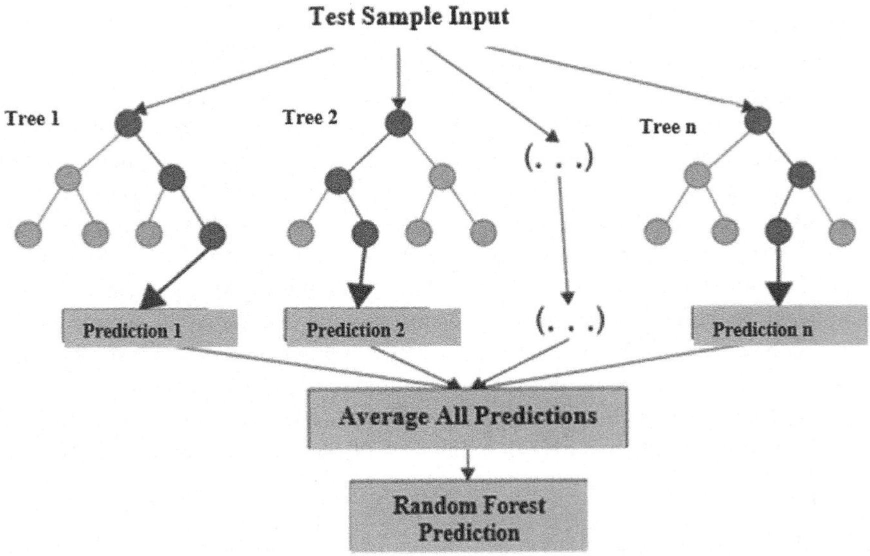

Fig. 4 Structure of Random Forest Containing Many Trees [28]

4 Artificial Intelligence and Machine Learning Applications

As also mentioned in more previous sections, machine learning, which is towards the usage of AI, can be carried out via very different ways, and smart systems, software, and tools can be developed in multiple study areas because of this variety. By this means, it takes the lead to performing of activities in many fields of daily life, scientific environments and even space as quicker, saving time and effort.

In this regard, a summary is presented in Table 2 intended for applications in several areas such as healthcare, technology and science, security etc. performed via AI technologies together with machine learning methods.

4.1 Prediction Applications on Structural Engineering Via Machine Learning Techniques

Engineering is one of the disciplines, where existing of pretty indeterminacies and that some situations are arranged via assumptions to stay on the safe side. In this respect, making required calculations must be done correctly for any design, which will be presented in engineering, controlling of all conditions by reviewing, and thus reaching of it to the fittest model, are some of main objectives. Nevertheless, in design process, one of the basic targets of engineering science is known as the most correct and efficient usage of the time and available sources. In this direction, the mentioned cases give task to an engineer, who will take responsibility in the present process, that it is to generate a result, which can provide the most relevant cost by using of time efficiently, besides making a design as safe, aesthetics, usable and appropriate.

When all these expressions are taken into consideration, it is required that the used classical computer and analysis programs have an optimal structure, and all these results are made possible to be accessed. In this respect, in the process of designing and operation of any engineering system, the most correct and proper methods, which can provide these advantages, should be consulted.

As mentioned, a waiting is also at stake for obtaining of outputs belonging to designing process due to that many engineering problems have long-timing analysis phases. Even, sometimes, this duration can last for months. However, just like a human, what would happen, if a smart and intelligent device exists that it can make comment by think, and can inference for different situations? Yes, as answer, studies from past times until today, and progressing made through these studies, are demonstrated to us that usage of such a type system is possible, and provides many advantages. In this scope, as follows, a literature summary is also given for structural engineering area. In the last section, six different structural design examples that are investigated via machine learning models (more intended for numerical prediction), are presented.

Table 2 Applications within AI with machine learning

Area	Applications	References
Science, technology and industry	• Determining of kind and location of mining beds in a site • Modelling of working mechanism of human brain • Modelling of different points within space and to determining of hierarchies between them • Describing of numerous genes in each new genome • Determining the construction of organic compounds • Monitoring of oil spill • Forecasting of the future electricity demand for power • Making maintenance plan for preventing of harms, which may occur in parts such as motors or generators	[14, 15, 19]
Pattern recognition	• Classification of images • Matching by finding of organs such as eye and nose belonging to a specific face • Recognition of complex patterns such a natural language text or chess position • Developing of intelligent monitoring and interference systems intended for natural disasters such as earthquakes, floods and droughts etc.	[1, 29]
Speech recognition and text mining	• Ensuring of searching of query • Modelling of sentences • Analyzing of handwriting • Detecting of keywords within the texts • Advising of documents that users can enjoy them • Transforming of data ensured with the combining of speeches, to voices • Using of speech recognition to be given of flight information such as flight numbers and city names	[1, 5, 16, 19]

(continued)

Table 2 (continued)

Area	Applications	References
Medicine and healthcare	• Detecting of benign and malignant tumors by making feature extraction automatically • Determining of whether be effective a specific medicine for patient • Diagnosing of disease from samples validated via biopsy • Recognition of cancerous cells in diseases such as breast cancer • Classification of images to certain diagnosis categories • Prediction of sensibility towards cancer, risk of relapse of cancer and survival situation during cancer disease • Separating of cancer patients according to risk levels • Developing of flexible, wearable, biocompatible monitoring systems as physiological • Generating of intelligent treatment programs	[16, 29–32]
Marketing and banking	• Determining of whether give to credit to customers by a company • Presenting of personalized gifts or offers to users after online shopping • Predicting what kind of people might buy which items • Estimating of likely customers in future	[15, 19]
Transportation	• Constructing of autonomous robots, which can work independent, besides robotic taxis • Detecting of pedestrians in traffic • Prediction of flowing of traffic	[7, 16]
Safety and Security	• Transforming of security tools to intelligent ones in significant public areas • Detection of criminal identity • Construction of intelligent monitoring, early warning systems and control system for public safety • Determining an action related with or intrusion or an unauthorized operation and preventing of these cases before not get damaged	[17, 29, 30]

In the study performed by Atici [33], ANNs and multiple regression analysis are utilized with the aim of forecasting of compressive strength for concrete mixtures according to properties and values of additive materials found by using of ribaunt numbers of non-destructive testing and values of impulse velocity within the changeable curing times where different rates fly ashes and blast furnace cinder exist. In the same way, both ANNs and multiple regression analysis are applied to foresee compressive strength of many different high-strength concrete samples (contain specific rates of additives as nano silica and copper slag) [34]. Additionally, Behnood and Golafshani [35] developed a model, which is designed with the use of both an optimization and machine learning method. In this regard, grey wolf optimization algorithm and ANNs are combined to determine of the compressive strength values of concrete mixtures, which are generated with a specific rate of silica fume instead of Portland cement.

On the other hand, structural members can be designed through obtained outputs from AI technologies, directly. According to this, in a study, besides support vector regression (SVR), also a kind of ANNs was used to observe its success about estimation for strength of shear of deep beams with two different creations as concrete and pre-stressed concrete [36]. Salazar et al. [37] handled a study connected with the determination of how arch dams behave depending on displacement and leakage. In here, five different techniques within machine learning as random forest, boosted regression trees, neural networks, support vector machines and multivariate adaptive regression splines are operated by comparing with classical statistic methods. On the other hand, two different techniques are applied as multiple linear regression and multivariate adaptive regression splines in a study, which is carried out with the aim of modelling the behavior of a rotary brace damper structure in a steel frame structure under different loading conditions, by detecting the strength of damper [38]. Asteris et al. [39] forecasted the values of the relative displacements belonging to storeys in any structure. To perform of this application, they used ANNs as feed-forward by trained via Krill-Herd optimization technique, besides the other ones such as linear regression and backpropagation neural networks to be compared with each other. Furthermore, Yang et al. [40] predicted the values of a parameter as a dynamic increase factor belonging to steel fiber-reinforced concrete structures by using random forest algorithm, which is combined of firefly algorithm. As differently, a study is realized to observe critical failure mode including as flexural, flexural with shear, and only shear for circular columns of reinforced concrete bridge structures by Mangalathu and Jeon [41]. With this aim, many machine learning methods are benefited like k-nearest neighbor, decision tree, random forest, ANNs and so. Load-carrying capacity with mode failure property for concrete of beam-column joint are established via extreme learning machine method [42].

Addition to these, also, design parameters and application decisions can be determined intended for fiber reinforced design for structural members such as beams, columns, masonry walls etc. with the help of machine learning techniques. For example, Lee and Lee [43] conducted a study that it includes the determination of shear strength of flexural members reinforced via FRP, which do not have any stirrups. By this means, it is provided that the effective parameters from all ones (used as input) such as effective depth, shear span-to-depth ratio, and compressive strength of concrete are found for prediction of strength, and finally, some formulas are developed intended for application by using ANNs. Furthermore, Cascardi et al. [44] are benefited from ANNs with the purpose of determination of compressive strength intended for circular concrete columns, which are reinforced by using fiber reinforced polymers. Yaseen et al. [45] also carried out a study related to estimation of shear strength of concrete beams reinforced with steel fibers. In this scope, they benefited from SVR, besides ANNs by both hybridized with particle swarm optimization algorithm.

Nevertheless, in structural engineering field, there is the other working branch and topic, which is related to management or operation of a construction, too. One of applied studies that is conducted for prediction of energy usage for a building according to properties consisting of material thickness and capability of thermal insulation, through 180 different data. With this objective, both extreme machine learning and ANNs methods are dealt, besides genetic programming [46]. On the other respect, also Yücel and Namli [47], in the year of 2017, analyzed four machine learning methods intended for estimation of building performance/efficiency classes, which represent the effectiveness level of heating and cooling energies with the usage of 127 real structures. Here, ANNs, classifiers as Bayesian and k-nearest neighbor and C4.5 algorithm were operated for heating energy; besides that only ANNs are handled for cooling energy.

5 An Overview for Structural Engineering Applications Via Machine Learning

5.1 Prediction of Optimum Tuned Mass Damper (TMD) Parameters

In this application developed by Yücel et al. [48], a shear building model, which has a TMD on top of structure, can be seen in Fig. 5. The mentioned structure is subjected to a ground motion acceleration ($\ddot{x}_g$) and so response of structure is indicated as x, which means displacement of floor, besides that TMD displacement is expressed as x_d.

On the other hand, this building is a multi-degree of freedom (MDOF) structure, cause of the addition of TMD. For this reason, structural design parameters (m, k and c) should be formulized via matrices of mass (M), stiffness (K) and damping (C) as Eqs. (1)–(3). Also, TMD optimum design parameters (T_d and ξ_d) are taken place in Eqs. (4) and (5), respectively.

$$M = \begin{bmatrix} m & 0 \\ 0 & m_d \end{bmatrix} \tag{1}$$

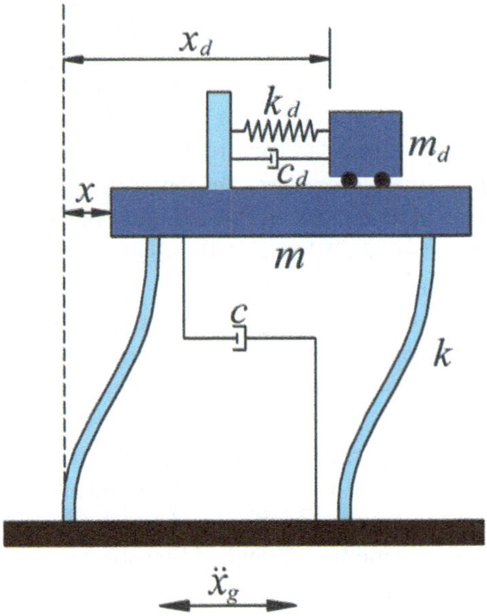

Fig. 5 The SDOF Structure with TMD [48]

$$K = \begin{bmatrix} k + k_d & -k_d \\ -k_d & k + k_d \end{bmatrix} \quad (2)$$

$$C = \begin{bmatrix} c + c_d & -c_d \\ -c_d & c + c_d \end{bmatrix} \quad (3)$$

$$T_d = 2\pi\sqrt{\frac{m_d}{k_d}} \quad (4)$$

$$\xi_d = \frac{c_d}{2m_d\sqrt{\frac{k_d}{m_d}}} \quad (5)$$

The building and TMD design constants are mass, stiffness and damping values expressed as m, k, c for structure and m_d, k_d and c_d for TMD, respectively.

On the other hand, the aim of structural optimization problem is the minimization of the maximum value belonging to transfer function regard to acceleration of structure through of determination of optimum TMD parameters. In this respect, transfer function (TF(ω) as a function of frequency-ω) is expressed with Eq. (6) and objective function (amplitude of TF as f) can be formulized as Eq. (7):

$$TF(\omega) = \begin{bmatrix} TF \\ TF_d \end{bmatrix} = \left[-M\omega^2 + C\omega j + K\right]^{-1} M\omega^2\{1\} \quad (6)$$

$$f = 20\mathrm{Log}_{10}|\max(TF)| \quad (7)$$

Table 3 Period and mass ratio combinations used in training

T_s (s)	0.1	0.3	0.5	0.7	1.0	1.5	2.0	3.0	4.0	5.0
μ	0.01	0.03	0.05	0.08	0.10	0.15	0.20	0.25	0.30	0.40

In the first process of prediction for optimum TMD parameters ensured via flower pollination algorithm (FPA) proposed by Yang [49], artificial neural networks (ANNs) were used and trained for detection of these values directly and without an optimization application. For this aim, a SDOF structure were handled by using different combinations for each period (T_s) and mass ratio (μ) of structure shown in Table 3.

Furthermore, in prediction process, input parameters are considered as T_s and μ, besides output parameters are T_d and ξ_d. After the training of ANNs, a test model, which has 10 new couples for structure period-mass ratio that they are not exist in training data and determined randomly, is proposed due to investigate the prediction success and performance. In here, objective function and error rates according to FPA were calculated by using ensured prediction results of optimum TMD parameters (Table 4).

In the second part of prediction application, a new second test model was generated via T_s values as Table 1 by dividing as 0.01 for μ by means of success of prediction and small error for model. In this stage, initial developed ANN model was used for prediction of new test couples and various graphics were created by benefiting from this model regard to the mentioned test data. Also, some equations are proposed via curve fitting method for optimum TMD parameters as T_d, $\xi_{d,\,opt}$ and f_{opt} (TMD frequency ratio) according to mass ratio and controlled for SDOF, besides MDOF by comparing with Sadek et al. [52] idealization method. These equations are linear

Table 4 Optimum results and predictions for design parameters of test samples

T_s (s)	μ	ANN Model			FPA			Error Rates (%)		
		T_d (s)	$\xi_{d,\,opt}$	f (TF) (dB)	T_d (s)	$\xi_{d,\,opt}$	f (TF) (dB)	T_d (s)	$\xi_{d,\,opt}$	f (TF) (dB)
1.750	0.030	1.7913	0.1104	13.5661	1.7904	0.1095	13.5447	0.0503	0.8222	0.1580
0.200	0.180	0.2266	0.2482	8.6806	0.2266	0.2478	8.6791	0.0147	0.1688	0.0179
0.100	0.065	0.1072	0.1623	12.0867	0.1049	0.1594	11.5984	2.2354	1.8230	4.2102
0.600	0.080	0.6394	0.1753	11.1307	0.6358	0.1705	11.0224	0.5534	2.8398	0.9824
2.300	0.100	2.4720	0.1896	10.3816	2.4668	0.1881	10.3377	0.2121	0.7968	0.4241
1.000	0.015	1.0117	0.0811	15.1436	1.0100	0.0836	15.0958	0.1691	2.9439	0.3164
3.500	0.200	4.0067	0.2495	8.3188	4.0001	0.2603	8.2997	0.1639	4.1322	0.2296
0.150	0.250	0.1787	0.2855	7.7992	0.1774	0.2837	7.7091	0.7372	0.6206	1.1685
0.280	0.400	0.3585	0.3349	6.3972	0.3360	0.3280	6.3404	6.7027	2.1077	0.8944
5.000	0.035	5.1272	0.1140	12.8309	5.0779	0.1084	12.7371	0.9718	5.1459	0.7367
							Mean	1.1811	2.1401	0.9138

(Eqs. 8–9), polynomial (Eqs. 10–11) and exponential (Eqs. 12–13) expressions for ξ_d and f_{opt} as follows:

$$\xi_{d,opt} = 0.5673\mu + 0.1235 \tag{8}$$

$$f_{opt} = -0.6438\mu + 0.9966 \tag{9}$$

$$\xi_{d,opt} = -54.673\mu^4 + 54.639\mu^3 - 19.274\mu^2 + 3.2302\mu + 0.0237 \tag{10}$$

$$f_{opt} = -249.91\mu^5 + 400.09\mu^4 - 208.03\mu^3 + 43.801\mu^2 - 4.1453\mu + 1.0675 \tag{11}$$

$$\xi_{d,opt} = 1.1258e^{2.8573\mu} \tag{12}$$

$$f_{opt} = 1.0038e^{-0.747\mu} \tag{13}$$

In the third stage, for test combinations belonging to SDOF structure, developed equations and some methods from literature were compared for objective function, optimum damping and frequency ratio as in Tables 5, 6 and 7, respectively.

On the other hand, two different MDOF structures were handled as 10 and 40 storey buildings (Fig. 6) and analyzed under earthquake recordings in FEMA P-695 [55] as far-field (Table 8).

In Tables 9 and 10, results for 10 storey structure, which has 360 tons, 650 MN/m stiffness and 6.2 MNs/m damping coefficient for each storey (Singh et al. [56]), were

Table 5 Minimum transfer function predictions for SDOF structures

		Den Hartog [50]	Warbrton [51]	Sadek et al. [52]	Leung and Zhang [53]	Linear Equation [48]	Polynomial Equation [48]	Exponential Equation [48]
T_s (s)	μ	*f* (TF)						
1.750	0.030	14.0587	14.2548	15.5528	14.9337	13.8870	13.5955	13.9493
0.200	0.180	9.6945	10.4208	10.9300	11.7059	8.7949	9.2198	8.9605
0.100	0.065	12.2607	12.6500	13.7799	13.3847	11.6518	12.0725	11.7025
0.600	0.080	11.7440	12.1871	13.2353	12.9593	11.0765	11.4098	11.1279
2.300	0.100	11.1464	11.6711	12.5920	12.5186	10.3910	10.5839	10.4817
1.000	0.015	15.4922	15.6105	16.8852	16.2817	16.1544	16.2026	16.1755
3.500	0.200	9.3472	10.1937	10.6105	11.5010	8.4678	8.6709	8.6402
0.150	0.250	8.8661	9.7376	9.9223	11.6586	7.9227	7.8941	8.0130
0.280	0.400	7.7161	8.8354	8.4328	12.8164	6.8316	6.5578	6.7719
5.000	0.035	13.3659	13.4803	14.7582	13.7873	13.3798	12.9310	13.3688

Table 6 Optimum damping ratio predictions for SDOF structures

		Den Hartog [50]	Warburton [51]	Sadek et al. [52]	Leung and Zhang [53]	Linear Equation [48]	Polynomial Equation [48]	Exponential Equation [48]
T_s (s)	μ	$\xi_{d,opt}$						
1.750	0.030	0.0853	0.0857	0.2192	0.0853	0.1405	0.1047	0.1371
0.200	0.180	0.1953	0.2001	0.4329	0.1977	0.2256	0.2419	0.2104
0.100	0.065	0.1235	0.1246	0.2940	0.1237	0.1604	0.1663	0.1515
0.600	0.080	0.1361	0.1375	0.3185	0.1364	0.1689	0.1845	0.1581
2.300	0.100	0.1508	0.1527	0.3470	0.1514	0.1802	0.2032	0.1674
1.000	0.015	0.0608	0.0609	0.1708	0.0607	0.1320	0.0680	0.1313
3.500	0.200	0.2041	0.2097	0.4499	0.2071	0.2370	0.2484	0.2228
0.150	0.250	0.2236	0.2315	0.4872	0.2281	0.2653	0.2668	0.2570
0.280	0.400	0.2673	0.2835	0.5702	0.2782	0.3504	0.3292	0.3945
5.000	0.035	0.0919	0.0924	0.2322	0.0919	0.1434	0.1154	0.1390

Table 7 Optimum frequency predictions for SDOF structures

		Den Hartog [50]	Warburton [51]	Sadek et al. [52]	Leung and Zhang [53]	Linear Equation [48]	Polynomial Equation [48]	Exponential Equation [48]
T_s (s)	μ	f_{opt}						
1.750	0.030	0.9709	0.9636	0.9626	0.9410	0.9773	0.9773	0.9816
0.200	0.180	0.8475	0.8084	0.8309	0.7469	0.8807	0.9000	0.8775
0.100	0.065	0.9390	0.9236	0.9274	0.8958	0.9548	0.9328	0.9562
0.600	0.080	0.9259	0.9072	0.9133	0.8769	0.9451	0.9253	0.9456
2.300	0.100	0.9091	0.8861	0.8954	0.8515	0.9322	0.9205	0.9315
1.000	0.015	0.9852	0.9815	0.9792	0.9624	0.9869	1.0145	0.9926
3.500	0.200	0.8333	0.7906	0.8163	0.7197	0.8678	0.8864	0.8645
0.150	0.250	0.8000	0.7483	0.7821	0.6500	0.8357	0.8371	0.8328
0.280	0.400	0.7143	0.6389	0.6952	0.4246	0.7391	0.7868	0.7445
5.000	0.035	0.9662	0.9577	0.9573	0.9343	0.9741	0.9677	0.9779

expressed according to 5% and 20% mass ratio. Also, for 40 storey (mass is 980 tons, stiffness and damping linearly decrease from initial to top storey as 2130-998 MN/m and 42.6-20 MNs/m, respectively (Liu et al. [57]), the results can be seen in Table 11 for only 5% mass ratio.

x_1, x_{10} and x_{40} are displacements with number/level that expresses the storey where displacements are observed. And also, TMD stroke expresses the measurement of moving of TMD, namely the capacity of lengthen of TMD spring.

Fig. 6 MDOF Structure with TMD [54]

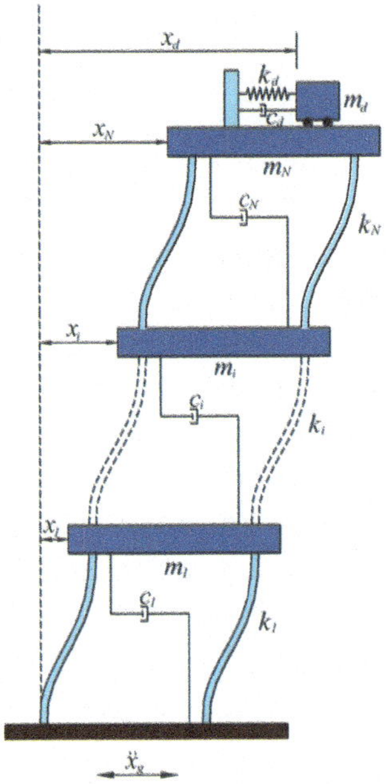

5.2 *Tubular Column Section Sizes Prediction (Case 1 and 2)*

The tubular column model is shown in Fig. 7. This study is related to the minimization of total cost of the column structure (by using FPA [49]) as in Eq. (14).

The current structure is subjected to an axial compression load indicated as P. Also, l, σ_y, ρ and E are length, yield stress, density and elasticity modulus of column, respectively. Additionally, d is diameter of central area and t is the thickness of column.

$$f(\mathbf{d}, \mathbf{t}) = 9.8\mathbf{dt} + 2\mathbf{d} \tag{14}$$

On the other hand, some limitations are applied as controlling of compression capacity of column (Eq. 15); capacity of buckling force (Eq. 16) and minimum and maximum ranges for design parameters (d and t) as in Eqs. (17)–(20).

$$g_1 = \frac{\mathrm{P}}{\pi \mathrm{dt}\sigma_y} - 1 \leq 0 \tag{15}$$

$$g_2 = \frac{8\mathrm{P}l^2}{\pi^3 E\mathrm{dt}\left(\mathrm{d}^2 + \mathrm{t}^2\right)} - 1 \leq 0 \tag{16}$$

$$g_3 = \frac{2.0}{\mathrm{d}} - 1 \leq 0 \tag{17}$$

Table 8 Far-field earthquake recordings to test and evaluate of equations [55]

Record number	Date	Name of record	1st direction	2nd direction
1	1994	Northridge	NORTHR/MUL009	NORTHR/MUL279
2	1994	Northridge	NORTHR/LOS000	NORTHR/LOS270
3	1999	Düzce, Turkey	DUZCE/BOL000	DUZCE/BOL090
4	1999	Hector Mine	HECTOR/HEC000	HECTOR/HEC090
5	1979	Imperial Valley	IMPVALL/H-DLT262	IMPVALL/H-DLT352
6	1979	Imperial Valley	IMPVALL/H-E11140	IMPVALL/H-E11230
7	1995	Kobe, Japan	KOBE/NIS000	KOBE/NIS090
8	1995	Kobe, Japan	KOBE/SHI000	KOBE/SHI090
9	1999	Kocaeli, Turkey	KOCAELI/DZC180	KOCAELI/DZC270
10	1999	Kocaeli, Turkey	KOCAELI/ARC000	KOCAELI/ARC090
11	1992	Landers	LANDERS/YER270	LANDERS/YER360
12	1992	Landers	LANDERS/CLW-LN	LANDERS/CLW-TR
13	1989	Loma Prieta	LOMAP/CAP000	LOMAP/CAP090
14	1989	Loma Prieta	LOMAP/G03000	LOMAP/G03090
15	1990	Manjil, Iran	MANJIL/ABBAR—L	MANJIL/ABBAR—T
16	1987	Superstition Hills	SUPERST/B-ICC000	SUPERST/B-ICC090
17	1987	Superstition Hills	SUPERST/B-POE270	SUPERST/B-POE360
18	1992	Cape Mendocino	CAPEMEND/RIO270	CAPEMEND/RIO360
19	1999	Chi-Chi, Taiwan	CHICHI/CHY101-E	CHICHI/CHY101-N
20	1999	Chi-Chi, Taiwan	CHICHI/TCU045-E	CHICHI/TCU045-N
21	1971	San Fernando	SFERN/PEL090	SFERN/PEL180
22	1976	Friuli, Italy	FRIULI/A-TMZ000	FRIULI/A-TMZ270

$$g_4 = \frac{\mathrm{d}}{14} - 1 \leq 0 \tag{18}$$

$$g_5 = \frac{0.2}{\mathrm{t}} - 1 \leq 0 \tag{19}$$

$$g_6 = \frac{\mathrm{t}}{0.9} - 1 \leq 0 \tag{20}$$

For prediction application, two different cases were considered. The first is to train of ANNs with the optimum design results together with minimum costs determined for 10,000 different P and l variations [58]. For this respect, P and l are handled in range of 100–5000 kgf and 100–800 cm as input parameters, besides d, t, and minimum cost are outputs.

Table 9 Optimum design results and maximum responses for 10-storey structure added TMD ($\mu = 5\%$)

	Den Hartog [50]	Warburton [51]	Sadek et al. [52]	Leung and Zhang [53]	Linear Equation [48]	Polynomial Equation [48]	Exponential Equation [48]
m_d (t)	152.627	152.627	152.627	152.627	152.627	152.627	152.627
T_d (s)	1.052	1.069	1.065	1.102	1.035	1.059	1.033
k_d (MN/m)	5.444	5.272	5.312	4.963	5.624	5.370	5.643
ξ_d (%)	13.827	13.916	33.690	13.832	19.246	18.188	18.391
c_d (MNs/m)	0.252	0.250	0.607	0.241	0.357	0.329	0.341
TF (dB)	−1.541	−1.084	0.378	−0.181	−2.405	−1.974	−2.432
x_{10} (m)	0.304	0.304	0.333	0.305	0.312	0.311	0.311
x_1 (m)	0.046	0.046	0.047	0.046	0.045	0.045	0.045
a (m/s^2)	15.164	15.296	15.526	15.551	15.157	15.284	15.128
TMD Stroke	2.891	2.955	1.664	3.084	2.387	2.545	2.442

Table 10 Optimum design results and maximum responses for 10-storey structure added TMD ($\mu = 20\%$)

	Den Hartog [50]	Warburton [51]	Sadek et al. [52]	Leung and Zhang [53]	Linear Equation [48]	Polynomial Equation [48]	Exponential Equation [48]
m_d (t)	610.506	610.506	610.506	610.506	610.506	610.506	610.506
T_d (s)	1.240	1.327	1.269	1.534	1.187	1.187	1.191
k_d (MN/m)	15.672	13.686	14.975	10.248	17.103	17.092	16.990
$\xi_{d,\,opt}$ (%)	25.869	26.578	57.018	26.242	30.030	31.482	28.232
c_d (MNs/m)	1.600	1.536	3.448	1.313	1.941	2.034	1.819
TF (dB)	−6.017	−5.075	−5.234	−2.872	−7.369	−7.534	−7.059
x_{10} (m)	0.233	0.233	0.274	0.240	0.242	0.245	0.238
x_1 (m)	0.033	0.034	0.037	0.037	0.033	0.034	0.033
a (m/s^2)	10.622	11.412	9.736	13.038	9.883	9.833	9.995
TMD Stroke	2.410	2.465	1.629	2.573	2.193	2.147	2.260

After, a test model was generated and the optimum values for design couples were determined. Tables 12, 13 and 14 present the optimization and ANN predictions for d, t and minimum costs with different error measurements belonging to test model.

Moreover, the proposed 1st and 2nd design constraints for column are observed by calculating via predictions for d, t and cost in order to control and provide of conditions (Table 15).

Table 11 Optimum design results and maximum responses for 40-storey structure added TMD ($\mu = 5\%$)

	Den Hartog [50]	Warburton [51]	Sadek et al. [52]	Leung and Zhang [53]	Linear Equation [48]	Polynomial Equation [48]	Exponential Equation [48]
m_d (t)	1541.809	1541.809	1541.809	1541.809	1541.809	1541.809	1541.809
T_d (s)	4.084	4.153	4.135	4.283	4.015	4.109	4.009
k_d (MN/m)	3.650	3.529	3.560	3.319	3.776	3.604	3.787
ξ_d (%)	14.442	14.535	35.188	14.447	20.102	18.996	19.209
c_d (MNs/m)	0.685	0.678	1.649	0.654	0.970	0.896	0.928
TF (dB)	−14.560	−14.007	−12.433	−13.099	−15.812	−15.295	−16.002
x_{40} (m)	1.749	1.734	1.793	1.705	1.776	1.756	1.775
x_1 (m)	0.052	0.052	0.053	0.051	0.053	0.052	0.053
a (m/s^2)	7.277	7.277	7.277	7.277	7.277	7.277	7.276
TMD Stroke	2.243	2.288	1.381	2.377	1.848	1.964	1.888

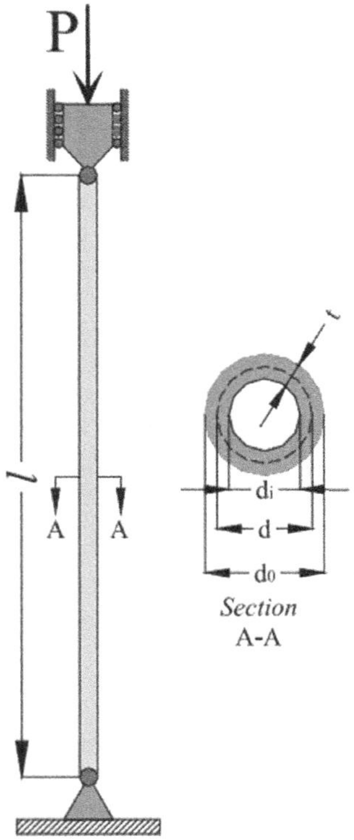

Fig. 7 Main Structure and Cross-section of Tubular Column Model [58]

Table 12 d results of new samples with ANN [58]

P (kgf)	l (cm)	FPA results	ANN Predictions	Error metric values for FPA		
		d (cm)		Absolute error	Absolute error %	Square error
171	551	4.2840	4.3687	0.0847	1.9780	0.0072
1373	306	5.7978	5.7872	0.0105	0.1819	0.0001
958	110	2.3883	2.4948	0.1064	4.4569	0.0113
749	411	5.7669	5.7955	0.0286	0.4954	0.0008
2043	701	11.5059	11.4767	0.0293	0.2542	0.0009
433	568	5.9607	5.8880	0.0727	1.2198	0.0053
3786	270	5.8814	5.7367	0.1447	2.4605	0.0209
2702	217	4.7244	4.7064	0.0180	0.3805	0.0003
2810	259	5.6466	5.6852	0.0386	0.6840	0.0015
870	686	8.5318	8.5527	0.0209	0.2452	0.0004
			Average	0.0554	1.2356	0.0049

Table 13 t results of new samples with ANN [58]

P (kgf)	l (cm)	FPA results	ANN predictions	Error metric values for FPA		
		t (cm)		Absolute error	Absolute error %	Square error
171	551	0.200	0.197	0.0026	1.3181	0.0000
1373	306	0.200	0.228	0.0288	14.3849	0.0008
958	110	0.255	0.265	0.0098	3.8375	0.0001
749	411	0.200	0.219	0.0198	9.8751	0.0004
2043	701	0.200	0.187	0.0128	6.3985	0.0002
433	568	0.200	0.209	0.0092	4.5756	0.0001
3786	270	0.409	0.479	0.0693	16.9100	0.0048
2702	217	0.364	0.408	0.0441	12.1022	0.0019
2810	259	0.316	0.373	0.0564	17.8151	0.0032
870	686	0.200	0.198	0.0011	0.5428	0.0000
			Average	0.0254	8.7760	0.0011

In case 2 [59], optimum design values were determined for a different test structure with the usage of developed ANN prediction model in case 1. In this case, test samples have values between in 500 and 2000 kgf by increasingly 500 kgf for P and 200–500 cm by increasingly 50 cm for l.

Table 14 Minimum cost results of new samples with ANN [58]

P (kgf)	l (cm)	FPA results	ANN predictions	Error metric values for FPA		
		Min f(d,t)		Absolute error	Absolute error %	Square error
171	551	16.9646	17.2501	0.2855	1.6829	0.0815
1373	306	22.9593	22.9529	0.0064	0.0279	0.0000
958	110	10.7535	11.2577	0.5042	4.6883	0.2542
749	411	22.8371	22.9484	0.1114	0.4878	0.0124
2043	701	45.5635	45.7778	0.2143	0.4703	0.0459
433	568	23.6043	23.3401	0.2642	1.1193	0.0698
3786	270	35.3832	35.4033	0.0200	0.0566	0.0004
2702	217	26.3062	26.3915	0.0854	0.3245	0.0073
2810	259	28.8245	28.8227	0.0018	0.0061	0.0000
870	686	33.7859	33.9598	0.1739	0.5147	0.0302
			Average	0.1667	0.9378	0.0502

Table 15 Controlling of design constraints for new samples within test model [58]

P (kgf)	l (cm)	g_1	g_2
171	551	−0.8737	−0.0443
1373	306	−0.3398	−0.1213
958	110	−0.0780	−0.1549
749	411	−0.6256	−0.1035
2043	701	−0.3946	0.0766
433	568	−0.7762	−0.0080
3786	270	−0.1231	−0.0802
2702	217	−0.1045	−0.0991
2810	259	−0.1570	−0.1694
870	686	−0.6744	−0.0019

Prediction results are shown in Figs. 8, 9, and 10 for d, t and minimum costs, respectively to prove the changing of values according to design constants (P and l) within test structure.

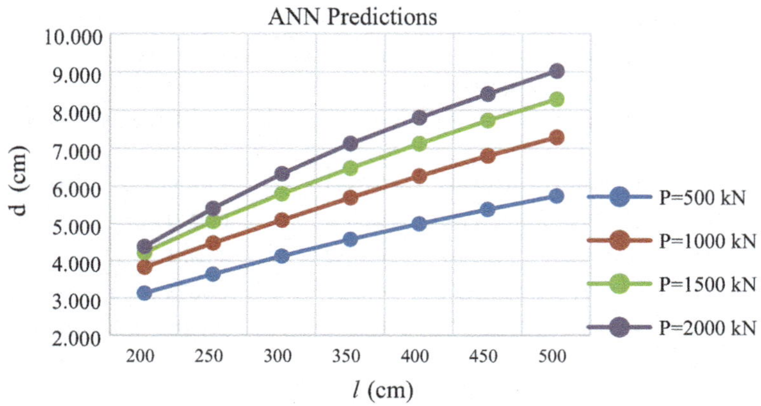

Fig. 8 Optimum d Results Obtained via ANN Prediction Model [59]

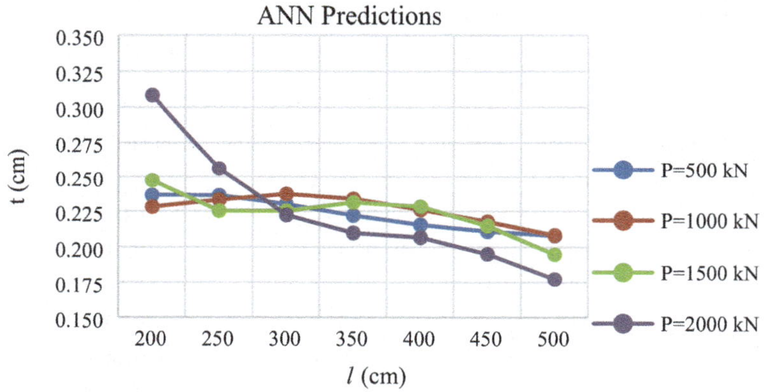

Fig. 9 Optimum t Results Obtained via ANN Prediction Model [59]

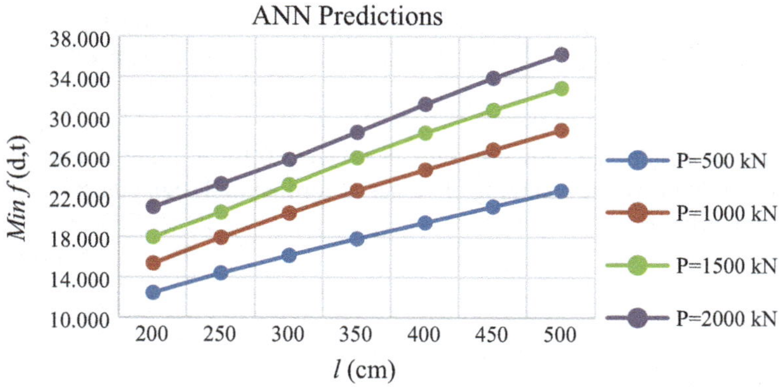

Fig. 10 Minimum Objective Function (Min Cost) Results Obtained via ANN Prediction Model [59]

5.3 Optimum Design and Estimation of Parameters for I-Beam (Case 1 and 2)

The I-beam structure used in the current application can be seen in Fig. 11. In here, optimization target is connected with determination of optimum section sizes, which can provide the minimum vertical deflection (Eq. 21), besides this process is conducted via FPA [49].

$$f(x) = \frac{\mathbf{PL}^3}{48\mathbf{EI}} \tag{21}$$

where, E means the elasticity modulus of beam and its value is 20,000 kN/cm^2. Also, I express the inertia moment as formulized below Eq. (22):

$$I = \frac{t_w(h - 2t_f)^3}{12} + \frac{bt_f^3}{6} + 2bt_f\left(\frac{h - t_f}{2}\right)^2 \tag{22}$$

1st design constraint is expressed in Eq. (23) respect to cross section of I-beam structure (this should be smaller than 300 cm^2), and also Eq. (24) for moment stress level cannot exceed 6 kN/cm^2. Besides that, design section parameters including h, b, t_w and t_f have minimum and maximum limits as in 10–100 cm, 10–60 cm, 0.9–6 cm and 0.9–6 cm, respectively.

$$g_1 = 2bt_f + t_w(h - 2t_f) \leq 300 \tag{23}$$

$$g_2 = \frac{1.5PLh}{t_w(h - 2t_f)^3 + 2bt_w\left(4t_f^2 + 3h(h - 2t_f)\right)} + \frac{1.5QLb}{t_w^3(h - 2t_f) + 2t_wb^3} \leq 6 \tag{24}$$

Moreover, for estimation applications, two independent cases were applied [61, 62]. Applications of the cases are related to determination of optimum design section

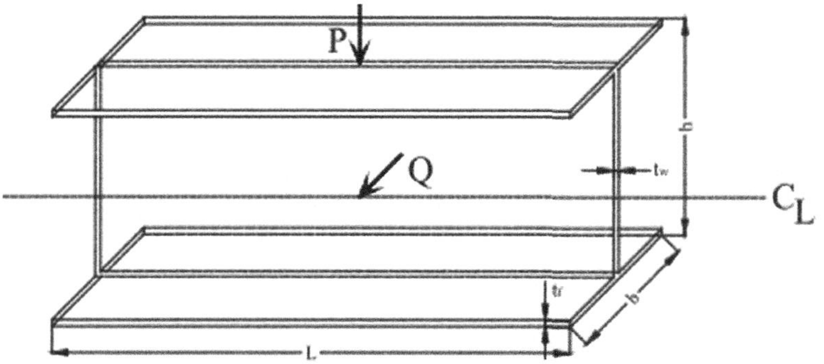

Fig. 11 I-beam Structural Model and Applied Load Conditions with Section Parameters [60]

Table 16 Optimum Values and Predictions for Test Couples Respect to *h* (Case 1) [61]

L (cm)	P (kN)	Design member	Optimum value	ANN estimation	Error calculations for FPA		
					Absolute error	Absolute error %	Square error
120	652	*h* (cm)	100.0000	99.9989	0.0011	0.0011	0.0000
350	520		100.0000	100.0121	0.0121	0.0121	0.0001
285	743		100.0000	100.0283	0.0283	0.0283	0.0008
150	200		100.0000	100.0045	0.0045	0.0045	0.0000
345	264		100.0000	100.0061	0.0061	0.0061	0.0000
100	690		100.0000	99.9984	0.0016	0.0016	0.0000
250	442		100.0000	99.9970	0.0030	0.0030	0.0000
310	675		100.0000	100.0279	0.0279	0.0279	0.0008
270	482		100.0000	100.0029	0.0029	0.0029	0.0000
220	355		100.0000	99.9942	0.0058	0.0058	0.0000
				Mean	0.0093	0.0093	0.0002

Table 17 Optimum values and predictions for test couples respect to *b* (Case 1) [61]

L (cm)	P (kN)	Design member	Optimum value	ANN estimation	Error calculations for FPA		
					Absolute error	Absolute error %	Square error
120	652	*b* (cm)	60.0000	59.9968	0.0032	0.0053	0.0000
350	520		60.0000	60.0180	0.0180	0.0300	0.0003
285	743		60.0000	60.0295	0.0295	0.0492	0.0009
150	200		60.0000	59.9652	0.0348	0.0580	0.0012
345	264		60.0000	60.0047	0.0048	0.0080	0.0000
100	690		60.0000	59.9933	0.0067	0.0112	0.0000
250	442		60.0000	59.9290	0.0710	0.1183	0.0050
310	675		60.0000	60.0200	0.0200	0.0333	0.0004
270	482		60.0000	59.9336	0.0664	0.1106	0.0044
220	355		60.0000	59.9266	0.0734	0.1223	0.0054
			Mean		0.0328	0.0546	0.0018

parameters and minimum deflections by using a trained ANNs. In here, P and Q are applied loads as vertical and horizontal located at center of beam span and web, respectively. In applied cases, these values and beam length (L) are not constant to handle in training process via ANNs. However, in case 1, Q is constant as 50 kN, besides P and L have values in 100–750 kN and 100–350 cm as random determined, respectively. As to case 2, Q is also changeable as 25–75 kN; P and L range in 250–750 kN and 150–400 cm, respectively.

In Tables 16, 17, 18 and 19, proposed values of design constants (L and P) for test model with estimations of design variables and calculated error rates are expressed respect to case 1 intended for h, b, t_w and t_f, respectively. Also, minimum deflections corresponding each test sample were calculated through estimation of design parameters by using the ANN model (Table 20). On the other hand, for case 2, estimations and error evaluations can be seen in Tables 21, 22, 23 and 24 respect to each design variable, besides that validation calculations for deflections can be seen in Table 25 respect to this case.

Table 18 Optimum values and predictions for test couples respect to t_w (Case 1) [61]

L (cm)	P (kN)	Design member	Optimum value	ANN estimation	Error calculations for FPA		
					Absolute error	Absolute error %	Square error
120	652	t_w (cm)	0.9000	0.8851	0.0149	1.652	0.0002
350	520		1.6303	1.5850	0.0454	2.782	0.0021
285	743		1.6833	1.7029	0.0196	1.163	0.0004
150	200		0.9000	0.9401	0.0401	4.457	0.0016
345	264		1.1171	1.0728	0.0443	3.965	0.0020
100	690		0.9000	0.9265	0.0265	2.948	0.0007
250	442		1.0635	1.0968	0.0333	3.133	0.0011
310	675		1.9552	1.7537	0.0420	2.451	0.0018
270	482		1.2077	1.2371	0.0295	2.447	0.0009
220	355		0.9000	0.9422	0.0422	4.684	0.0018
				Mean	0.0338	2.9681	0.0012

Table 19 Optimum values and predictions for test couples respect to t_f (Case1) [61]

L (cm)	P (kN)	Design member	Optimum value	ANN estimation	Error calculations for FPA		
					Absolute error	Absolute error %	Square error
120	652	t_f (cm)	1.7766	1.7867	0.0101	0.5669	0.0000
350	520		1.1733	1.2128	0.0395	3.3666	0.0000
285	743		1.1289	1.1206	0.0083	0.7344	0.0000
150	200		1.7766	1.7410	0.0356	2.0065	0.0000
345	264		1.5989	1.6336	0.0347	2.1718	0.0000
100	690		1.7766	1.7486	0.0280	1.5764	0.0000
250	442		1.6428	1.6088	0.0340	2.0714	0.0000
310	675		0.9000	1.0812	0.0238	2.1558	0.0000
270	482		1.5244	1.4951	0.0293	1.9223	0.0000
220	355		1.7766	1.7341	0.0426	2.3957	0.0000
			Mean		0.0286	1.8968	0.0000

Table 20 Minimum deflection calculations via ANN estimations for test model (Case 1) [61]

L (cm)	P (kN)	Design member	Optimum value	Calculations via ANN estimations	Error calculations for FPA		
					Absolute error	Absolute error %	Square error
120	652	*Min f* (*x*) (cm)	0.002018	0.0020	0.0000	0.277	0.0000000
350	520		0.049381	0.0486	0.0008	1.613	0.0000006
285	743		0.038774	0.0388	0.0000	0.070	0.0000000
150	200		0.001209	0.0012	0.0000	1.225	0.0000000
345	264		0.020572	0.0203	0.0002	1.143	0.0000001
100	690		0.001236	0.0012	0.0000	1.006	0.0000000
250	442		0.012915	0.0131	0.0002	1.357	0.0000000
310	675		0.049937	0.0461	0.0003	0.653	0.0000001
270	482		0.018465	0.0187	0.0002	1.191	0.0000000
220	355		0.006771	0.0069	0.0001	1.614	0.0000000
			Mean		0.0002	1.015	0.0000001

Table 21 Optimum values and predictions for test couples respect to *h* (Case 2) [62]

L (cm)	P (kN)	Q (kN)	Design member	Optimum value	ANN estimation	Error calculations for FPA		
						Absolute error	Absolute error %	Square error
223	432	55	*h* (cm)	100.0000	99.9840	0.016	0.016	0.000
155	289	28		100.0000	99.9972	0.003	0.003	0.000
387	540	38		100.0000	99.9742	0.026	0.026	0.001
250	265	64		100.0000	99.9816	0.018	0.018	0.000
345	566	56		100.0000	99.9695	0.031	0.031	0.001
165	746	70		100.0000	100.0294	0.029	0.029	0.001
171	480	68		100.0000	99.9871	0.013	0.013	0.000
320	675	26		100.0000	100.3811	0.381	0.381	0.145
276	715	30		100.0000	100.2504	0.250	0.250	0.063
160	345	48		100.0000	99.9900	0.010	0.010	0.000
				Mean		0.078	0.078	0.021

5.4 Prediction Application for Linear Base Isolation Systems

In this application, seismically isolated benchmark structure, which has only one storey, is handled. Superstructure is three-dimensional shear building and has base isolation system as rubber bearing under of base floor. Also, building plan is symmetrical and bays are 5 m in each direction [63]. Besides, mass of each storey and the stiffnesses are assumed as equal and modal damping ratio of superstructure is 5% for all modes.

Table 22 Optimum values and predictions for test couples respect to b (Case 2) [62]

L (cm)	P (kN)	Q (kN)	Design member	Optimum value	ANN estimation	Error calculations for FPA		
						Absolute error	Absolute error %	Square error
223	432	55	b (cm)	60.0000	59.9995	0.000	0.001	0.000
155	289	28		60.0000	60.0022	0.002	0.004	0.000
387	540	38		60.0000	59.9927	0.007	0.012	0.000
250	265	64		60.0000	60.0036	0.004	0.006	0.000
345	566	56		60.0000	59.9967	0.003	0.006	0.000
165	746	70		60.0000	59.9934	0.007	0.011	0.000
171	480	68		60.0000	60.0015	0.001	0.002	0.000
320	675	26		60.0000	59.9981	0.002	0.003	0.000
276	715	30		60.0000	60.0014	0.001	0.002	0.000
160	345	48		60.0000	60.0012	0.001	0.002	0.000
				Mean		0.003	0.005	0.001

Table 23 Optimum values and predictions for test couples respect to t_w (Case 2) [62]

L (cm)	P (kN)	Q (kN)	Design member	Optimum value	ANN estimation	Error calculations for FPA		
						Absolute error	Absolute error %	Square error
223	432	55	t_w (cm)	0.9758	0.9781	0.002	0.233	0.000
155	289	28		0.9000	0.8991	0.001	0.104	0.000
387	540	38		1.6835	1.7165	0.033	1.959	0.001
250	265	64		0.9341	0.9540	0.020	2.128	0.000
345	566	56		1.7645	1.7801	0.016	0.885	0.000
165	746	70		1.1016	1.2296	0.128	11.615	0.016
171	480	68		0.9000	0.9370	0.037	4.114	0.001
320	675	26		1.5065	1.4971	0.009	0.625	0.000
276	715	30		1.4027	1.3936	0.009	0.650	0.000
160	345	48		0.9000	0.9089	0.009	0.990	0.000
					Mean	0.026	2.330	0.002

In here, base isolation system stiffness, total weight with gravitational acceleration are expressed via K_I, W and g, respectively and superstructure period is T_I can be calculated with Eq. (25).

$$T_I = 2\pi\sqrt{\frac{W}{K_I g}} \tag{25}$$

Table 24 Optimum values and predictions for test couples respect to t_f (Case 2) [62]

L (cm)	P (kN)	Q (kN)	Design member	Optimum value	ANN estimation	Error calculations for FPA		
						Absolute error	Absolute error %	Square error
223	432	55	t_f (cm)	1.7147	1.7116	0.000	0.003	0.179
155	289	28		1.7766	1.7711	0.000	0.006	0.311
387	540	38		1.1287	1.0896	0.002	0.039	3.469
250	265	64		1.7488	1.7237	0.001	0.025	1.439
345	566	56		1.0608	1.0407	0.000	0.020	1.897
165	746	70		1.6116	1.5117	0.010	0.100	6.199
171	480	68		1.7766	1.7416	0.001	0.035	1.974
320	675	26		1.2766	1.2788	0.000	0.002	0.174
276	715	30		1.3629	1.3657	0.000	0.003	0.205
160	345	48		1.7766	1.7658	0.000	0.011	0.613
				Mean		0.024	1.646	0.001

Table 25 Minimum deflection calculations via ANN estimations for test model (Case 2) [62]

L (cm)	P (kN)	Q (kN)	Design member	Optimum value	Calculations via ANN estimations	Error calculations for FPA		
						Absolute error	Absolute error %	Square error
223	432	55	$Min\, f(x)$ (cm)	0.0088	0.0088	0.000	0.154	0.000
155	289	28		0.0019	0.0019	0.000	0.277	0.000
387	540	38		0.0706	0.0719	0.001	1.909	0.000
250	265	64		0.0075	0.0076	0.000	0.986	0.000
345	566	56		0.0539	0.0544	0.001	1.064	0.000
165	746	70		0.0063	0.0065	0.000	3.295	0.000
171	480	68		0.0043	0.0044	0.000	1.222	0.000
320	675	26		0.0471	0.0467	0.000	0.837	0.000
276	715	30		0.0310	0.0308	0.000	0.578	0.000
160	345	48		0.0025	0.0025	0.000	0.422	0.000
				Mean		0.000	1.074	0.000

Also, C_I and M is viscous damping coefficient and total mass for base isolation system and damping ratio as ζ_I can be determined via Eq. (26).

$$\xi_I = \frac{C_I}{2M\omega_I} \tag{26}$$

By respect to the prediction stage, isolator periods and damping ratios are considered as various in the dynamic analysis in order to decide with ANN model directly

[64]. So, periods have values between in 2 and 4 s by increasingly 0.1 s and damping ratio is ranged in 0.1–0.3 by increasingly 0.01, respectively. Also, it is handled that ground excitations have different pulse periods (TP_s) as 2, 3 and 4 s. According to this information, the isolator period (T_I), damping ratio (ζ_I) and pulse periods are inputs, and maximum displacement (x) and acceleration (a) results are considered as outputs during ANN training.

Also, after training, test model was generated to observe the success of ANN prediction model. Test model is evaluated for three different cases by including of variety values for isolator combinations and pulse periods as in Tables 26, 27 and 28 according to training of whole design as case 1, handling of T_I considered as 2.1, 2.8, 3.3 and 3.9 s for case 2 and, taking couples of T_I as 2.4 and 3.6 s for case 3.

5.5 Estimation of Optimal Section Areas and Minimum Volume of 3-Bar Truss

The structural model of 3-bar truss, which is dealt for determination of minimum structural volume (by benefiting from harmony search (HS) [65]), can be seen in Fig. 12 [66]. Truss model has three steel bars with 20,000 kN/cm^2 elasticity modulus and section areas (A) of 1st and 3rd are same due to existing of symmetry. Also, there is an applied load at the junction point to southwest direction as P, besides l is span for supports as 100 cm.

In here, P is set in range of 0.1–2.8 kN by the way of 0.1 kN increasingly intended for training of machine learning technique as ANNs [67]. So, optimum results for A are handled as output data, besides that P values are input data.

On the other side, objective function and design limit conditions (depending on that bar stresses cannot violate $\sigma = 2$ kN/cm^2) for the mentioned optimization problem considered can be expressed via Eqs. (27)–(30), respectively.

$$f(v) = \left(2\sqrt{2}A_1 + A_2\right)l \tag{27}$$

$$g_1 = \frac{\sqrt{2}A_1 + A_2}{\sqrt{2}(A_1)^2 + 2A_1A_2}\mathrm{P} - \sigma \leq 0 \tag{28}$$

$$g_2 = \frac{A_2}{\sqrt{2}(A_1)^2 + 2A_1A_2}\mathrm{P} - \sigma \leq 0 \tag{29}$$

$$g_3 = \frac{1}{A_1 + \sqrt{2}A_2}\mathrm{P} - \sigma \leq 0 \tag{30}$$

The mentioned estimation process is realized for a created test design and ensured values together with the optimization results and error calculations are shown in Table 29 for 1st and 3rd, Table 30 for 2nd bars and Table 31 for minimum volume.

Table 26 Prediction results for new isolator combinations (Case 1) [64]

T_I	ζ_I	TP_s	x (cm)	Absolute error	Error calculations for ANN predictions		a (cm/s^2)	Error calculations for ANN predictions		
					Absolute error %	Square error		Absolute error	Absolute error %	Square error
2	0.21	2	0.9553	0.01	0.98	0.0001	9.9323	1.95	0.0391	0.20
2.2	0.1	3	1.3167	0.00	0.10	0.0000	10.6556	3.13	0.1186	0.34
2.9	0.25	4	1.0950	0.01	0.46	0.0000	5.5215	1.05	0.0034	0.06
3.1	0.3	4	1.1503	0.04	3.63	0.0016	5.1141	0.12	0.0000	0.01
2.5	0.28	2	1.0558	0.04	3.91	0.0016	7.5203	1.31	0.0099	0.10
4	0.15	3	2.0370	0.04	1.85	0.0014	5.2976	0.52	0.0008	0.03
2	0.19	2.5	0.9027	0.02	2.52	0.0005	9.4782	3.77	0.1382	0.37
2.8	0.27	3.2	1.0741	0.06	5.54	0.0040	6.0389	5.05	0.1031	0.32
2.4	0.16	2.2	1.2130	0.05	3.73	0.0022	9.0538	0.62	0.0032	0.06
3.7	0.12	3.8	2.1044	0.10	4.35	0.0091	6.3499	3.79	0.0625	0.25
4.3	0.21	2	1.2612	0.03	2.54	0.0010	3.4093	3.63	0.0142	0.12
1.8	0.22	3	0.6338	0.04	7.60	0.0020	7.3997	1.86	0.0197	0.14

(continued)

Table 26 (continued)

T_I	ζ_I	TP_s	x (cm)	Absolute error	Error calculations for ANN predictions		a (cm/s^2)	Error calculations for ANN predictions		
					Absolute error %	Square error		Absolute error	Absolute error %	Square error
2.2	0.05	4	0.9278	0.01	1.08	0.0001	6.9785	9.13	0.4920	0.70
3.8	0.4	2	0.9817	0.02	1.83	0.0003	4.2006	6.03	0.0726	0.27
3.4	0.18	2.5	1.5338	0.06	3.53	0.0032	5.8129	2.30	0.0188	0.14
2.5	0.26	3.2	0.9082	0.07	6.76	0.0043	6.5724	2.12	0.0203	0.14
2.1	0.15	2	1.1170	0.01	0.97	0.0001	10.5583	0.20	0.0005	0.02
2.8	0.18	3	1.3965	0.01	0.47	0.0000	7.2780	1.69	0.0156	0.12
3.3	0.25	4	1.3655	0.01	0.93	0.0002	5.3439	0.13	0.0000	0.01
3.9	0.28	3	1.5347	0.03	1.87	0.0009	4.8824	0.46	0.0005	0.02
			Average	0.03	2.73	0.0016	***Average***	0.17	2.4400	0.05
			Root mean square error	0.0404			***Root mean square error***	0.2380		

Table 27 Prediction results for new isolator combinations (Case 2) [64]

T_I	ζ_I	TP_s	x (cm)	Absolute error	Error calculations for ANN predictions		a (cm/s^2)	Error calculations for ANN predictions		
					Absolute error %	Square error		Absolute error	Absolute error %	Square error
2	0.21	2	0.9553	0.04	4.05	0.0015	9.9323	0.10	0.98	0.0098
2.2	0.1	3	1.3167	0.03	2.11	0.0008	10.6556	0.09	0.80	0.0077
2.9	0.25	4	1.0950	0.02	1.95	0.0005	5.5215	0.02	0.42	0.0005
3.1	0.3	4	1.1503	0.04	4.01	0.0020	5.1141	0.02	0.44	0.0005
2.5	0.28	2	1.0558	0.03	2.98	0.0009	7.5203	0.07	0.89	0.0046
4	0.15	3	2.0370	0.02	0.85	0.0003	5.2976	0.05	0.88	0.0022
2	0.19	2.5	0.9027	0.10	10.39	0.0093	9.4782	0.73	7.41	0.5331
2.8	0.27	3.2	1.0741	0.07	5.82	0.0044	6.0389	0.44	6.96	0.1960
2.4	0.16	2.2	1.2130	0.01	0.48	0.0000	9.0538	0.16	1.74	0.0252
3.7	0.12	3.8	2.1044	0.02	0.95	0.0004	6.3499	0.31	4.73	0.0974
4.3	0.21	2	1.2612	0.39	32.06	0.1555	3.4093	2.35	71.34	5.5083
1.8	0.22	3	0.6338	0.16	26.99	0.0253	7.3997	0.36	4.74	0.1277

(continued)

Table 27 (continued)

T_I	ζ_I	TP_s	x (cm)	Absolute error	Error calculations for ANN predictions		a (cm/s^2)	Error calculations for ANN predictions		
					Absolute error %	Square error		Absolute error	Absolute error %	Square error
2.2	0.05	4	0.9278	0.40	42.34	0.1577	6.9785	0.10	1.34	0.0106
3.8	0.4	2	0.9817	0.10	10.15	0.0103	4.2006	0.04	0.82	0.0013
3.4	0.18	2.5	1.5338	0.05	2.88	0.0021	5.8129	0.37	6.15	0.1337
2.5	0.26	3.2	0.9082	0.00	0.37	0.0000	6.5724	0.10	1.44	0.0093
2.1	0.15	2	1.1170	0.03	2.24	0.0006	10.5583	0.08	0.78	0.0067
2.8	0.18	3	1.3965	0.06	4.22	0.0034	7.2780	0.50	6.79	0.2529
3.3	0.25	4	1.3655	0.02	1.59	0.0005	5.3439	0.13	2.52	0.0181
3.9	0.28	3	1.5347	0.04	2.40	0.0014	4.8824	0.08	1.63	0.0062
			Average	0.08	7.94	0.0188	***Average***	0.30	6.140	0.3476
			Root mean square error	0.1373			***Root mean square error***	0.5896		

Table 28 Prediction results for new isolator combinations (Case 3) [64]

T_I	ζ_I	TP_s	x (cm)	Error calculations for ANN predictions			a (cm/s^2)	Error calculations for ANN predictions		
				Absolute error	Absolute error %	Square error		Absolute error	Absolute error %	Square error
2	0.21	2	0.9553	0.01	1.56	0.0002	9.9323	0.06	0.59	0.0036
2.2	0.1	3	1.3167	0.00	0.01	0.0000	10.6556	0.03	0.26	0.0008
2.9	0.25	4	1.0950	0.05	4.54	0.0025	5.5215	0.03	0.45	0.0006
3.1	0.3	4	1.1503	0.03	2.74	0.0009	5.1141	0.16	3.11	0.0253
2.5	0.28	2	1.0558	0.02	2.13	0.0005	7.5203	0.01	0.13	0.0001
4	0.15	3	2.0370	0.01	0.29	0.0000	5.2976	0.03	0.65	0.0012
2	0.19	2.5	0.9027	0.38	41.49	0.1476	9.4782	0.73	7.40	0.5317
2.8	0.27	3.2	1.0741	0.21	18.48	0.0441	6.0389	0.46	7.27	0.2137
2.4	0.16	2.2	1.2130	0.28	21.86	0.0758	9.0538	0.79	8.65	0.6216
3.7	0.12	3.8	2.1044	0.58	26.42	0.3378	6.3499	1.40	21.25	1.9663
4.3	0.21	2	1.2612	0.11	9.28	0.0130	3.4093	0.12	3.67	0.0146
1.8	0.22	3	0.6338	0.00	0.06	0.0000	7.3997	0.14	1.87	0.0198

(continued)

Table 28 (continued)

T_I	ζ_I	TP_s	x (cm)	Error calculations for ANN predictions			a (cm/s^2)	Error calculations for ANN predictions		
				Absolute error	Absolute error %	Square error		Absolute error	Absolute error %	Square error
2.2	0.05	4	0.9278	0.05	5.54	0.0027	6.9785	0.08	1.09	0.0070
3.8	0.4	2	0.9817	0.06	5.82	0.0034	4.2006	0.16	3.60	0.0259
3.4	0.18	2.5	1.5338	0.15	9.64	0.0235	5.8129	0.35	5.88	0.1225
2.5	0.26	3.2	0.9082	0.18	18.84	0.0337	6.5724	0.26	3.88	0.0679
2.1	0.15	2	1.1170	0.06	5.11	0.0033	10.5583	0.05	0.43	0.0021
2.8	0.18	3	1.3965	0.04	2.61	0.0013	7.2780	0.32	4.37	0.1046
3.3	0.25	4	1.3655	0.02	1.36	0.0003	5.3439	0.03	0.63	0.0011
3.9	0.28	3	1.5347	0.03	2.21	0.0012	4.8824	0.08	1.69	0.0068
			Average	0.11	9.00	0.03	***Average***	0.27	3.84	0.19
			Root mean square error	0.1860			***Root mean square error***	0.4323		

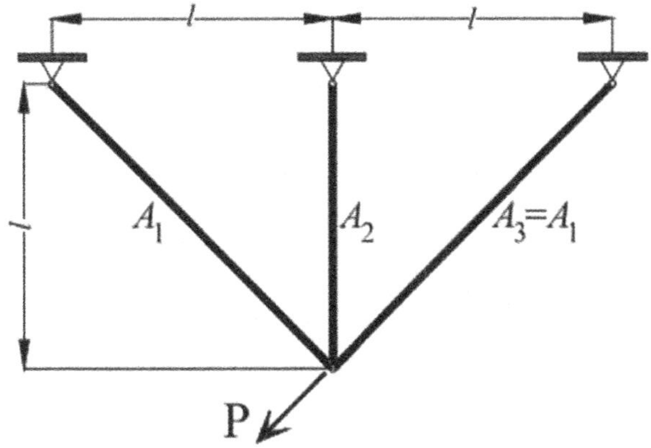

Fig. 12 3-bar Truss Model and Section Sizes [66]

Table 29 Estimations with error measurements for test model ($A_1 = A_3$) [67]

P (kN)	HS Results	Predictions with ANN	Errors for HS		
	$A_1 = A_3$		Error	Absolute Error	Squared Error
1.74	0.6866	0.6880	−0.00138	0.00138	0.0000019
2.42	0.9533	0.9446	0.00872	0.00872	0.0000760
0.85	0.3350	0.3378	−0.00280	0.00280	0.0000079
0.07	0.0277	0.0230	0.00474	0.00474	0.0000224
2.81	1.0000	0.9932	0.00678	0.00678	0.0000460
		Average		0.00488	0.000031
		Root mean square error		0.00555	

Table 30 Estimations with error measurements for test model (A_2) [67]

P (kN)	HS Results	Predictions with ANN	Errors for HS		
	A_2		Error	Absolute Error	Squared Error
1.74	0.3540	0.3497	0.00430	0.00430	0.0000185
2.42	0.4969	0.5214	−0.02445	0.02445	0.0005979
0.85	0.1741	0.1662	0.00797	0.00797	0.0000635
0.07	0.0140	0.0261	−0.01211	0.01211	0.0001465
2.81	0.9626	0.9791	−0.01650	0.01650	0.0002722
		Average		0.01306	0.0002197
		Root mean square error		0.01482	

Table 31 Estimations with error measurements for test model ($\boldsymbol{Minf}(\boldsymbol{v})$) [67]

P (kN)	HS Results	Predictions with ANN	Errors for HS		
	$Minf_{(v)}$		Error	Absolute Error	Squared Error
1.74	229.5904	229.5731	0.01737	0.01737	0.0003015
2.42	319.3153	319.2859	0.02940	0.02940	0.0008646
0.85	112.1577	112.1616	−0.00392	0.00392	0.0000154
0.07	9.2368	9.0810	0.15587	0.15587	0.0242942
2.81	379.1043	378.8521	0.25223	0.25223	0.0636211
		Average		0.09176	0.0178194
		Root mean square error		0.13349	

5.6 Determination of Optimum Section Parameters Regard to Fiber Reinforced Polymer Design

In this application, a reinforced concrete T-beam design with carbon fiber reinforced polymer (CFRP) is realized for optimization of sizing parameters by using Jaya algorithm [68] according to conditions of regulation of ACI 318-Building [69] in order to increase the capacity of shear force. This beam model and CFRP design are shown in Fig. 13. Also, the aim is the minimization of required CFRP unit area ($\boldsymbol{A}$) expressed as follows:

$$\boldsymbol{A} = \frac{\mathbf{w}_f\left(\frac{2\mathbf{d}_f}{\sin\beta} + \mathbf{b}_w\right)}{\mathbf{s}_f}\boldsymbol{x}1000 \tag{31}$$

s_f is spacing between CFRP strips, w_f is CFRP width, β is bonding angle of strips, b_w is width of beam and d_f is the depth of strips from tensile reinforcement. Also, some structural design limitations from ACI Regulation are considered for design.

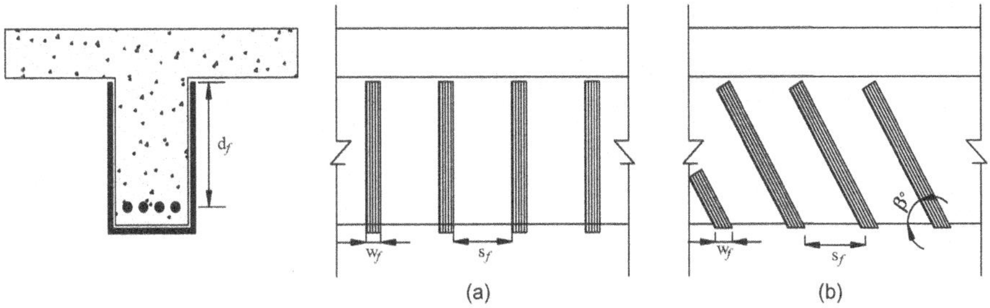

Fig. 13 Beam Model with the Application Design of CFRP [70]

In the prediction process, a dataset is generated including design parameters (beam height (h), breadth of beam (b_w) and thickness of slab (h_f), additive shear force (V_{add}) as input and target results (bonding angle of strips (β) and usage rate of CFRP (r)) as output data [71]. In here, r is developed by Eq. (32):

$$r = \frac{w_f}{w_f + s_f} \tag{32}$$

In prediction for this structural design, various machine learning methods and hybridizations of them (Random tree, ANNs and usage of both with Bagging) are benefited. By this respect, the results of training performance can be seen for β and r in Table 32.

Following of this process, a test data is generated through that the prediction capability of main prediction models is extremely high as more than 98%. For this respect, input data of new test samples with optimum results, and predictions are presented in Tables 33 and 34, respectively.

Table 32 Performance evaluations of used techniques corresponding training data [71]

Outputs	Success indicators	Prediction techniques			
		ANNs (%)	Bagging with ANNs (%)	Random Tree (%)	Bagging with random tree (%)
β (°)	Correlation coefficient (R)	99.69	99.77	98.89	99.38
r		99.70	99.49	99.36	99.59

Table 33 Input couples and optimum values of output parameters for test model [71]

Inputs				Outputs			Calculations via predictions
h (mm)	b_w (mm)	h_f (mm)	V_{add} (kN)	β (°)	w_f (mm)	s_f (mm)	r
350	200	120	70,000	65.016	35.170	61.271	0.365
530	420	85	55,000	63.324	18.639	54.053	0.256
750	370	100	120,000	63.593	83.198	59.184	0.584
680	500	95	11,0000	62.608	58.911	53.272	0.525
420	450	120	70,000	62.753	61.816	110.000	0.360
300	300	100	10,0000	64.259	99.541	104.068	0.489
770	220	90	80,000	64.974	58.013	93.878	0.382
400	510	85	90,000	62.683	76.496	106.484	0.418
550	200	80	50,,000	64.839	31.058	102.775	0.232
800	300	100	12,0000	64.217	97.372	68.646	0.587

Table 34 β and r results for design couples within test model [71]

Method	β (°)	Error Values for JA		r	Error Values for JA	
		Absolute error	Square error		Absolute error	Square error
ANNs	63.657	1.359	1.847	0.604	0.239	0.057
	63.644	0.320	0.102	0.240	0.016	0.000
	64.902	1.309	1.714	0.377	0.207	0.043
	63.930	1.322	1.747	0.372	0.153	0.023
	61.858	0.895	0.801	0.456	0.096	0.009
	61.944	2.315	5.358	0.910	0.421	0.177
	65.969	0.995	0.991	0.220	0.162	0.026
	61.508	1.175	1.382	0.545	0.127	0.016
	65.498	0.659	0.435	0.212	0.020	0.000
	65.531	1.314	1.726	0.353	0.234	0.055
	Average	1.166	1.610		0.168	0.041
	Root mean square error		1.269		0.256	0.202
Bagging with ANNs	63.556	1.460	2.132	0.621		0.066
	63.634	0.310	0.096	0.236	0.020	0.000
	64.917	1.324	1.754	0.370	0.214	0.046
	63.880	1.272	1.618	0.367	0.158	0.025
	61.780	0.973	0.947	0.460	0.100	0.010
	61.846	2.413	5.822	0.945	0.456	0.208
	65.977	1.003	1.007	0.210	0.172	0.030
	61.506	1.177	1.386	0.551	0.133	0.018
	65.487	0.648	0.420	0.208	0.024	0.001
	65.558	1.341	1.798	0.343	0.244	0.059
	Average	1.192	1.698		0.178	0.046
	Root mean square error		1.303			0.214
Random Tree	64.239	0.777	0.604	0.353	0.012	0.000
	63.597	0.273	0.075	0.237	0.019	0.000
	64.944	1.351	1.826	0.277	0.307	0.094
	63.940	1.332	1.774	0.323	0.202	0.041
	61.143	1.610	2.592	0.353	0.007	0.000
	61.733	2.526	6.380	0.987	0.498	0.248
	66.158	1.184	1.402	0.279	0.103	0.011
	61.753	0.930	0.866	0.607	0.189	0.036
	65.762	0.923	0.852	0.191	0.041	0.002

(continued)

Table 34 (continued)

Method	β (°)	Error Values for JA		r	Error Values for JA	
		Absolute error	Square error		Absolute error	Square error
	65.458	1.241	1.539	0.279	0.308	0.095
	Average	1.215	1.791		0.169	0.053
	Root mean square error		1.338			0.230
Bagging with Random Tree	64.335	0.681	0.464	0.346	0.019	0.000
	63.595	0.271	0.073	0.235	0.021	0.000
	64.892	1.299	1.688	0.282	0.302	0.091
	63.942	1.334	1.779	0.326	0.199	0.040
	61.227	1.526	2.328	0.372	0.012	0.000
	61.707	2.552	6.512	0.955	0.466	0.217
	66.115	1.141	1.302	0.290	0.092	0.008
	61.818	0.865	0.749	0.605	0.187	0.035
	65.708	0.869	0.756	0.191	0.041	0.002
	65.468	1.251	1.564	0.285	0.302	0.091
	Average	1.179	1.722		0.164	0.049
	Root mean square error		1.312			0.221

References

1. McCarthy, J.: What is Artificial Intelligence?, Computer Science Department, Stanford University, Stanford (2004). http://cse.unl.edu/~choueiry/S09-476-876/Documents/whatisai.pdf. Accessed: 9 July 2020
2. Grebow, D.: The Short Definition of "Real Learning" (2018). https://knowledgestar.blog/2018/07/15/the-short-definition-of-real-learning/. Accessed: 10 July 2020
3. Nilsson, N.J.: Introduction to Machine Learning an Early Draft of a Proposed Textbook. Robotics Laboratory Department of Computer Science, Stanford University (1998). http://ai.stanford.edu/~nilsson/MLBOOK.pdf. Accessed: 1 July 2020
4. Oppermann, A.: Artificial Intelligence vs. Machine Learning vs. Deep Learning (2019). https://towardsdatascience.com/artificial-intelligence-vs-machine-learning-vs-deep-learning-2210ba8cc4ac. Accessed: 8 July 2020
5. Tutorialspoint: Artificial Intelligence Simply Easy Learning. Tutorials Point (I) Pvt. Ltd., ISBN: TP00057 (2015)
6. A Brief History of AI: AI Topics (Association for the Advancement of Artificial Intelligence (AAAI)). https://aitopics.org/misc/brief-history. Accessed: 5 July 2020
7. Ertel, W.: Introduction to Artificial Intelligence. In: Mackie, I. (ed.) Undergraduate Topics in Computer Science, 2nd edn. Springer International Publishing AG, Cham, Switzerland, ISBN: 978-3-319-58486-7 (2017)
8. Russell, S.J., Norvig, P.: Introduction. Artificial Intelligence: A Modern Approach, Prentice Hall, Eaglewood Cliffs, New Jersey, USA, 18–23 (1995)

9. Skansi, S.: Introduction to deep learning: from logical calculus to artificial intelligence. In: Mackie, I. (ed.) Undergraduate Topics in Computer Science. Springer International Publishing AG, Cham, Switzerland, ISBN: 978-3-319-73003-5 (2018)
10. Jones, M.T.: Artificial Intelligence: A Systems Approach. Infinity Science Press LLC, ISBN: 978-0-9778582-3-1 (2008)
11. Han, J., Kamber, M.: Classification and Prediction, Data Mining: Concepts and Techniques, 2nd edn, pp. 285–368. Morgan Kaufmann, California, USA (2006)
12. Kubat, M.: An Introduction to Machine Learning, 2nd edn. Springer International Publishing AG, Cham, Switzerland (2017). ISBN 978-3-319-63912-3
13. Feigenbaum, E., Buchanan, B.: Dendral and meta-dendral: roots of knowledge systems and expert system applications. Artif. Intell. **59**(i1-2), 233–240
14. Luger, G.F.: Artificial Intelligence: Structures and Strategies for Complex Problem Solving, 6th edn. Pearson Education, Boston (2009). ISBN 978-0-321-54589-3
15. Witten, I.H., Hall, M.A., Frank, E.: Data Mining: Practical Machine Learning Tools and Techniques. Morgan Kaufmann Series, Burlington, MA, USA (2011). ISBN 978-0-12-374856-0
16. Şeker, A., Diri, B., Balık, H.H.: Derin Öğrenme Yöntemleri ve Uygulamaları Hakkında Bir Inceleme. Gazi Mühendislik Bilimleri Dergisi **3**(3), 47–64 (2017)
17. Introduction to Machine Learning Algorithms. https://www.educba.com/machine-learning-algorithms/
18. James, G., Witten, D., Hastie, T., Tibshirani, R.: Support Vector Machines, An Introduction to Statistical Learning, vol. 112. Springer, New York, USA, p. 18, 337 (2013)
19. Nagy, Z.: Artificial Intelligence and Machine Learning Fundamentals: Develop Real-world Applications Powered by the Latest AI Advances. Packt Publishing Ltd., Birmingham, UK (2018). ISBN 978-1-78980-165-1
20. Ping-Feng, P., Ming-Fu, H.: An enhanced support vector machines model for classification and rule generation. In: Koziel, S., Yang, X.S. (eds.) Computational Optimization, Methods and Algorithms, vol. 356, pp. 241–243. Springer, Heidelberg (2011)
21. Bramer, M.: Principles of Data Mining (Vol. 180). In: Mackie, I. (ed.) Undergraduate Topics in Computer Science, 3rd edn. Springer, London, UK, ISBN: 978-1-4471-7306-9 (2016)
22. Taneja, S., Gupta, C., Goyal, K., Gureja, D.: An enhanced k-nearest neighbor algorithm using information gain and clustering. In: 2014 Fourth International Conference on Advanced Computing & Communication Technologies, 8–9 February 2014 Rohtak-India, IEEE, pp. 325–329 (2014)
23. Loh, W.Y.: Classification and regression tree methods. In: Ruggeri, Kenett and Faltin (eds.) Encyclopedia of Statistics in Quality and Reliability, John Wiley & Sons, Ltd., Department of Statistics, University of Wisconsin, Madison, USA, pp. 315–323 (2008)
24. Steinberg, D.: CART: classification and regression trees. In: Kumar, V., Wu, X. (Eds.), The Top Ten Algorithms in Data Mining, Chapman & Hall/CRC Press, USA, pp. 179–181 (2009)
25. Shafer, J., Agrawal, R., Mehta, M.: SPRINT: a scalable parallel classifier for data mining. In: Proceedings of the 22nd International Conference Very Large Data Bases, 1996 Mumbai-India, IBM Almaden Research Center, pp. 544–555 (1996)
26. Shih, Y.S.: QUEST Classification Tree (version 1.10.1) (2020). https://discovery.ccu.edu.tw/Site/nu26786/Document/article_5YvQJM.html?group_login=-1. Accessed 15 July 2020
27. Rokach, L., Maimon, O.: Decision Forests, Data Mining with Decision Trees: Theory and Applications, 2nd edn. World Scientific Publishing Co. Pte. Ltd., Singapore, pp. 121–126 (2010)
28. https://corporatefinanceinstitute.com/resources/knowledge/other/random-forest/
29. Jia, Y., Liu, S., Jiang, S.: Analysis of the development status of artificial intelligence technology at home and abroad. In: 2019 International Conference on Virtual Reality and Intelligent Systems (ICVRIS), 14–15 September 2019 Jishou-China, IEEE, pp. 195–198 (2010)
30. Hurwitz, J., Kirsch, D.: Machine Learning for Dummies, IBM Limited. John Wiley & Sons Inc, New Jersey, USA (2018). ISBN 978-1-119-45495-3

31. Kourou, K., Exarchos, T.P., Exarchos, K.P., Karamouzis, M.V., Fotiadis, D.I.: Machine learning applications in cancer prognosis and prediction. Comput. Struct. Biotechnol. J. **13**, 8–17 (2015)
32. Nichols, J.A., Chan, H.W.H., Baker, M.A.: Machine learning: applications of artificial intelligence to imaging and diagnosis. Biophys. Rev. **11**(1), 111–118 (2019)
33. Atici, U.: Prediction of the strength of mineral admixture concrete using multivariable regression analysis and an artificial neural network. Expert Syst. Appl. **38**(8), 9609–9618 (2011)
34. Chithra, S., Kumar, S.S., Chinnaraju, K., Ashmita, F.A.: A comparative study on the compressive strength prediction models for high performance concrete containing nano silica and copper slag using regression analysis and artificial neural networks. Constr. Build. Mater. **114**, 528–535 (2016)
35. Behnood, A., Golafshani, E.M.: Predicting the compressive strength of silica fume concrete using hybrid artificial neural network with multi-objective grey wolves. J. Cleaner Prod. **202**, 54–64 (2018)
36. Pal, M., Deswal, S.: Support vector regression based shear strength modelling of deep beams. Comput. Struct. **89**(13–14), 1430–1439 (2011)
37. Salazar, F., Toledo, M.A., Oñate, E., Morán, R.: An empirical comparison of machine learning techniques for dam behaviour modelling. Struct. Safety **56**, 9–17 (2015)
38. Mansouri, I., Safa, M., Ibrahim, Z., Kisi, O., Tahir, M.M., Baharom, S., Azimi, M.: Strength prediction of rotary brace damper using MLR and MARS. Struct. Eng. Mech. **60**(3), 471–488 (2016)
39. Asteris, P.G., Nozhati, S., Nikoo, M., Cavaleri, L., Nikoo, M.: Krill Herd algorithm-based neural network in structural seismic reliability evaluation. In: Mechanics of Advanced Materials and Structures, pp. 1–8 (2018)
40. Yang, L., Qi, C., Lin, X., Li, J., Dong, X.: Prediction of dynamic increase factor for steel fibre reinforced concrete using a hybrid artificial intelligence model. Eng. Struct. **189**, 309–318 (2019)
41. Mangalathu, S., Jeon, J.S.: Machine learning–based failure mode recognition of circular reinforced concrete bridge columns: comparative study. J. Struct. Eng. **145**(10), 04019104 (2019)
42. Alwanas, A.A.H., Al-Musawi, A.A., Salih, S.Q., Tao, H., Ali, M., Yaseen, Z.M.: Load-carrying capacity and mode failure simulation of beam-column joint connection: application of self-tuning machine learning model. Eng. Struct. **194**, 220–229 (2019)
43. Lee, S., Lee, C.: Prediction of shear strength of FRP-reinforced concrete flexural members without stirrups using artificial neural networks. Eng. Struct. **61**, 99–112 (2014)
44. Cascardi, A., Micelli, F., Aiello, M.A.: An artificial neural networks model for the prediction of the compressive strength of FRP-confined concrete circular columns. Eng. Struct. **140**, 199–208 (2017)
45. Yaseen, Z.M., Tran, M.T., Kim, S., Bakhshpoori, T., Deo, R.C.: Shear strength prediction of steel fiber reinforced concrete beam using hybrid intelligence models: a new approach. Eng. Struct. **177**, 244–255 (2018)
46. Naji, S., Keivani, A., Shamshirband, S., Alengaram, U.J., Jumaat, M.Z., Mansor, Z., Lee, M.: Estimating building energy consumption using extreme learning machine method. Energy **97**, 506–516 (2016)
47. Yücel, M., Namli, E.: Yapay Zekâ Modelleri ile Betonarme Yapılara Ait Enerji Performans Sınıflarının Tahmini. Uludağ Univ. J. Faculty Eng. **22**(3), 325–346 (2017)
48. Yucel, M., Bekdaş, G., Nigdeli, S.M., Sevgen, S.: Estimation of optimum tuned mass damper parameters via machine learning. J. Build. Eng. 100847 (2019)
49. Yang, X.S.: Flower pollination algorithm for global optimization. In: Unconventional Computation and Natural Computation 2012, Lecture Notes in Computer Science, vol. 7445, pp. 240–249 (2012)
50. Den Hartog, J.P.: Mechanical Vibrations, 3rd edn. McGraw-Hill, New York, ISBN: 978-1443725361

51. Warburton, G.B.: Optimum absorber parameters for various combinations of response and excitation parameters. Earthquake Eng. Struct. Dyn. **10**, 381–401 (1982)
52. Sadek, F., Mohraz, B., Taylor, A.W., Chung, R.M.: A method of estimating the parameters of tuned mass dampers for seismic applications. Earthquake Eng. Struct. Dyn. **26**(6), 617–635 (1997)
53. Leung, A.Y.T., Zhang, H.: Particle swarm optimization of tuned mass dampers. Eng. Struct. **31**(3), 715–728 (2009)
54. Bekdaş, G., Nigdeli, S.M.: Mass ratio factor for optimum tuned mass damper strategies. Int. J. Mech. Sci. **71**, 68–84 (2013)
55. FEMA P-695: Quantification of Building Seismic Performance Factors. Federal Emergency Management Agency, Washington DC (2009)
56. Singh, M.P., Singh, S., Moreschi, L.M.: Tuned mass dampers for response control of torsional buildings. Earthquake Eng. Struct. Dyn. **31**, 749–769 (2002)
57. Liu, M.Y., Chiang, W.L., Hwang, J.H., Chu, C.R.: Wind-induced vibration of high-rise building with tuned mass damper including soil–structure interaction. J. Wind Eng. Industrial Aerodyn. **96**(6), 1092–1102 (2008)
58. Yucel, M., Bekdaş, G., Nigdeli, S.M., Sevgen, S.: Artificial neural network model for optimum design of tubular columns. Int. J. Theor. Appl. Mech. **3**, 82–86 (2018)
59. Yucel, M., Nigdeli, S.M., Bekdaş, G.: Optimum design of tubular columns for different cases by using prediction model generated via artificial neural networks. In: IV. Eurasian Conference on Civil and Environmental Engineering (ECOCEE), 17–18 June 2019 Istanbul, Turkey (2019)
60. Yang, X.-S., Bekdaş, G., Nigdeli, S.M. (eds.): Metaheuristics and Optimization in Civil Engineering. Springer, Switzerland (2016). ISBN: 9783319262451
61. Yucel, M., Nigdeli, S.M., Bekdaş, G.: Generation of an artificial neural network model for optimum design of I-beam with minimum vertical deflection. In: 12th HSTAM International Congress on Mechanics, 22–25 September 2019 Thessaloniki, Greece (2019)
62. Yucel, M., Nigdeli, S.M., Bekdaş, G.: Estimation model for generation optimization of design variables for I-beam vertical deflection minimization. In: IV. Eurasian Conference on Civil and Environmental Engineering (ECOCEE), 17–18 June 2019 Istanbul, Turkey (2019)
63. Alhan, C., Sürmeli, M.: Necessity and adequacy of near-source factors for not-so-tall fixed-base buildings. Earthquake Eng. Eng. Vib. **14**, 13–26 (2015)
64. Yucel, M., Öncü Davas, S., Nigdeli, S.M., Bekdaş, G., Sevgen, S.: Estimating of analysis results for structures with linear base isolation systems using artificial neural network model. Int. J. Control Syst. Robot. **3**, 50–56 (2018)
65. Geem, Z.W., Kim, J.H., Loganathan, G.V.: A new heuristic optimization algorithm: harmony search. Simulation **76**(2), 60–68 (2001)
66. Nowcki, H.: Optimization in pre-contract ship design. In: Fujita, Y., Lind, K., Williams, T.J. (eds.) Computer Applications in the Automation of Shipyard Operation and Ship Design, vol. 2, pp. 327–338. Elsevier, New York (1974)
67. Yucel, M., Bekdaş, G., Nigdeli, S.M.: Prediction of optimum 3-bar truss model parameters with an ANN model. In: Proceeding of 6th International Conference on Harmony Search, Soft Computing and Applications (ICHSA 2020), 16–17 July 2020 Istanbul-Turkey, Springer (2020)
68. Rao, R.V.: Jaya: a simple and new optimization algorithm for solving constrained and unconstrained optimization problems. Int. J. Industrial Eng. Comput. **7**(1), 19–34 (2016)
69. ACI 318 M-05: Building Code Requirements for Structural Concrete and Commentary, American Concrete Institute (2005)
70. Khalifa, A., Nanni, A.: Improving shear capacity of existing RC T-section beams using CFRP composites. Cement Concr. Compos. **22**(3), 165–174 (2000)

71. Yucel, M., Kayabekir, A.E., Nigdeli, S.M., Bekdaş, G.: Optimum design of Carbon Fiber Reinforced Polymer (CFRP) beams for shear capacity via machine learning methods: optimum prediction methods on advance ensemble algorithms – bagging combinations. In: Bekdaş, G., Nigdeli, S.M., Yucel, M. (eds.) Artificial Intelligence and Machine Learning Applications in Civil, Mechanical, and Industrial Engineering, pp. 84–102. IGI Global Publications, New York, USA (2019)

Design Optimization of Multi-objective Structural Engineering Problems Via Artificial Bee Colony Algorithm

Serdar Carbas, Deniz Ustun, and Abdurrahim Toktas

Abstract The construction sector constitutes a significant portion of global gross national expenditures with huge financial budget requirements and provides employment for more than one hundred million people. Besides, considering that people spend more than 80% of their time indoors today, it is necessary to make optimal structure designs. This requirement stems from the inadequacy of existing structures in the face of today's changing conditions. Indeed, realistic design optimization of the structures can be done not only by taking into account a single objective but also considering a number of structural criteria. It means that there is inherent multi-purpose in most structural design optimization problems. Thus, it is very difficult engineering task to solve these kinds of problems, as it is necessary to optimize multiple purposes simultaneously to obtain optimal designs. With the help of the improvisation in optimization techniques used for multi-objective structural engineering design, algorithms are provided to achieve the optimal designs by creating a strong synergy between the structural requirements and constraints mentioned in the design specifications. The recent addition to this trend is so-called Artificial Bee Colony (ABC) algorithm which simulates the nectar searching ability of the bees in nature for nutrition. In this chapter, an optimal design algorithm via ABC is proposed in order to obtain the optimum design of multi-objective structural engineering design

S. Carbas (✉)
Department of Civil Engineering, Faculty of Engineering, Karamanoglu Mehmetbey University, Karaman, Turkey
e-mail: scarbas@kmu.edu.tr

D. Ustun
Department of Computer Engineering, Faculty of Engineering, Tarsus University, Tarsus, Mersin, Turkey
e-mail: denizustun@tarsus.edu.tr

A. Toktas
Department of Electrical and Electronics Engineering, Faculty of Engineering, Karamanoglu Mehmetbey University, Karaman, Turkey
e-mail: atoktas@kmu.edu.tr

S. M. Nigdeli et al. (eds.), *Advances in Structural Engineering—Optimization*, Studies in Systems, Decision and Control 326,
https://doi.org/10.1007/978-3-030-61848-3_3

problems. The applications in design examples have shown the robustness, effectiveness, and reliability of ABC in attaining the design optimization of multi-objective constrained structural engineering design problems.

Keywords Structural engineering design · Artificial bee colony algorithm · Multi-objective optimization · Metaheuristics

1 Introduction

The task of a structural engineer is to size and construct buildings so that they can safely withstand various loads thought to be exposed throughout their service life, and also having minimum cost as much as possible. It is only possible for the structural engineer to guarantee that the structure dimensioned has sufficient strength, but only if the behaviour of the building provides defined limits. Consequently, the structural optimization problem can be defined as both choosing the sectional dimensions of the members forming the structure and determining its structural behaviour simultaneously in accordance with certain limitations under the effects of loads that the structure may be exposed to throughout its life. With this definition, in fact, it provides the safest structure in terms of structural characteristic and at the same time, it is possible to obtain the structure with the minimum cost as much as possible by choosing the section sizes in the most appropriate way. Until recent developments in numerical and optimization methods have been achieved, the solution of such problems was made by repeated analysis based on the experience and intuition of the structural engineer. Advances in computer technologies and accordingly optimization methods made it possible to formulate and solve this complex structural design problem as multiple decision-making problems [1]. The methods developed to obtain solutions of multi-objective optimization problems are generally techniques that find solutions for all objectives to make a decision. There is more than one objective function that should be minimized (or maximized) and also the constraints that should be provided in such decision-making problems. Structural engineering design problems are also the kind of problems that exactly fit this definition [2, 3].

In this context, many of the structural engineering design problems encountered in real life have multiple, complex and intricate objective functions. As a simple example, while construct a structure, the main objectives are to maximize both the usage area and structural strength while also minimize the cost. In such multi-objective optimization problems, it is aimed to solve all objectives at the same time by considering the objective functions, constraints and decision variables. The most popular of the methods developed for yielding these solutions is the Pareto optimal approach [4]. This approach can be summarized briefly as a set of solutions representing the best balances among the objectives of the design problems. By adapting it into multi-objective optimization algorithms, it effectively searches the solution space and provides optimal solutions to problems at wider ranges. Many metaheuristic-based optimization algorithms developed to obtain the solutions of multi-objective design optimization problems are available in the literature [5–12].

In this chapter, the Artificial Bee Colony (ABC) algorithm, which was developed by imitating the natural nectar (nutrient)-finding behaviour of bees living in the colony [13], is utilized to obtain the optimum solutions for multi-objective structural engineering design problems. A natural bee colony has much intelligence to organize the tasks to be done and to share tasks related to the things to be done in the hive. In nature, a bee colony contains one queen bee, hundreds of drones (male bees) and thousands of female (employed) bees. Each of the bees in the hive and/or colony has different tasks. The task of drones is to mate with queen bees. The task of queen bees is to mate with drones and produce new offspring. The task of employed bees is to meet the basic needs of the hive. The most important of these basic needs is that employed bees find food. When any of the employed bees finds a new food source (nectar), it is reported to the other onlooker bees in the hive with a waggle dance. After this information sharing, the onlooker bees waiting in the hive move to the food source where the amount of nectar is higher. Depending on the quantity and quality of the food source (nectar), the employed bee that consumes the food source starts to look for a new nectar and turns into a scout bee or waits for new information in the hive and turns into an onlooker bee [14–16].

The multi-objective (MO) engineering design optimization problems chosen from structural engineering design scope are executed to ensure the potency of proposed mo-ABC algorithm with Pareto optimal approach. As a first MO structural engineering design example, a two-bar truss is selected where the structural volume (fabrication cost) and the stresses in the structural members are minimized simultaneously. The second design example is taken as the design of an I-beam. It is aimed to find the optimal dimensions of the beam while minimizing the cross-sectional area and static deflection of the beam at the same time. The last MO structural engineering design problem is treated as a steel welded beam which is optimized for synchronously minimizing the cost of welding and point the tip deflection of the beam. The acquired final optimal designs clearly exhibit that the proposed mo-ABC algorithm based on Pareto optimal approach come into possession of an effective algorithmic ability and computational capability on attaining the design optimization of MO structural engineering design problems.

2 General Construction for Pareto Optimal Approach Based Multi-objective Optimization

MO optimization is a concept that allows multiple decision makers to optimize MO functions simultaneously. Making optimal decisions between two or more intricate objectives is important not only in engineering design problems, but also in social sciences (game, welfare, production and team theories), logistics and economics, as well. In MO optimization problems, it is essential to optimize multiple objectives systematically and synchronously. Especially, while trying to maximize one of the objectives; and trying to minimize another one increases the complexity and difficulty of the problem even more. Therefore, in MO optimization, it is more difficult to create a final combined objective function when establishing the model of decision problem than single-objective optimization. The algorithms used in single-objective optimization may not work best to obtain the solutions for MO problems where multiple solutions are in intricate with each other. Hence, in order to achieve the optimal results in solving MO problems, MO optimization methods having different algorithmic structures should be utilized. In MO optimization where the objectives are combined as a single purpose, various solution methods can be employed. One of the most popular MO optimization methods applied to recognize the solution is Pareto optimal approach [17].

In Pareto optimal approach, instead of finding a single solution point, a set of optimum solutions is obtained. Optimal solutions are dominant solutions in the entire solution set and occur in the region called Pareto optimal front or shortly Pareto front (PF). The mathematical expression of the PF is significant for the healthy start and progress of a solution a procedure. So, Pareto optimal and dominative solutions can be explained together [18]. In the Pareto optimal set, the concept of dominance can be explained as the difference between superior and non-superior solutions. The solution points, which can be defined as superior or dominant, are the best solution points that form the Pareto-optimal set.

The optimal solution to a MO problem is to investigate a set of solutions that each fulfill the objectives at an acceptable level, without being dominated by the other solutions. For solution points A, B, and D on the Pareto-optimal front, shown in Fig. 1, one cannot be said to be better or worse than the other, and that point C contains a worse solution to the minimization optimization problem. Therefore, the solutions at points A, B and D dominate the solution at point C. Since none

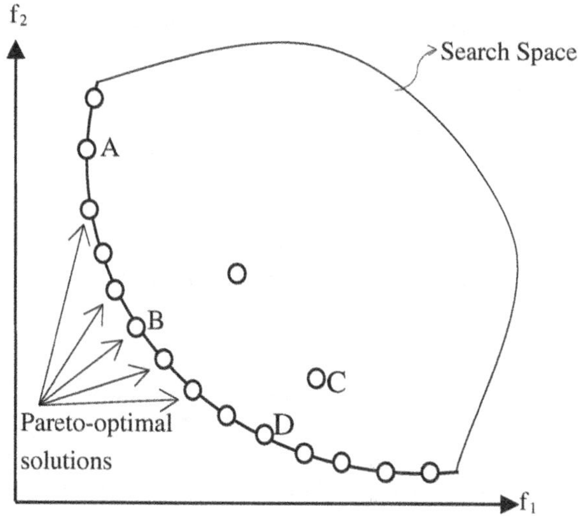

Fig. 1 An ideal Pareto-optimal set

of the Pareto-optimal solutions can be classified as better than other solutions, the primary aim in MO optimization is to achieve as many Pareto-optimal solutions as possible [19]. MO optimization has two main purposes; (i) Finding closer solutions (convergence) to Pareto-optimal front (if known before) as much as possible, (ii) Ensuring that the non-dominated (Pareto-optimal) solutions obtained are distributed as smoothly as possible on the Pareto-optimal front (distribution or diversity) [20]. These two objectives, namely convergence and diversity, which construct the basis of multi-purpose optimizations, form the Pareto-optimal front in the solution set. The results obtained for the purpose of convergence should produce a maximized or minimized front depending on the problem. Again, considering the purpose of distribution in MO optimization, the distance between the solutions on the resulting front and their placement to form a single surface line should be come into the picture. Achievement (performance) criteria of MO optimization algorithms are calculated for these two main purposes. In this context, along with MO optimization algorithms, many performance metrics have been developed in order to determine the success of these algorithms such as C-measure, Inverted Generational Distance (IGD), Hypervolume (HV), Epsilon (ε-) Indicator, Generational Distance (GD), Maximum Spread, and so forth [4, 18, 21, 22].

2.1 *Metrics for Algorithmic Performance Measurement*

To quantify the significance of the attained Pareto-optimal fronts through the proposed mo-ABC, some metrics are applied. That is to say Hypervolume, Maximum Spread, and Spacing metrics are engaged to specify particular aspects of the PFs [23].

2.1.1 Hypervolume

The Hypervolume can be considered as one of the most famous performance metrics which keeps in the view both the convergence and diversity of the optimal designs (solutions) [21]. This metric can be described as the whole area limited within the hypercubes produced by the points on the non-dominated PF and a chosen reference point. For a minimization optimization problem, the Hypervolume is desired for a higher value since it is indicant of a smooth convergence and well distribution of solutions. If an optimization problem is considered with two objectives, the hypervolume can be defined as the area of the search space limited by the yielded PF. To apprehend this concept, one can take into account a rectangle delimited by one solution point which is pertained to PF and the reference point. So, each solution placed on PF generates a rectangle in the search space. Hence, the hypervolume comply with the area shaped by the combining of all these rectangles. The hypervolume can be applied to the optimization problems using Eq. (1) as follows;

$$HV = \{U_i a_i | x_i \in PF\} \tag{1}$$

here, x_i stands for a non-dominated vector in PF and a_i presents hypervolume identified by the constituents of x_i and origin.

2.1.2 Maximum Spread

The Maximum Spread describes the assessment of the maximum expansion encircled by the non-dominated solutions on the PF. So, Maximum Spread taking into account the distribution of the designs (solutions) [18, 24]. Thus, this metric measure how better the PF is enveloped by the convergence set. In the case of an optimization problem having two objectives, Maximum Spread, matches up with Euclidean distance between two successive solutions as defined in Eq. (2).

$$MS = \sqrt{\sum_{k=1}^{K} \left(\max_{i=1}^{ns} f_k^i - \min_{i=1}^{ns} f_k^i \right)^2} \tag{2}$$

here, the ns represents the number of solutions on the PF and K stands for the total objectives number in the specified optimization problem. It should be worthy to mention that the higher the MS, the better the algorithmic performance.

2.1.3 Spacing

The Spacing evaluates the spread (distribution) of non-dominated solutions along the Pareto-optimal front [25]. In fact, it quantifies the disparity of the distance between contiguous non-dominated solutions and can be assessed by Eq. (3).

$$S = \sqrt{\frac{1}{n-1} \sum_{i=1}^{n} (\overline{d} - d_i)^2} \tag{3}$$

here, the $d_i = \min_j(|f_1^i(x) - f_1^j(x)| + |f_2^i(x) - f_2^j(x)|),\ i, j = 1, \ldots, n$. $\overline{d}$ is the mean gap between all the contiguous solutions and n is the non-dominated solutions number. The Spacing metric needs less amount of computational expenses.

3 Artificial Bee Colony Algorithm

3.1 Bees in Habitat

Bees living in their own habitats establish a social order among themselves and act accordingly. Bees ensure that the food (pollen and/or nectar) is found, brought to the nest, stored and distributed thanks to the hierarchy among them. The works in the bee colony (hive) are done by bees with different tasks specific to that work. Bees can job-share among themselves and organize themselves. Intrinsically, bees aim to find the maximum amount of nectar by spending minimum energy in nature. There are three different types of bees in a hive. These are named as queen bee, male bees (drones) and employed (female) bees [26]. Each hive has a single queen bee, and its task is to ensure the functioning of employed bees and drones and to check the integrity of the colony. Thanks to a chemical substance excreted by the queen bee, it also guarantees the protection of the colony by noticing the foreign bees entering the hive. The queen is the only bee that produces offspring in the colony, and the amount of offspring directly depends on the food (pollen) supply of the colony. In this way, it keeps the size of the colony under control, that is, in balance. There are many drones in a hive that do not have needles for defense. The only task of drones that do not have organs to collect pollen is to mate with the queen bee. Female bees have the most outnumber in the hive. These employed bees are responsible for not only collecting and storing the pollens/nectars, but also securing the colony and expelling dead bees from the hive. Although these employed bees are female, the only difference from the queen bee is that they are not fertile. Another important feature of these bees is that they are responsible for searching for food in the second half of their lives. At this stage of their lives, they observe the environment and determine the location of nectars.

The most important task in a bee colony is to search for food sources. After a bee leaves the hive, the food search process begins. When the bee finds a nectar, it stores the pollens collected, thanks to its related organs, to the abdomen. The process of collecting pollen depends on the distance of the nectar to the hive. Meanwhile, the bee begins to make the food it stores in the abdomen and after returning to the hive, it transfers the honey it forms to the empty comb. Then, after the honey is left in the comb, this bee informs other bees in the hive with a special dance (waggle dance), about the distance of the food source from the hive, the quality and quantity of the food source, as well. During this dance, other bees in the hive get this information by touching the antenna of the dancing bee. The distance of the food source and dance time is directly proportional. By monitoring the available food sources, the most suitable food source with the highest probability value is determined [27].

3.2 Concept of Artificial Bee Colony Algorithm

As explained in the previous section, Artificial Bee Colony (ABC) algorithm is modeled by examining that the bees living in the swarm in nature having their own organization and division of labor. ABC, which is one of the favorite swarm intelligence algorithms with this perspective, has emerged as a result of the examination of foraging behavior of bees [28].

The algorithm contains some principal assumptions. The first of these assumptions is that there is only one employed bee per food source. So, the total number of food sources is equal to the number of employed bees. The second assumption is that the number of onlooker bees is equal to the number of employed bees. Also, when an employed bee consumes the food supply, this bee becomes a scout bee. In another assumption, while the locations of food sources coincide with possible solutions in the optimization problem, the nectar richness in this source corresponds to the quality of the solutions. In the ABC algorithm, there are three different types of bees in the colony such as the employed bees, onlooker bees, and the scout bees.

3.2.1 Production of Food Sources (Phase of Initialization)

First of all, the algorithm starts by randomly generating food source locations that can correspond to the solutions in the search space (Eq. (4)).

$$x_{i,j} = x_{\min_j} + rand(0, 1)(x_{\max_j} - x_{\min_j}) \tag{4}$$

Here, i indicates the food source ($i = 1, \ldots, NEB$), *NEB* stands for number of employed bees, and j indicates the parameter ($j = 1, \ldots, D$) to be optimized. $x_{\min_j}$ and $x_{\max_j}$ are the lower and upper limits of this parameter, respectively. *rand* is a uniformly distributed random number between [0, 1].

3.2.2 Sending Employed Bees to Food Sources (Phase of Employed Bees)

The employed bee determines a new food source close to the current one, and then assesses its quality and if it is better than current one, memorizes it. Since there is only one employed bee responsible for each food source, each employed bee head towards to the food source they are responsible for, evaluate the result according to the objective of the problem and update and store their information. The better solutions that improve the objective value are kept in memory and the worse ones are discarded. The following Eq. (5) is used to determine the new food source.

$$v_{ij} = x_{ij} + \phi_{ij}(x_{ij} - x_{kj}) \tag{5}$$

here, $x_{ij}$is the current food source. v_{ij}new food source, $x_{ij}$is found by changing the j parameter of the current source. The j parameter value is a random number between zero and the upper limit of the parameter. When changing the randomly selected parameter j, the difference between the randomly selected x_k neighborhood solution's j_{th} parameter and the jth parameter of the current source is taken. This difference is added to the jth parameter of the current source after it is weighted by the number of ϕ_{ij} that takes a random value between $[-1, 1]$. As the difference between x_{ij} and x_{kj} decreases, the change of x_{ij} parameter value decreases. Greedy selection is applied between candidate solution v_i and current solution x_i to determine the individual with the highest fitness value.

$$fitness(x_i) = \begin{cases} \frac{1}{1+f(x_i)} & \text{if } f(x_i) \geq 0 \\ 1 + abs(f(x_i)) & \text{if } f(x_i) < 0 \end{cases} \tag{6}$$

here, $f(x_i)$ is the objective function value of the food source. The calculation method of the fitness value may differ depending on the type of the optimization problem. The objective function solution is computed according to the calculated fitness value. If the fitness value is better than the previously found, it is replaced with the old one. When the fitness value does not change, the counter is incremented with one, otherwise, if the new food source is memorized, the counter is reset as zero.

3.2.3 Calculation of Probability (Phase of Onlooker Bees)

After the employed bees complete their research in the environment and return to the hive, they start to waggle dance in the dance area. Thanks to this dance, they convey information about the nectar amount of the food resources they found to the onlooker bees. Then, onlooker bees go towards to these food sources with the probability determined according to their fitness values. This selection is engaged utilizing the roulette wheel method in the standard artificial bee colony algorithm. The angle of each slice on the wheel is proportional to the fitness value. In other words, the ratio of the fitness value of a food resource to the sum of the fitness value of all resources gives the probability of the relative selection of that food resource relative to other resources. This probability calculation is given in Eq. (7).

$$P_i = \frac{fitness_i}{\sum_{i=1}^{NEB} fitness_i} \tag{7}$$

Here, the value $fitness_i$ indicates the fitness value of the food resource, while the value NEB stands for the number of employed bees. As the fitness value of the food resource increases, the probability of being selected raises.

3.2.4 Selection of Food Sources by Onlooker Bees

After calculating the probability values, a random number in the range of [0, 1] is generated for each food resource using the roulette wheel method. If the probability value (P_i) of the food source is greater than this randomly generated number, a new food source is produced by Eq. (5). Then the fitness value of the new solution is calculated and compared to the old one. Likewise, if the new solution is better $failure_i$ counter is reset, otherwise counter is increased by 1 (one).

3.2.5 Determination of Abandoned Food Source and Scout Bee Production (Phase of Scout Bees)

At the end of each cycle, the failure to develop solutions ($failure_i$) is checked. If this counter reaches a higher value than a previously determined value (limit), this indicates that the resource is exhausted. The employed bee, whose resource is exhausted, leaves this food source and start searching new sources so that it becomes a scout bee [29]. If a scout bee finds a good food source in somewhere while doing research without any guidance, this scout bee is employed for this food source.

The best amount of nectar found is considered the value of the objective function, and the location of this food source is considered the value of the design variables. Thus, the value of the design variables found gives global optimum results.

A pseudo code of the artificial bee colony algorithm, whose steps are summarized above, is given Fig. 2.

4 Pareto-Based Multi-objective ABC (Mo-ABC) Algorithm

In the proposed mo-ABC algorithm, at first the initial solutions are generated. Amid these solutions, non-dominated solutions are retained in an archive. The current solutions improvements are supplied by the employed bee phase. New solutions are obtained by Eq. (8):

$$v_{ij} = x_{ij} + \delta(x_{ij} - x_{kj}) \tag{8}$$

where x_{ij} indicates the current solution, the x_{kj} stands for a neighbor solution ($i \neq k$) and v_{ij} defines a new candidate solution. The δ is a random value generated in range of [−1, 1].

A new solution is chosen by executing a greedy selection method between the candidate and current solutions. The failure counter is increased, when the current solution is selected. Otherwise, the counter is reset, when the new solution is selected. The archive is updated when the attained solutions are compared with each archive (AR) elements. This guarantees that the updates solutions are kept in the AR. The

Step 1. *Assigning random initial values to x_{ij} solutions, the initial food sources, and resetting the failure to develop solutions ($failure_i$).*
Step2. *Calculation of fitness values ($fitness_i$) for each food source.*
Step 3. *Repeat*
for i=1 to NEB do
- Calculating the fitness value by generating new food source v_i for each x_i food source using Eq. (5).
- Choosing the one with a high fitness value by applying greedy selection between x_i and v_i.
- If the v_i solution is better, resetting the non-solution counter ($failure_i$=0), if not, increasing the value of the counter ($failure_i$=$failure_i$+1).
end for
Calculation of P_i probability values using Eq. (7)
k=0, i=1.
repeat
if** P_i >random **then
- Using the Eq. (5), to create a new v_i solution in the neighborhood of x_i and calculate the fitness value of this solution.
- Greedy selection between v_i and x_i solutions and choosing the best.
- If the v_i solution is better, resetting the non-solution counter ($failure_i$=0), otherwise increasing the value of the counter ($failure_i$=$failure_i$+1).
-k=k+1
end if
***until** k=NEB*
if** max($failure_i$)>limit **then
-Omitting x_i and generating a new solution using Eq. (4)
end if
Keeping the best solution in memory.
Step 4. ***Until** (NC=maximum cycle number)*

Fig. 2 Pseudocode of standard ABC algorithm

Fig. 3 presents the employed bees phase and the AR update process, respectively [30].

The onlooker bees in the mo-ABC are received as archive members conversely to the standard ABC algorithm, and some other archive member is utilized to upgrade an archive member. This procedure is executed as a similar formulation as presented in Eq. (8). The candidate solution v_{ij} in the onlooker bees phase is thus developed by utilizing AR_{ij} which is an archive member, and AR_{kj} is a neighbor archive member ($i \neq k$). If a neighbor archive member is selected, the crowding distance values of all archive members are computed [31] and the member AR_{kj} with the smallest crowding distance value is chosen.

The current solution is an archive member, and the candidate solution is developed by utilizing the archive members as in Eq. (8). The archive upgrade process illustrated in Fig. 3 makes use of between these two solutions. Associated with this procedure, it is intended to raise the local search capability of the algorithm in the archive

```
for i=1 to NEB/2
Chose a random dimension j and neighbor k from the colony (i≠k)
vij= xij + δ * (xij − xkj)
    if vi dominates xi
    xi=vi
    failurei = 0
    else if xi dominates vi
    failurei = failurei + 1
    else if vi and xi are non-dominated solutions
    if rand< 0.5
    xi=vi
    failurei = 0
    else
    failurei = failurei + 1
    end if
    end if
Upgrade archive (vi and AR)
    for i=1 to length(AR)
      if v dominates ARi
       Add v to the archive
      else if ARi dominates v
      Do nothing
      else if ARi and v are non-dominated solutions
      Add v to the archive
      end if
    end for
  for i=1 to length(AR)
  Chose a random dimension j
  Chose neighbor k from the archive (i≠k) consistent with crowding
   distance values
  vij = ARij+ δ (ARij − ARkj)
  Upgrade Archive (vi and AR)
  end for
  Decide only one xi solution with maximum failure in the colony
  if failurei ≥ limit
  Produce initial position for xi
  failurei=0
  end if
end for
```

employed bee phase

archive upgrading process

onlooker bee phase

scout bee phase

Fig. 3 Pseudocode of Pareto-based mo-ABC algorithm

members. The onlooker bees phase of the mo-ABC algorithm is tabulated in Fig. 3, as well.

A constant-sized archive is utilized in the mo-ABC algorithm. If the archive is upgraded, the archive size is checked. If the archive size arrives at a pre-decided value, elite archive members are retained in the archive. Crowding distance values are utilized in the chosen of elite members.

Likewise, in the standard one, also in the mo-ABC algorithm, the failure counters of the food sources (amount of nectars) are checked in the scout bees phase. If there is a food source that arrives at a pre-decided limit value, the new position is settled. There can be only one scout bee in each cycle. The termination criterion is taken as the evaluation number. If the termination criterion is met, the algorithm is ended, and the current archive is reported as a final solution (design).

5 Design Examples

In this chapter, three structural engineering design problems are resolved to show the efficacy of the proposed mo-ABC for the MO design optimization problems. Namely, these problems are a two-bar truss design problem, an I-beam design problem, and a welded beam design problem. All the way through the Pareto-based MO optimization process, the functional control parameters of mo-ABC algorithm are taken as follows; the bee colony size as 30, the maximum cycle number (MCN) as 1000, and the limit as 30. With these selections the total objective function evaluations become is 30,000. Also, the Pareto population number defines the amount of non-dominated solution values on the Pareto-front is taken as 150. The abovementioned values for control parameters of proposed Pareto-based mo-ABC are taken into account with aforesaid reported studies [14, 16, 32–34].

Essentially due to the natural formation features of the structural engineering design problems are constrained in real practices. To manage these constraints, the easiest way is turning those kind of design problems into the unconstrained problems with the aid of the penalty function method. Here, the Eq. (9) is applied to the mo-ABC to operate this transition [35].

$$F_p = F(1 + C)^{\xi} \tag{9}$$

where, F_p stand for the penalized objective function. F shows the objective function of the design problem. C presents the total violations of the constraints. ζ is a penalty coefficient which may be selected as 2.0 [35–37] as calculated in Eqs. (10) and (11).

$$C = \sum_{i=1}^{cn} c_i \tag{10}$$

$$c_i = \begin{cases} 0 & \text{if } g_k \leq 0 \\ g_k & \text{if } g_k > 0 \end{cases} \quad k = 1, 2, 3, \ldots, cn \tag{11}$$

where, g_k is the kth constraint function and cn is the number of constraints in the MO structural design problem. It is worth to specify that before utilizing in the design algorithm entire of the constraints are needed to be normalized as less or equal than zero.

5.1 Two-Bar Truss Design Problem

As a first design example, a two-bar truss design problem is chosen in which it has 3 nodes, and 2 structural members as shown in Fig. 4 [38, 39]. The objectives of this structural design problem are that both minimizing the volume of the truss $f_1(x)$ and the stresses in each of the two structural members $f_2(x)$, synchronously. The vertical distance y between B and C (in meters), length x_1 of AC (in meters), and length x_2 of BC (in meters) are treated as the design variables of this structural engineering design problem. The mathematical formation of the problem is stated through Eqs. (12) and (13).

$$\min f_1(x) = x_1 * \sqrt{16 + y^2} + x_2 * \sqrt{1 + y^2}) \tag{12}$$

$$\min f_2(x) = \max(\sigma_{AC}, \sigma_{BC}) \tag{13}$$

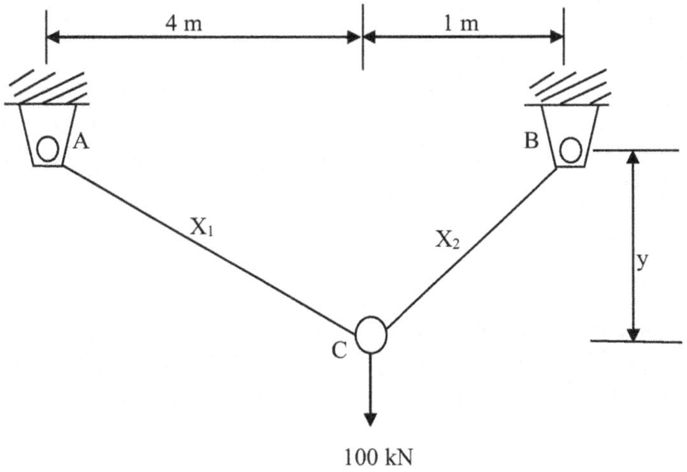

Fig. 4 Two-bar truss

Subjected to

$$g = \frac{\max(\sigma_{AC}, \sigma_{BC})}{10^5} - 1 \leq 0.0 \tag{14}$$

$$1 \leq y \leq 3 \quad \text{and} \quad \mathrm{x} \geq 0 \tag{15}$$

where the stresses can be computed as,

$$\sigma_{AC} = \frac{20 * \sqrt{16 + y^2}}{y * x_1} \quad \text{and} \quad \sigma_{BC} = \frac{80 * \sqrt{1 + y^2}}{y * x_2} \tag{16}$$

In order to implement the proposed mo-ABC technique, the lower and the upper bounds on x_1 and x_2 are considered as $0 \leq x_1$ and $x_2 \leq 0.01$.

The Pareto-front curve of four-bar planar truss problem attained through the mo-ABC is demonstrated in Fig. 5. From this figure, it is obvious that for this MO structural engineering design optimization problem, the Pareto-based mo-ABC algorithm produces evenly spread out non-dominated solutions which comprise a pulchritudinous convergence on the Pareto-front curve. The mo-ABC finalizes in a Pareto-front curve between (0.04039 m^3 (maxf_2), 99779.7406 kPa (minf_1)) and (0.052057 m^3 (minf_2), 8432.74043 kPa (max f_1)). This edge values were obtained as (0.004445 m^3, 89,983 kPa) and (0.004833 m^3, 83,268 kPa) using a ε-constraint MO design algorithm. Furthermore, the final design solutions attained via EM-MOPSO are distributed in the interval of (0.004026 m3, 99,996 kPa) and (0.05273

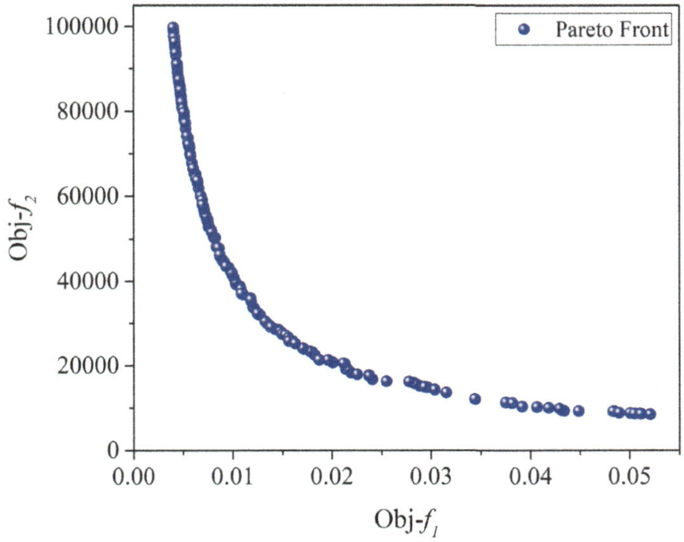

Fig. 5 Pareto-front curve obtained through mo-ABC algorithm for two-bar truss design problem

m^3, 8434.493 kPa), and those accomplished by NSGA-II are distributed in the interval of (0.00407 m^3, 99,755 kPa) and (0.05304 m^3, 8439 kPa) [39]. These solutions are tabulated in Fig. 6, as well. If these extremum point values of objective functions are compared, it is clearly seen that proposed mo-ABC finds out near-optimal border tail designs.

In Table 1, the design variables and constraint belong to the optimal border tail points objective functions are tabulated. Also, from the same table, it can be concluded that there is no any constraint violation. This proves the algorithmic performance supremacy of the proposed mo-ABC algorithm on this structural engineering design problem.

The performance metrics obtained from the Pareto-optimal set of mo-ABC and previously declared EM-MOPSO and NSGA-II [39] methods are exhibited in Table 2. Investigating by this table, it is obvious that the mo-ABC achieves a wide variety of final design solutions having uniform spread and the smooth convergence rate. For this MO structural engineering design optimization problem, the mo-ABC ensures its effectiveness over the other reported optimizers. If the values of Hypervolume and Maximum Spread metrics are yielded as higher as possible and the Spacing is attained as lower as possible, these prove the algorithmic superiority and validity of mo-ABC.

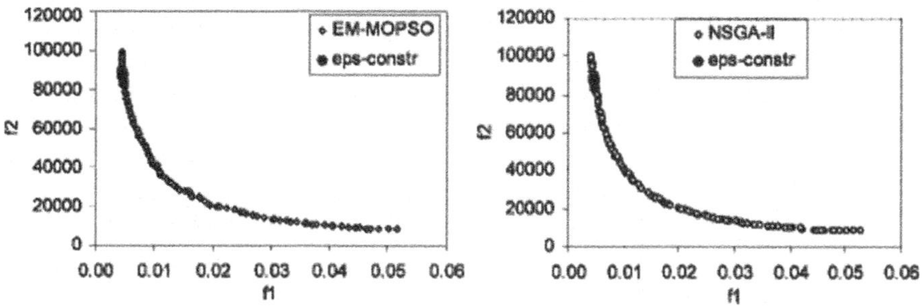

Fig. 6 Pareto-front curve obtained through ε-constraint, EM-MOPSO, and NSGA-II algorithms for two-bar planar truss design problem [39]

Table 1 Optimal designs for two-bar truss design problem attained by mo-ABC

Design variables and constraint		x_1	x_2	y	g
Border tail points Objective Function values	$\max f_2 = 0.04039$	0.004087	0.01	3.0	−91567.2596
	$\min f_1 = 99779.7406$				
	$\min f_2 = 0.052057$	0.000439	0.000904	2.0543	−220.2594
	$\max f_1 = 8432.74043$				

Table 2 Performance metrics of two-bar truss design problem

Two-bar truss design			
Metrics	ABC	EM-MOPSO	NSGA-II
Hypervolume	0.7766	0.7679	0.7737
Spacing	0.0151	0.0203	0.0154
Maximum Spread	1.1063	1.0717	1.1025

5.2 I-Beam Design Problem

The I-beam structural engineering design problem is selected as the second design example of this chapter [39, 40]. The main objective of the problem is to obtain the optimal dimensions of the beam as illustrated in Fig. 7. The objective functions of this problem are both minimizing the cross-sectional area and displacement under affected loads while satisfying the geometric dimensions and strength constraints treated as design variables.

The mathematical model of this structural engineering design problem is as follows:

$$\text{minimizing cross - sectional area; } f_1(x) = 2 * x_2 x_4 + x_3 * (x_1 - 2 * x_4) \quad (17)$$

$$\text{minimizing dispalcement; } f_2(x) = \frac{P * L^3}{48 * E * I} \quad (18)$$

Subjected to

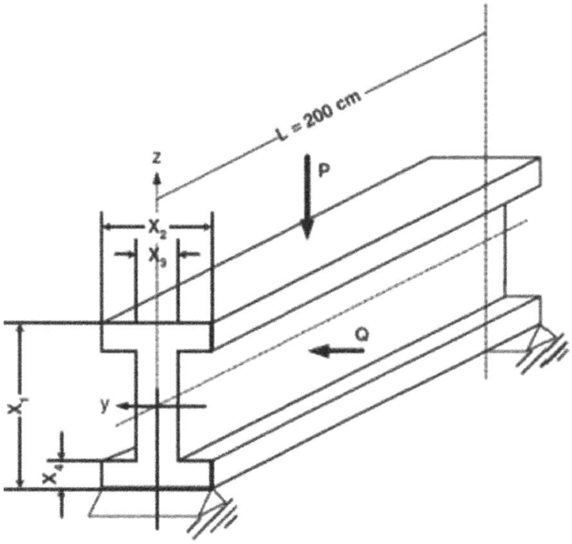

Fig. 7 I-beam

$$g(x) = \left(\frac{M_y}{Z_y} + \frac{M_z}{Z_z}\right) - \sigma_a \leq 0.0 \tag{19}$$

where,

$$I = \frac{1}{12} * \{x_3 * (x_1 - 2 * x_4)^3 + 2 * x_2 * x_4[4 * x_4^2 + 3 * x_1 * (x_1 - 2 * x_4)]\} \tag{20}$$

$$Z_y = \frac{1}{6 * x_1} * \{x_3 * (x_1 - x_4)^3 + 2 * x_2 * x_4[4 * x_4^2 + 3 * x_1 * (x_1 - 2 * x_4)]\} \tag{21}$$

$$Z_z = \frac{1}{6 * x_2} * \{x_3^3 * (x_1 - x_4) + 2 * x_2^3 * x_4\} \tag{22}$$

$$M_y = \frac{P * L}{4} \text{ and } \mathrm{M}_z = \frac{Q * L}{4} \tag{23}$$

$$10 \leq x_1 \leq 80,\ 10 \leq x_2 \leq 50,\ 0.9 \leq x_3 \leq 5,\ 0.9 \leq x_4 \leq 5 \tag{24}$$

Figure 8 presents the non-dominated Pareto-optimal solutions attained by mo-ABC, and Fig. 9 represents both EM-MOPSO and NSGA-II Pareto-optimal solutions for the same structural design problem [39]. It can be understood that mo-ABC brings off with a fine uniform spread final design than other optimizers. The

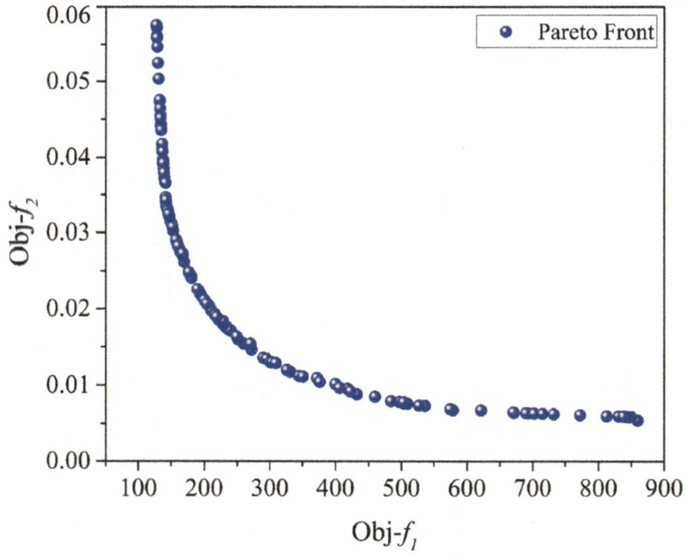

Fig. 8 Pareto-front curve obtained through mo-ABC algorithm for I-beam design problem

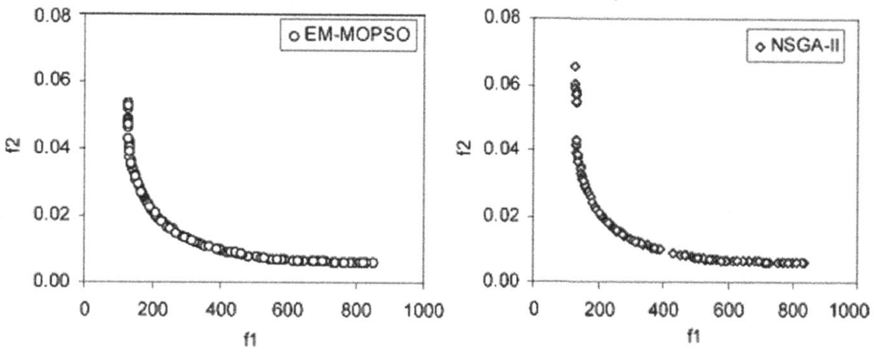

Fig. 9 Pareto-front curve obtained through EM-MOPSO and NSGA-II algorithms for I-beam design problem [39]

mo-ABC accomplishes the minimum cross-sectional area as 127.2510 versus the minimum displacement as 0.057634. Also, mo-ABC yields the minimum displacement as 0.005961 versus the minimum cross-sectional area as 829.5524. The EM-MOPSO preserve the minimum cross-sectional area as 127.9508 versus a displacement as 0.05368, and for the minimum displacement as 0.005961 versus the minimum cross-sectional area as 829.5748. Also, the same problem is resolved using NSGA-II with resulting a minimum cross-sectional area as 127.2341 versus the displacement as 0.0654, and a minimum displacement as 0.0060 versus the minimum cross-sectional area as 829.8684. Thus, the proposed mo-ABC method is capable of finding a relatively broad distribution of Pareto-optimal solutions.

Table 3, not only contains the edge point designs on the Pareto-front curve, but also comprises the design variables and constraint values for MO structural engineering problem of I-beam design. It is clearly seen from this table that regarding to the optimal design variables, there is no any constraint violation. This guarantees the solution ability of mo-ABC for this design problem.

The performance metrics are collected in Table 4 for the mo-ABC, EM-MOPSO, and NSGA_II optimizers concerning to Pareto-optimal sets. According to the Hypervolume and the Maximum Spread metric values, the proposed mo-ABC having higher metric values over EM-MOPSO and NSGA-II, and having lower Spacing metric value strengthen the characteristics of fine convergence and diversity, and a uniform distribution, as well.

Table 3 Optimal designs for two-bar truss design problem attained by mo-ABC

design variables and constraint		x_1	x_2	x_3	x_4	$g(x)$
Border tail points Objective Function values	$\max f_2 = 0.057634$	62.99	40.10	0.90	0.90	−3.242443
	$\min f_1 = 127.2510$					
	$\min f_2 = 0.005961$	79.99973	50.00	4.70803	4.99991	−14.30745
	$\max f_1 = 829.5524$					

Table 4 Performance metrics of I-beam design problem

I-Beam Design			
Metrics	ABC	EM-MOPSO	NSGA-II
Hypervolume	0.7992	0.7655	0.7606
Spacing	0.0124	0.0315	0.0255
Maximum Spread	1.0141	0.945	1.0123

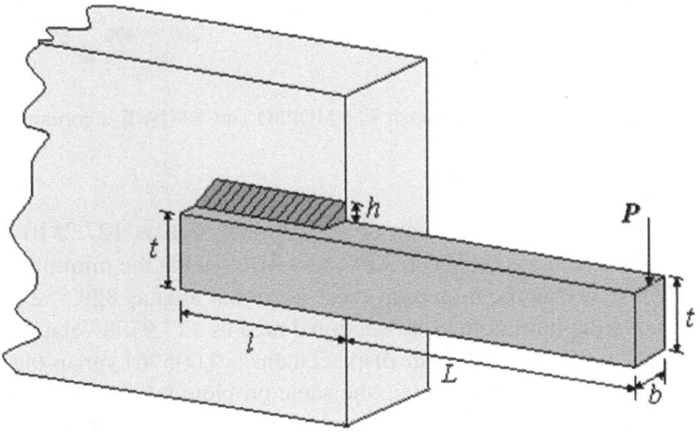

Fig. 10 Steel welded beam

5.3 Steel Welded Beam Design Problem

A steel welded beam is selected as third MO structural engineering design problem. The 3-D view and the structural dimensions of the beam are illustrated on Fig. 10. The MO design optimization of this beam is executed to bear a certain concentrated load by synchronously meeting the minimum overall fabrication cost and the smallest deflection under the tip (load applied) point. This MO structural engineering design optimization problem has four design variables which are thickness of weld $h = x_1$, length of weld $l = x_2$, beam width $t = x_3$, and beam thickness $b = x_4$ [39, 41].

The mathematical formula of the problem consisting two objective functions and five design constraints (the shear and bending stresses in the beam, the end deflection of the beam, the buckling load on the bar, and a side constraint,) are exhibited in Eq. (25) to Eq. (31).

Minimize;

$$f_1(x) = 1.10471\, x_1^2\, x_2 + 0.04811\, x_3\, x_4\, (14.0\; + x_2) \tag{25}$$

$$f_2(x) = \delta(x) = \frac{4\, P\, L^3}{E\, x_3^3\, x_4} = \frac{2.1952}{x_4 x_3^3} \tag{26}$$

subjected to

$$g_1(x) = \tau(x) - \tau_{\max} \leq 0 \tag{27}$$

$$g_2(x) = \sigma(x) - \sigma_{\max} \leq 0 \tag{28}$$

$$g_3(x) = P - P_c(x) \leq 0 \tag{29}$$

$$g_4(x) = x_1 - x_4 \leq 0 \tag{30}$$

Also, the design variable interval of the MO steel welded beam design problem is restricted as in Eq. (31).

$$0.125 \leq x_1 \leq 5.0,\ 0.1 \leq x_2 \leq 10,\ 0.1 \leq x_3 \leq 10,\ 0.125 \leq x_4 \leq 5.0 \tag{31}$$

where;

$$\tau(x) = \sqrt{(\tau')^2 + \frac{2\tau'\tau'' x_2}{2R} + (\tau'')^2},\ \sigma(x) = \frac{6PL}{x_4 x_3^2},$$

$$\tau' = \frac{P}{\sqrt{2}x_1 x_2},\ \tau'' = \frac{MR}{J},\ M = P\left(L + \frac{x_2}{2}\right),\ R = \sqrt{\frac{x_2^2}{4} + \left(\frac{x_1 + x_3}{2}\right)^2},$$

$$J = 2\left\{x_1 x_2 \sqrt{2}\left[\frac{x_2^2}{12} + \left(\frac{x_1 + x_3}{2}\right)^2\right]\right\},\ P_c(x) = \frac{4.013E\sqrt{\frac{(x_3^2 x_4^6)}{36}}}{L^2}\left(1 - \frac{x_3}{2L}\sqrt{\frac{E}{4G}}\right)$$

in which,

$$P = 6000 \text{ lb},\ L = 14 \text{ in},\ E = 30 \times 10^6 \text{ psi},\ G = 12 \times 10^6 \text{ psi}$$
$$\tau_{\max} = 13600 \text{ psi},\ \sigma_{\max} = 30000 \text{ psi}.$$

Figure 11 portrays the Pareto-front curve formed by non-dominated solutions through mo-ABC developed with the aim of this chapter. Also, the Pareto-front curves yielded by EM-MOPSO and NSGA-II methods are depicted in Fig. 12.

From these figures, the border tail points achieved by mo-ABC are (0.008261711 (maxf_2), 2.304837339 (minf_1)) and (0.000549 (minf_2), 29.312096 (maxf_1)). The border tail point values for minimum fabrication cost and related deflection (minf_1, maxf_2) are reported as (0.0157, 2.382) accomplished utilizing EM-MOPSO, and as (0.0101, 3.443) attained using NSGA-II [39]. It can be concluded from these edge values that proposed mo-ABC evidently generates better designs for minf_1 than EM-MOPSO and NSGA-II. Similarly, the minimized deflection (minf_2) associated

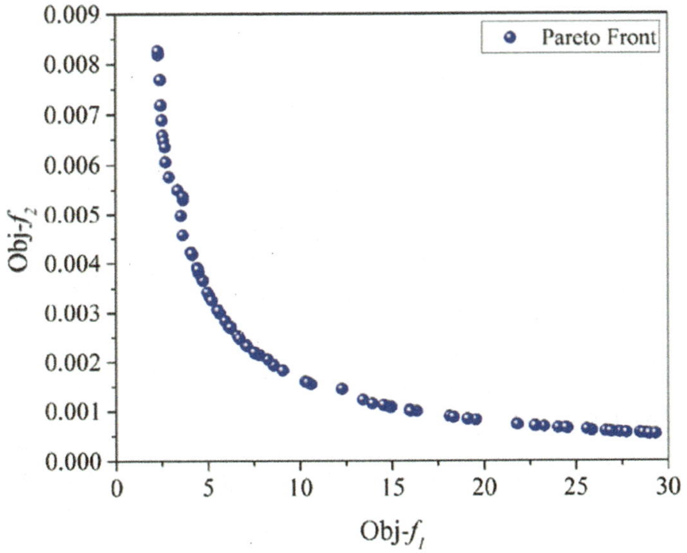

Fig. 11 Pareto-front curve achieved through Pareto-based mo-ABC algorithm for steel welded beam design problem

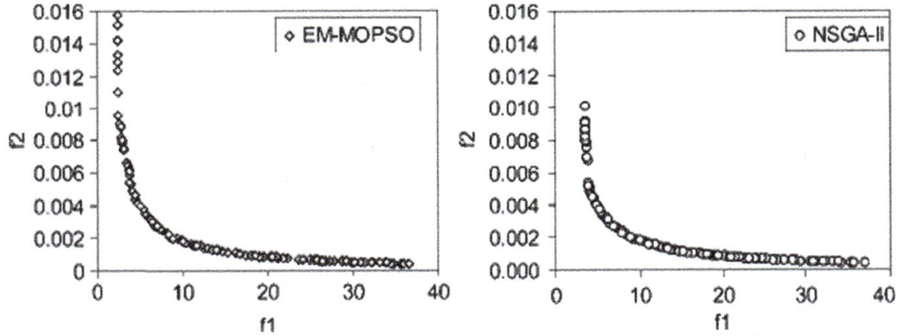

Fig. 12 Pareto-front curve attained through EM-MOPSO and NSGA-II algorithms for steel welded beam design problem [39]

to cost of fabrication (maxf_1) obtained by EM-MOPSO and NSGA-II are (0.000439, 36.4836) and (0.004, 36.9121). These results, also, guarantee that the proposed mo-ABC is a promising method in outcome with a great variety of Pareto-optimal non-dominated solutions.

The design variables and constraints values, by which the so-called global optimum trade-off MO design is attained, are tabulated in Table 5. It is apparently comprehended from this table that there is no any constraint violation proving the hegemony of proposed mo-ABC over this structural engineering design problem.

Table 5 Optimal designs for steel welded beam design problem achieved by the mo-ABC

Design variable values		x_1	x_2	x_3	x_4	g_1	g_2	g_3	g_4
Border tail points Objective Function values	$\max f_2 = 0.008262$ $\min f_1 = 2.304837$	0.2333	2.7035	9.983664	0.26701	−8.49	−11062.77	−0,0337	−7972.35
	$\min f_2 = 0.000549$ $\max f_1 = 29.31209$	0.52398	1.06409	10.0	4.0	−6.33	−28,740.0	−3.4760	−47013504

Table 6 Performance metrics of steel welded beam design problem

Welded beam design			
Metrics	ABC	EM-MOPSO	NSGA-II
Hyper volume	0.8649	0.8564	0.8396
Spacing	0.0104	0.0264	0.0178
Maximum spread	0.8292	1.2934	1.0424

In Table 6, the performance metrics of the mo-ABC, EM-MOPSO, and NSGA-II optimizers are represented for the steel welded beam MO structural engineering design problem. From this table, it can easily be reckoned that the proposed mo-ABC surpassed the EM-MOPSO and NSGA-II algorithms in terms of the Hypervolume, Spacing, and Maximum Spread. So, it is apparent that mo-ABC attained not only a better convergence and diversity, but also a good solution distribution. This verifies that the mo-ABC arrives at the best non-dominated solutions set.

6 Conclusions

In this chapter, for hard to solve MO structural engineering design optimization problems, a Pareto-based MO artificial bee colony (mo-ABC) algorithm is submitted. The principal idea of the proposed mo-ABC mimics the food search characteristics of honey bees to be used for breeding. For illustrating the algorithmic capacity of the proposed mo-ABC, three MO structural engineering design optimization problems are implemented. These problems are a two-bar truss design, an I-beam design, and a steel welded beam design. The algorithm includes a constraint handling strategy for MO problems with discrete design variables without any restrictions on the number of design variables, constraints, and objective functions. Moreover, the proposed mo-ABC does not require any additional gradient information. The algorithm uses a Pareto-optimal strategy to acquire the intricate designs via non-dominated solutions. The algorithm process is effective to obtain border tail points of the Pareto-front curve while preserving a fine spread over the curve. The yielded Pareto-front curves in terms of performance metrics having well convergence and diversity in non-dominated Pareto-optimal solutions sets. The metrics resolutions show that the mo-ABC has enough quality and quantity for solving MO structural engineering design problems. Besides, the accomplished optimal designs are compared with previously announced ones to verify the efficacy and robustness of mo-ABC. The optimum designs acquired in this chapter reveal that the proposed mo-ABC has a powerful tool to generate an optimal Pareto-front curve and provide illustrious non-dominated MO designs for structural engineering design optimization problems.

References

1. Zavala, G., Nebro, A.J., Luna, F., Coello Coello, C.A.: Structural design using multi-objective metaheuristics. Comparative study and application to a real-world problem. Struct. Multidiscip. Optim. **53**, 545–566 (2016). https://doi.org/10.1007/s00158-015-1291-3.
2. Zhou-Kangas, Y., Miettinen, K.: Decision making in multiobjective optimization problems under uncertainty: balancing between robustness and quality. OR Spectr. **41**, 391–413 (2019). https://doi.org/10.1007/s00291-018-0540-4
3. Padhye, N., Deb, K.: Multi-objective optimisation and multi-criteria decision making in SLS using evolutionary approaches. Rapid Prototyp. J. **17**, 458–478 (2011). https://doi.org/10.1108/13552541111184198
4. Zitzler, E., Thiele, L.: Multiobjective evolutionary algorithms: a comparative case study and the strength Pareto approach. IEEE Trans. Evol. Comput. **3**, 257–271 (1999). https://doi.org/10.1109/4235.797969
5. Deb, K., Agrawal, S., Pratap, A., Meyarivan, T.: A fast elitist non-dominated sorting genetic algorithm for multi-objective optimization: NSGA-II. In: Schoenauer, M., et al. (eds.) International Conference on Parallel Problem Solving from Nature, pp. 849–858. Springer, Berlin, Heidelberg (2000). https://doi.org/10.1007/3-540-45356-3_83
6. Zitzler, E., Laumanns, M., Thiele, L.: SPEA2: Improving the Strength Pareto Evolutionary Algorithm. Zurich (2001)
7. Coello Coello, C.A., Pulido, G.T., Lechuga, M.S.: Handling multiple objectives with particle swarm optimization. IEEE Trans. Evol. Comput. **8**, 256–279 (2004). https://doi.org/10.1109/TEVC.2004.826067
8. Mirjalili, S., Jangir, P., Saremi, S.: Multi-objective ant lion optimizer: a multi-objective optimization algorithm for solving engineering problems. Appl. Intell. **46**, 79–95 (2017). https://doi.org/10.1007/s10489-016-0825-8
9. Mirjalili, S.: Dragonfly algorithm: a new meta-heuristic optimization technique for solving single-objective, discrete, and multi-objective problems. Neural Comput. Appl. **27**, 1053–1073 (2016). https://doi.org/10.1007/s00521-015-1920-1
10. Mirjalili, S.Z., Mirjalili, S., Saremi, S., Faris, H., Aljarah, I.: Grasshopper optimization algorithm for multi-objective optimization problems. Appl. Intell. **48**, 805–820 (2018). https://doi.org/10.1007/s10489-017-1019-8
11. Sadollah, A., Eskandar, H., Kim, J.H.: Water cycle algorithm for solving constrained multi-objective optimization problems. Appl. Soft Comput. J. **27**, 279–298 (2015). https://doi.org/10.1016/j.asoc.2014.10.042
12. Jangir, P., Indrajit Trivedi, N., Engineer, J., Rajya Vidyut Prasaran Nigam Losal-Sikar, R.: Non-dominated sorting moth flame optimizer: a novel multi-objective optimization algorithm for solving engineering design problems. Eng. Technol. Open. Acc. **2** (2018)
13. Karaboga, D., Basturk, B.: A powerful and efficient algorithm for numerical function optimization: Artificial bee colony (ABC) algorithm. J. Glob. Optim. **39**, 459–471 (2007). https://doi.org/10.1007/s10898-007-9149-x
14. Carbas, S.: Optimum design of low-rise steel frames made of cold-formed thin-walled steel sections (2013)
15. Sonmez, M.: Artificial Bee Colony algorithm for optimization of truss structures. Appl. Soft Comput. J., pp. 2406–2418 (2011).https://doi.org/10.1016/j.asoc.2010.09.003
16. Aydogdu, I., Carbas, S., Akin, A.: Effect of Levy Flight on the discrete optimum design of steel skeletal structures using metaheuristics. Steel Compos. Struct. **24** (2017). https://doi.org/10.12989/scs.2017.24.1.093
17. De Weck, O.L.: Multiobjective optimization: history and promise. In: The Third China-Japan-Korea Joint Symposium on Optimization of Structural and Mechanical Systems. Kanazawa (2004).
18. Deb, K.: Multi-objective optimization using evolutionary algorithms. Wiley, New York (2001)
19. Deb, K.: Evolutionary algorithms for multi-criterion optimization in engineering design. Evol. Algorithms Eng. Comput. Sci. **8** (1999)

20. Deb, K.: Multi-Objective Optimization Using Evolutionary Algorithms: An Introduction. , Kanpur (2011).
21. Zitzler, E.: Evolutionary Algorithms for Multiobjective Optimization: Methods and Applications (1999)
22. Zitzler, E., Thiele, L., Laumanns, M., Fonseca, C.M., Da Fonseca, V.G.: Performance assessment of multiobjective optimizers: an analysis and review (2003). https://doi.org/10.1109/TEVC.2003.810758
23. Santana, R.A., Pontes, M.R., Bastos-Filho, C.J.A.: A multiple objective particle swarm optimization approach using crowding distance and roulette wheel. In: ISDA 2009—9th International Conference on Intelligent Systems Design and Applications, pp. 237–242 (2009). https://doi.org/10.1109/ISDA.2009.73.
24. Zitzler, E., Deb, K., Thiele, L.: Comparison of multiobjective evolutionary algorithms: empirical results. Evol. Comput. **8**, 173–195 (2000)
25. Schott, J.R.: Fault tolerant design using single and multicriteria genetic algorithm optimization (1995)
26. Karaboga, D., Akay, B.: A survey: algorithms simulating bee swarm intelligence. Artif. Intell. Rev. **31**, 61–85 (2009). https://doi.org/10.1007/s10462-009-9127-4
27. Karaboga, D.: An idea based on Honey Bee swarm for numerical optimization. Kayseri (2005)
28. Karaboga, D.: Artificial bee colony algorithm. https://www.scholarpedia.org/article/Artificial_bee_colony_algorithm. Last accessed 3 May 2020
29. Teodorović, D.: Bee colony optimization (BCO). Springer, Berlin, Heidelberg (2009). https://doi.org/10.1007/978-3-642-04225-6_3
30. Yilmaz Acar, Z., Aydemir, F., Basciftci, F.: A new multi-objective artificial bee colony algorithm for multi-onjective optimization problems. J. Selcuk. 144–152 (2018)
31. Deb, K., Pratap, A., Agarwal, S., Meyarivan, T.: A fast and elitist multiobjective genetic algorithm: NSGA-II. IEEE Trans. Evol. Comput. **6**, 182–197 (2002). https://doi.org/10.1109/4235.996017
32. Saka, M.P., Carbas, S., Aydogdu, I., Akin, A.: Use of swarm intelligence in structural steel design optimization (2016). https://doi.org/10.1007/978-3-319-26245-1_3
33. Saka, M.P., Carbas, S., Aydogdu, I., Akin, A., Geem, Z.W.: Comparative study on recent metaheuristic algorithms in design optimization of cold-formed steel structures (2015). https://doi.org/10.1007/978-3-319-18320-6_9
34. Carbas, S.: Artificial Bee Colony Based optimum design of reticulated domes To LRFD-AISC. Int. J. Adv. Mech. Civ. Eng. **2**, 68–74 (2015)
35. Coello Coello, C.A.: Theoretical and numerical constraint-handling techniques used with evolutionary algorithms: a survey of the state of the art. Comput. Methods Appl. Mech. Eng. **191**, 1245–1287 (2002). https://doi.org/10.1016/S0045-7825(01)00323-1
36. Hasançebi, O., Çarbaş, S., Doğan, E., Erdal, F., Saka, M.P.: Performance evaluation of metaheuristic search techniques in the optimum design of real size pin jointed structures. Comput. Struct. **87** (2009). https://doi.org/10.1016/j.compstruc.2009.01.002
37. Hasançebi, O., Çarbaş, S., Doğan, E., Erdal, F., Saka, M.P.: Comparison of non-deterministic search techniques in the optimum design of real size steel frames. Comput. Struct. **88** (2010). https://doi.org/10.1016/j.compstruc.2010.06.006
38. Deb, K., Pratap, A., Moitra, S.: Mechanical component design for multiple objectives using elitist non-dominated sorting GA. In: Lecture Notes in Computer Science (LNCS), pp. 859–868. Springer, Berlin, Heidelberg (2000). https://doi.org/10.1007/3-540-45356-3_84
39. Reddy, M.J., Kumar, D.N.: An efficient multi-objective optimization algorithm based on swarm intelligence for engineering design. Eng. Optim. **39**, 49–68 (2007). https://doi.org/10.1080/03052150600930493
40. Yang, B.S., Yeun, Y.S., Ruy, W.S.: Managing approximation models in multiobjective optimization. Struct. Multidiscip. Optim. **24**, 141–156 (2002). https://doi.org/10.1007/s00158-002-0224-0
41. Ray, T., Liew, K.M.: A swarm metaphor for multiobjective design optimization. Eng. Optim. **34**, 141–153 (2002). https://doi.org/10.1080/03052150210915

Optimal Parameter Identification of Fuzzy Controllers in Nonlinear Buildings Based on Seismic Hazard Analysis Using Tribe-Charged System Search

Siamak Talatahari and Mahdi Azizi

Abstract Tabriz is one of the ancient cities in the North-West of Iran that has experienced many severe earthquakes for a long time. Implementing control strategies for buildings encountering severe earthquakes can prevent damages in structural members. Smart control strategies are capable of reducing ground motion effects by utilizing efficient control algorithms. Fuzzy logic controller is one of the most effective control algorithms that is mostly formulated based on the human knowledge and expertise. In many cases, the expert knowledge does not yield optimal control responses for structures under strong ground motions so the optimization of these controllers is concerned. The main aim of this paper is to optimize a fuzzy controller implemented in a 20-story steel structure with nonlinear behavior. In most cases, this problem is formulated based on the linear behavior of the structure; however, in this paper, the objective function and the performance criteria are selected based on the nonlinear characteristics of the structure. The Tribe-Charged System Search algorithm is proposed and utilized for optimization of membership functions and rule base of fuzzy controller. The seismic inputs for nonlinear dynamic analysis is selected thorough the energy based ground motion selection and modification method by utilizing the probabilistic seismic hazard analysis for Tabriz. The performance of the proposed algorithm is compared with the standard Charged System Search Algorithm and eight different metaheuristic algorithms. The obtained results prove that the upgraded method is capable of providing competitive results in reducing building responses and damages due to the destructive earthquake records.

Keywords Fuzzy logic controller · Seismic hazard analysis · Optimization · Nonlinear structure · Tribe-charged system search · Tabriz

S. Talatahari (✉) · M. Azizi
Department of Civil Engineering, University of Tabriz, Tabriz, Iran
e-mail: talatahari@tabrizu.ac.ir

M. Azizi
e-mail: mehdi.azizi@tabrizu.ac.ir

S. Talatahari
Engineering Faculty, Near East University, Mersin 10, North Cyprus, Turkey

S. M. Nigdeli et al. (eds.), *Advances in Structural Engineering—Optimization*, Studies in Systems, Decision and Control 326,
https://doi.org/10.1007/978-3-030-61848-3_4

1 Introduction

Over the past few decades, there has been an increasing interest in Seismic Hazard Analysis (SHA) that results in some realistic models for describing seismic sources, seismic attenuation, earthquake recurrence frequency, and occurrence probability of earthquakes at a site. There are two kinds of quantitative approaches for SHA as Deterministic SHA (DSHA) and Probabilistic SHA (PSHA) while in both approaches, seismological records of past earthquakes are needed. DSHA is an approach that is formulated based on some specific earthquake scenarios. PSHA is an approach that estimates the probability of exceeding various levels of ground motion at a site (or a map of sites) for all possible earthquakes. In this approach, some specific components such as magnitude, source to site distance and the deviation measure of the ground motion from a predicted or median value based on seismological activities are utilized for hazard analysis. The results of the PSHA are presented as estimated probabilities or frequencies per unit time or as expected number of events per year. In order to identify the individual earthquake scenarios that mostly contribute to the hazard, a deaggregation process should be conducted. In this process, the exceedance probability of earthquakes based on the contributions of different magnitude-distance pairs is represented as a guide for selection of response spectra or acceleration time histories for dynamic analysis.

The PSHA has been developed over decades and the empirical statistical approaches [1] have been largely renewed by some analytical and numerical attempts [2]. Cornell [3] conducted the first analytical method for PSHA in 1968. The most comprehensive research in this field was presented by the Senior Seismic Hazard Analysis Committee (SSHAC) report [4], which covers many methodological issues. Many PSHA researches have been conducted around the world especially for Iran [5–10] while some of them discussed the city of Tabriz [11–14] which is located in North-West (N-W) of Iran. The N-W of Iran is a region of intense seismicity and deformation situated between two thrust belts of the Caucasus and the Zagros Mountains from north to south. The north Tabriz fault, which is one of the well-known seismogenic faults in this region, is passing within striking distance of urban area and has a major history of intense seismic activity [15].

One of the possible options in designing engineering structures is to install a smart system to provide the stiffness or damping required to withstand severe dynamic loads. In recent years, using control systems to reduce the response of structures under different types of dynamic loads such as earthquake, wind and shock loads is a worldwide concern. Active control systems are a kind of intelligent systems that use an external energy source to generate control power for structural response control. An important issue in the field of active control is how to determine the control force with respect to external excitation and structural response in such a way that the behavior of the structure is within the predetermined range. In other words, the choice of an effective control algorithm has a significant role on the efficiency of the control system. Many control algorithms have been proposed for active structural control applications each of which has certain advantages, depending on the specific applications and desired objectives. The Fuzzy Logic Controller (FLC) has been investigated as a control algorithm for the active control of civil engineering structures for many years. In order to design an FLC, knowledge or experience is required for construction of control rules and stating the membership functions. Using human knowledge and experience for this purpose does not yield optimal control responses for a given structure and tuning of these parameters is necessary [16]. A major drawback of FLC is that the tuning process of control rules and membership functions becomes more difficult and very time consuming when the number of the system inputs and outputs are increased. Nevertheless, their performance in practice has proven to be quite good, and they are quite versatile [17].

In recent years, there is an increasing interest to optimize the FLC with heuristic, metaheuristic and nature inspired approaches. The term "optimization" refers to the study of problems in which minimization or maximization of a function is required. Kaveh [18] developed some effective and efficient optimization algorithms and discussed the application of these algorithms in real projects. Kaveh [19] also presented the applications of metaheuristic optimization algorithms in civil engineering. Kaveh and Ghazaan [20] presented the utilizing of metaheuristic algorithms for optimal design of real-size structures. Fu et al. [21] utilized the Genetic Algorithm (GA) for optimization of a nonlinear fuzzy controller to use in the magnetorheological elastomer isolators. Soltani et al. [22] utilized Particle Swarm Optimization (PSO) to design a fuzzy sliding mode controller based on parallel-distributed compensator in order to overcome the problem caused by an inappropriate selection of sliding surface parameters. Chen et al. [23] discussed parameter estimation of fuzzy sliding mode controller for hydraulic turbine regulating system with Imperialist Competitive Algorithm (ICA). Olivas et al. [24] generalized the type-2 fuzzy logic approach for dynamic parameter adaptation with Ant Colony Optimization (ACO) algorithm in order to optimize the membership functions and the rule base. Amador-Angulo and Castillo [25] utilized Bee Colony Optimization (BCO) method for optimal distribution of the membership functions in the design of fuzzy controllers for complex nonlinear plants. Chamorro et al. [26] proposed the application of Differential Evolution (DE) algorithm for optimal tuning of a fuzzy controller in order to improve the synthetic inertia control in power systems. Ray et al. [27] discussed the parameter tuning of fuzzy Proportional-Integral-Derivative (PID) based power system stabilizer under different operating conditions by Firefly Algorithm (FA). Chrouta et al. [28] presented optimal control of an irrigation station process based on the FLC with Takagi–Sugeno (TS) model using Cuckoo Search Algorithm (COA). Giri and Bera [29] developed a fuzzy proportional-integral (PI) controller for damping the oscillation of system frequency of distributed energy generation in wind turbines and optimized the gains of this controller with Grey Wolf Optimization (GWO) algorithm. Chao et al. [30] utilized the Harmony Search (HS) algorithm to find optimal network parameters for better performances of fuzzy cerebellar model articulation controller. Li et al. [31] designed a fuzzy PID controller for a nonlinear hydraulic turbine governing system by using the Gravitational Search Algorithm (GSA). In addition, among these researches, some comparative studies based on utilizing different metaheuristic algorithms has been widely studied [32–36].

In most cases, the optimization of the FLC with metaheuristic algorithms is formulated based on the linear behavior of the structure [37]. In addition, it can be found that the mathematical models of the structures used in researches are not the accurate model and the objective functions in the optimization problems are chosen based on the imprecise models of the structure [38]. The main contribution of this paper is to optimize the FLC with the Charged System Search (CSS) algorithm and propose an improved version of this algorithm as the Tribe-Charged System Search (Tribe-CSS) in order to improve the performance of the standard CSS algorithm. The CSS, recently proposed by Kaveh and Talatahari [39], is a novel metaheuristic algorithm that that is formulated based on some principles of mechanics and physics that mimics the Newtonian and Coulomb laws of mechanics and electrostatics to formulate a search technique. This algorithm has very fast convergence rate with a special adaptive mechanism that strikes a balance between exploration and exploitation. It could be noted that this algorithm is capable of finding the global solution and is simple in implementation. The proposed upgrading process is based on the "continuous-time" concept in which it's assumed that time changes continuously in the optimization process and all updating processes are performed after creating just one solution. In this paper, the Tribe-CSS will be applied for optimization of the FLC implemented in a 20-story steel structure, while the mathematical model of the structure is some kind of precise model and the nonlinearity of the structure is taken into account. The seismic inputs for dynamic analysis are considered based on the PSHA information from the Moghaddam et al. [14] research for Tabriz city. The earthquake records in optimization problem are selected based on the Energy-Based Ground Motion Selection and Modification (EBGMSM) method proposed by Marasco and Cimellaro [40]. The EBGMSM is an efficient method with low variability of parameters and accuracy in preserving the median demand for a given hazard scenario. The selected earthquake records based on the mentioned method are utilized as seismic input for the 20-story structure and the capability of the Tribe-CSS is examined with respect to the destructive effects of these ground motions. Also, the performance and capability of the Tribe-CSS in FLC optimization is compared with the standard CSS and various classical and advanced optimization algorithms. In the implementation of these algorithms for an FLC tuning process, the latest and developed versions of these algorithms are utilized.

2 Seismic Hazard Analysis

The north Tabriz fault is an active fault with lateral strike-slip displacement and a vertical displacement on the north side. This fault has an average strike over a length of about 150 km similar to the vertical direction in the dip [41]. Some of the most destructive historical earthquakes generated by this fault are the 1042, 1721, 1780 and 1786 earthquakes with magnitudes of 7.6, 7.7, 7.7 and 6.3 respectively [42]. A summary of historical and instrumental earthquakes that have occurred in Tabriz is displayed in Fig. 1.

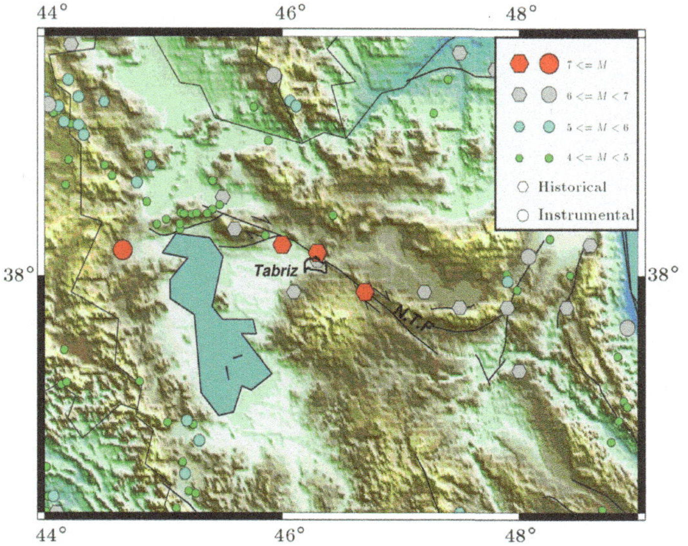

Fig. 1 Historical and instrumental earthquakes around north Tabriz fault [14]

As discussed by Masson et al. [43], based on the seismic events during the 700 years, the recurrent time intervals for Tabriz is about 250 years. According to the paleoseismological studies over the past 3600 years, Hessami et al. [15] suggested 821 ± 176 years for this parameter. Moghaddam et al. [14] conducted a complete study of the PSHA and the seismic hazard deaggregation for Tabriz. Based on a maximum likelihood method, they estimated the earthquake magnitude and source to site distance for a 10% probability of exceedance within 50 years (475-year return period). The seismic hazard deaggregation plot for Tabriz is displayed in Fig. 2.

Based on the PSHA for Tabriz [14] and the seismic hazard deaggregation plot for a 10% probability of exceedance within 50 years (475-year return period), the corresponding Uniform Hazard Spectrum (UHS) are depicted in Fig. 3. It should be noted that based on the magnitude (M), source to site distance (R) and the shear wave velocity (VS) for different periods, the UHS values are calculated.

In order to perform a dynamic analysis, selection and modification of earthquake records is concerned. In this study, the earthquake records in optimization problem are selected based on the EBGMSM method. This method emerges from comparing a set of horizontal ground motions at various ranges of frequency with a target spectrum. As mentioned by Marasco and Cimellaro [40], the Predicted Mean Spectrum (PMS), Conditional Mean Spectrum (CMS), Design Spectrum (DS) or Uniform Hazard Spectrum (UHS) are some of options for considering as target spectrum and the records with compelling contribution to the target spectrum are selected. In this purpose, the UHS for Tabriz is selected as the target spectrum. In the EBGMSM method, a set of ground motions with close seismic activity that matches the target spectrum at the period of interest (Tref) are considered. Two of the most common

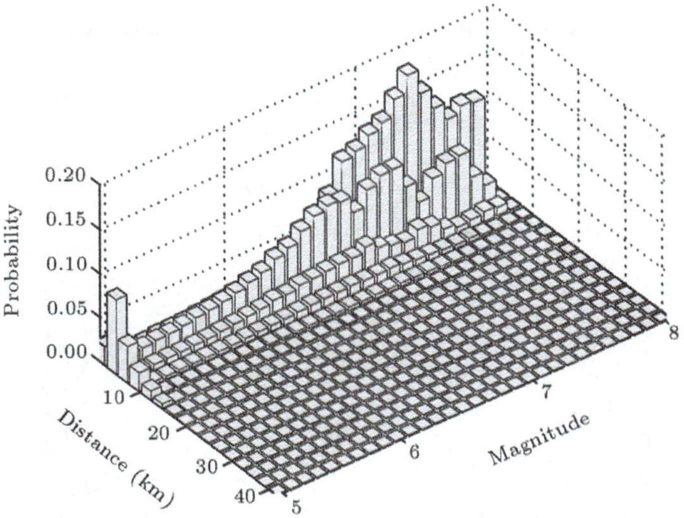

Fig. 2 Tabriz deaggregation plot for 10% exceedance probability in 50 years [14]

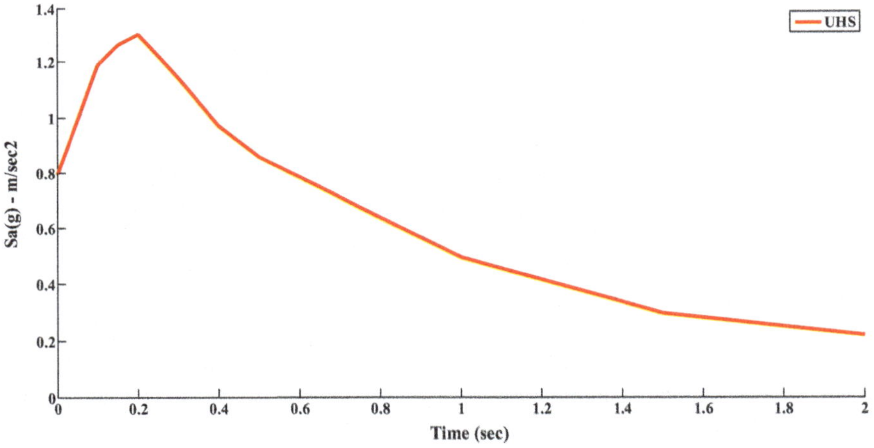

Fig. 3 Uniform Hazard Spectrum for Tabriz

ground motion databases as the Pacific Earthquake Engineering Research Center (PEER) and the European Strong-Motion Data (ESMD) are considered for data selection. The PEER database includes 600 shallow crustal events with 21,336 three-component records that has a magnitude range of 3–7.9, a rupture distance range of 0.05–1533 km, and the shear wave velocity at the site (in the top 30 m) range of 94 to 2100 m/s. The ESMD database includes 462 triaxial ground motion records from 110 earthquakes and 261 stations in Europe and the Middle East. The scaled mean spectrum of the ground motions should have an equivalent Housner Intensity

(HI) for the periods of $0.2T_{ref}$ to $2T_{ref}$. By utilizing a specific index in terms of the energy-frequency trend's shape and its scattering degree around the mean value, the horizontal components for each band of frequency are obtained which results in a set of spectrum-compatible records with almost identical severity and low dispersion of the structural response parameters [40].

By performing the EBGMSM method with UHS for Tabriz as the target spectrum and utilizing the OPENSIGNAL [44, 45] as a computer-based software for signal processing, the specifications and principles of the selected earthquake records are presented in Tables 1 and 2.

The selected ground motion records cover a magnitude range from 6 to 7.62 and a distance range from 0.27 to 8.92 km to the fault. In Fig. 4, the acceleration and mean spectra for the selected earthquakes are compared with the target spectrum

Table 1 Characteristics of the selected earthquake records

EQ	Earthquake—Date	M	R (km)	Fault mechanism
EQ1	South Iceland—2000	6.5	6	Strike Slip
EQ2	Chi Chi—1999	7.62	2.2	Reverse Oblique
EQ3	North Palm Springs—1986	6.06	6.04	Reverse Oblique
EQ4	Imperial Valley—1979	6.53	2.66	Strike Slip
EQ5	Tabas—1978	7.35	2.05	Reverse
EQ6	Umbria Marche—1997	6	8.92	Normal
EQ7	Kobe—1995	6.9	0.27	Strike Slip

Table 2 Specific details of the selected earthquake records

EQ	Component	Scale Factor	PGA (g)	PGV (cm/s)
EQ1	006263ya	1.49	0.9330	67.52
	006263za	1.20	0.5783	55.63
EQ2	WNT90	0.92	0.8866	63.64
	WNT0	0.94	0.5907	39.63
EQ3	WWT180	1.72	0.8478	59.70
	WWT270	1.62	0.9971	51.24
EQ4	BCR140	1.17	0.6869	52.76
	BCR230	1.11	0.8623	51.09
EQ5	TABL1	0.69	0.5832	68.18
	TABT1	0.75	0.6391	90.92
EQ6	NCR000	1.63	0.9487	55.48
	NCR270	1.78	0.8928	49.26
EQ7	TAZ000	0.82	0.5668	55.76
	TAZ090	0.95	0.6588	80.99

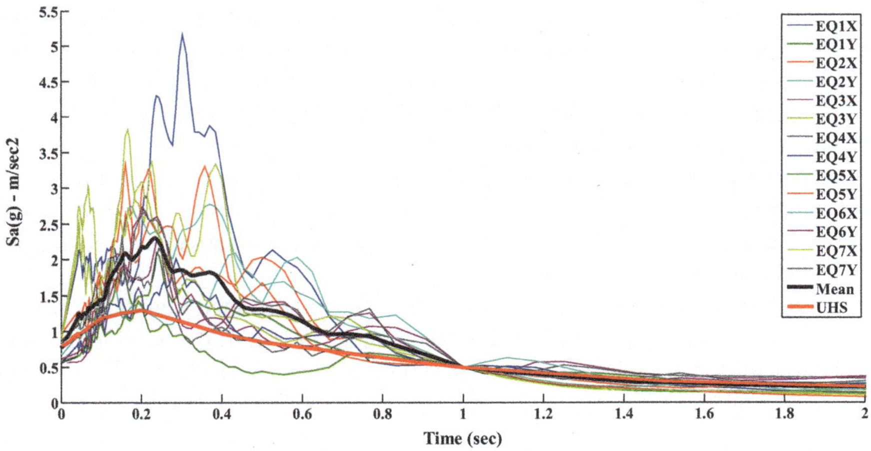

Fig. 4 Comparison of acceleration and mean spectra of selected earthquakes with the UHS

(UHS). It should be mentioned that the mean spectrum and the UHS are encouraged to be compatible for a wide range of periods especially for periods larger than one second that is concerned in high-rise buildings. The scale factor in EBGMSM method is utilized for this purpose. Both the North-South (N-S) and the East-West (E-W) components of the selected earthquake records can be considered for dynamic analysis. Because of the fact that the N-S components of the earthquake records have higher Peak Ground Acceleration (PGA) and Peak Ground Velocity (PGV) values and the numerical study is conducted as a 2-Dimentional (2-D) approach so the N-S components of the selected earthquake records are considered for dynamic analysis. The acceleration time histories of these components are depicted in Fig. 5.

In Fig. 6, the acceleration, velocity and displacement spectra of the N-S components of the selected earthquake records along with the mean spectra are illustrated for 5% of damping. In this illustration, the severity of the selected ground motions is in perspective.

3 Fuzzy Logic Controller

By moving human to the information age, human knowledge and expertise become increasingly important. For many years, humankind has been looking for a model for natural intelligence and human deduction. Human uses quantitative and qualitative information for their processing. Qualitative information is expressed in terms of verbal expressions (linguistic). Conventional processors are unable to analyze the qualitative information and human language. Therefore, a hypothesis that can formulate the human language or generally human knowledge in a systematic way and put it along with other mathematical models in engineering systems are needed.

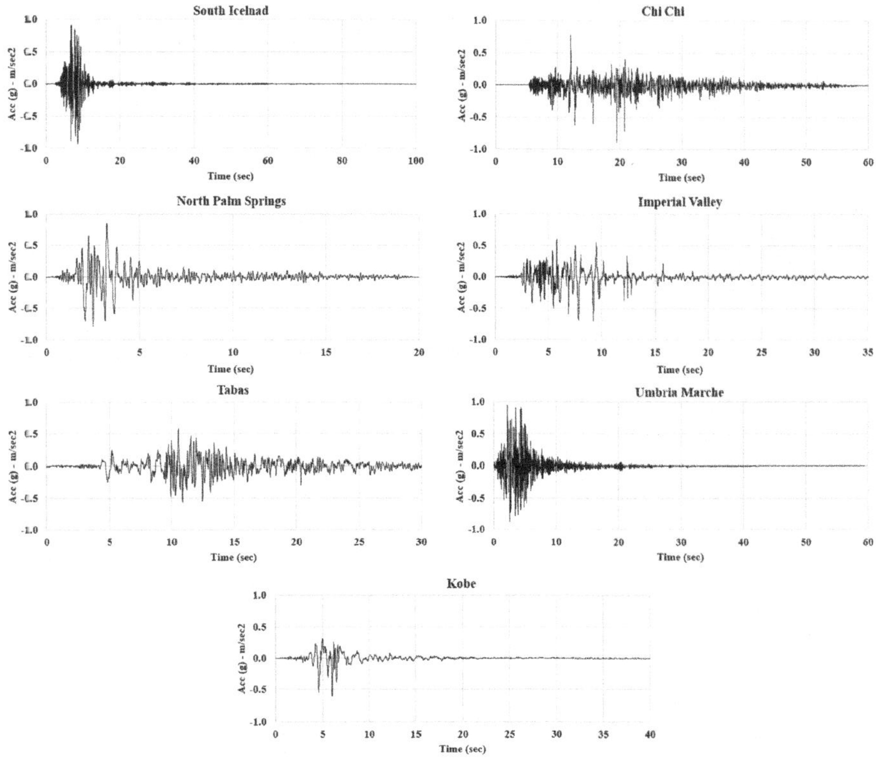

Fig. 5 Acceleration time histories of the selected earthquake records (N-S components)

The theory of fuzzy sets can model verbal expressions and approximate deductions and respond to these needs.

Many commonly used control methods are based on a model, which means that the controller design and parameter adjustment are based on the mathematical model of the system. Linear and PID controllers are formulated based on a model. However, in most cases, the mathematical model of the systems are not accurate and the control methods based on these models are not efficient. In such cases, a fuzzy system based on the expert knowledge can be designed to effectively control the system even if the mathematical model is completely inaccessible. In fact, one of the main applications of fuzzy systems is to control the closed loop of nonlinear systems where the mathematical models are unknown. In Fig. 7, a closed-loop control system with FLC as control algorithm is demonstrated.

The fuzzy control system described in Fig. 7 is comprised of the following components:

1. Fuzzification (the process of fuzzifying any input variables of control system, into linguistic terms);
2. Rule Base (containing fuzzy IF–THEN rules);

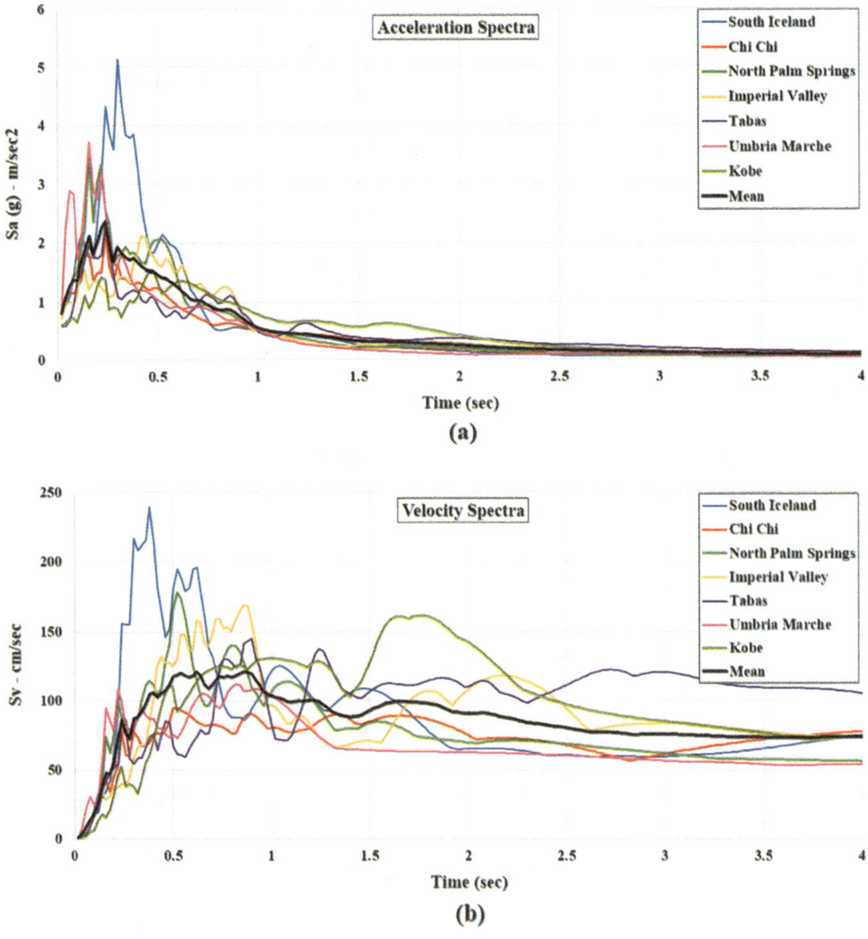

Fig. 6 Acceleration, velocity and displacement spectra of the N-S components

3. Inference Mechanism (extracts the resulting fuzzy outputs from rules);
4. Defuzzification (delivers the crisp signals of control).

As presented by Al-Dawod et al. [16], human knowledge and expertise do not provide the optimal membership functions and rule base. A human designed fuzzy controller is one of the possible configurations of the controller. In this section, an FLC based on the expert knowledge is presented in regard to prepare the FLC configuration for utilizing in optimization problems. In order to define a fuzzy space, 11 linguistic variables are selected. These variables are presented in Table 3.

The fuzzy controller has two input and one output variables. Each input variables have eight membership functions while the output variable has eleven membership functions. The shapes of the membership functions are assumed to be triangular for

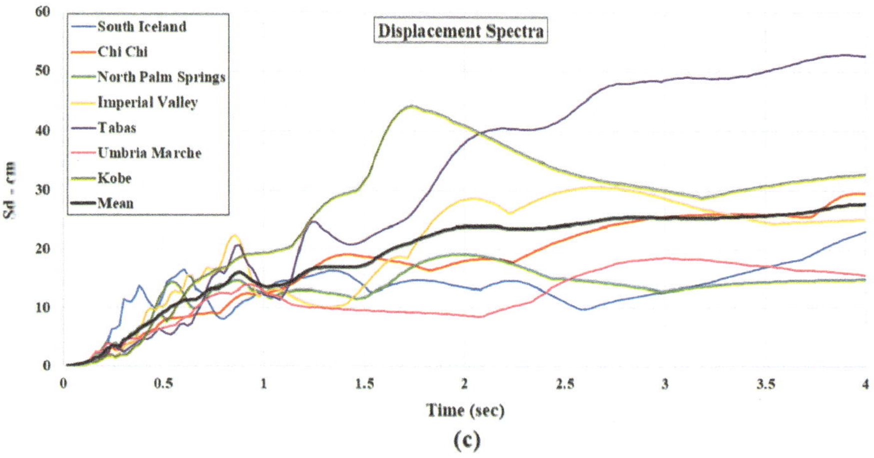

Fig. 6 (continued)

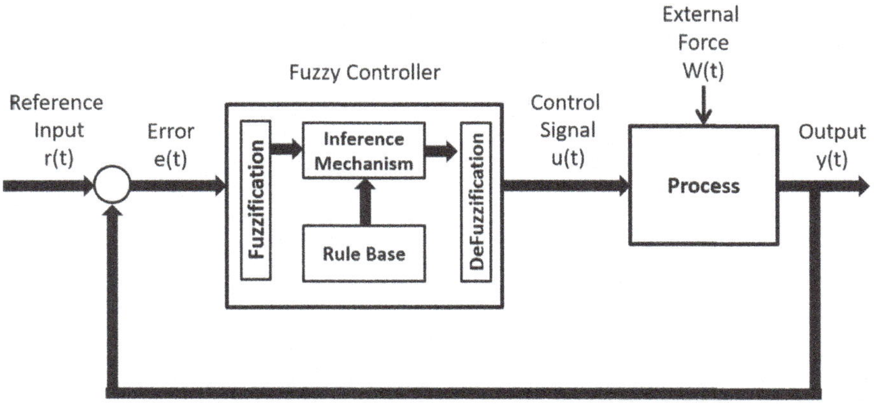

Fig. 7 Fuzzy logic control system

the input and output variables as depicted in Fig. 8. The rule base of FLC is also presented in Table 4.

In this paper, the fuzzy logic controller is implemented in a building while the first two input variables are utilized for indicating the maximum acceleration of the adjacent stories in building and the only output fuzzy variable is utilized for representing the maximum demanded control force in different stories of the building. As an example when PVL is considered as the fuzzy input or output variable, it means that the maximum story acceleration or the maximum story control force have a positive and very large value. It also should be mentioned that based on the inputs and output of the FLC, the defuzzification is implemented thorough the "Centroid" method provided by the fuzzy logic toolbox in MATLAB.

Table 3 Fuzzy Variables

Variable	Description
PVL	Positive and very Large
PL	Positive and Large
PM	Positive and Medium
PS	Positive and Small
PVS	Positive and very Small
ZR	Zero
NVS	Negative and very Small
NS	Negative and Small
NM	Negative and Medium
NL	Negative and Large
NVL	Negative and very Large

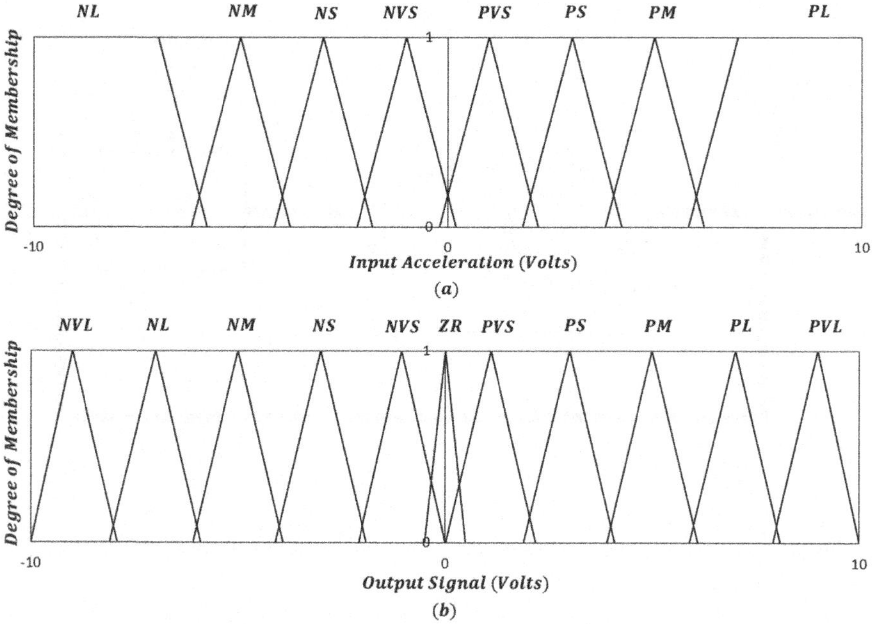

Fig. 8 Non-optimized membership functions for the inputs (**a**) and the output (**b**) of the FLC

4 Optimization Algorithms

As an influential instrument, optimization procedures are capable to solve many problems constructively. Among various classes of optimization methods, metaheuristic algorithms have proved their robust proficiency in resolving engineering issues. The

Table 4 Non-optimized rule base of the FLC

Control force								
Second input	First input							
	NL	NM	NS	NVS	PVS	PS	PM	PL
NL	PVL	PL	PM	PS	PVS	ZR	NVS	NS
NM	PL	PM	PS	PS	PVS	ZR	NVS	NS
NS	PM	PS	PS	PVS	PVS	ZR	NVS	NS
NVS	PM	PS	PVS	PVS	ZR	NVS	NS	NM
PVS	PM	PS	PVS	ZR	NVS	NVS	NS	NM
PS	PS	PVS	ZR	NVS	NVS	NS	NS	NM
PM	PS	PVS	ZR	NVS	NS	NS	NM	NL
PL	PS	PVS	ZR	NVS	NS	NM	NL	NVL

CSS can be categorized as a recently proposed metaheuristic optimization algorithms which has been used successfully for many engineering problems. In this paper, based on this algorithm, an improved CSS called Tribe-CSS is developed as a modified form to solve the problem. Summarizing the standard CSS as well as stating the tribe-CSS, the paper is followed.

4.1 *Standard CSS*

In 2010, Kaveh and Talatahari [39] established an algorithm founded on Coulomb and Gauss laws of electrostatics with the addition of physics' Newtonian law. Their proposed algorithm includes some agents, named charged particles (CPs), investigating search space for optimal solution. Moving over the search space is due to impact of CPs on each because of their electrical charges. In other words, CPs attract or repel each other because of their electrical charge. The magnitude of attraction or repulsion is measured using the Coulomb's law. Furthermore, the direction within the search spaces is determined by the Newtonian law. What's more, a memory, called charged memory (CM), saves the so far best CPs and their related fitness function. Illuminating the method, the flowchart of this algorithm is illustrated in Fig. 9.

4.2 *Presentation of the Tribe-CSS*

A premature convergence could be a probable event for many optimization algorithms. With the intention of eluding this occasion, several researchers have suggested various procedures. Tribe-CSS can be outlined as the boosted version of the standard CSS developed to overcome the likely early convergence of the CSS. Generally, the

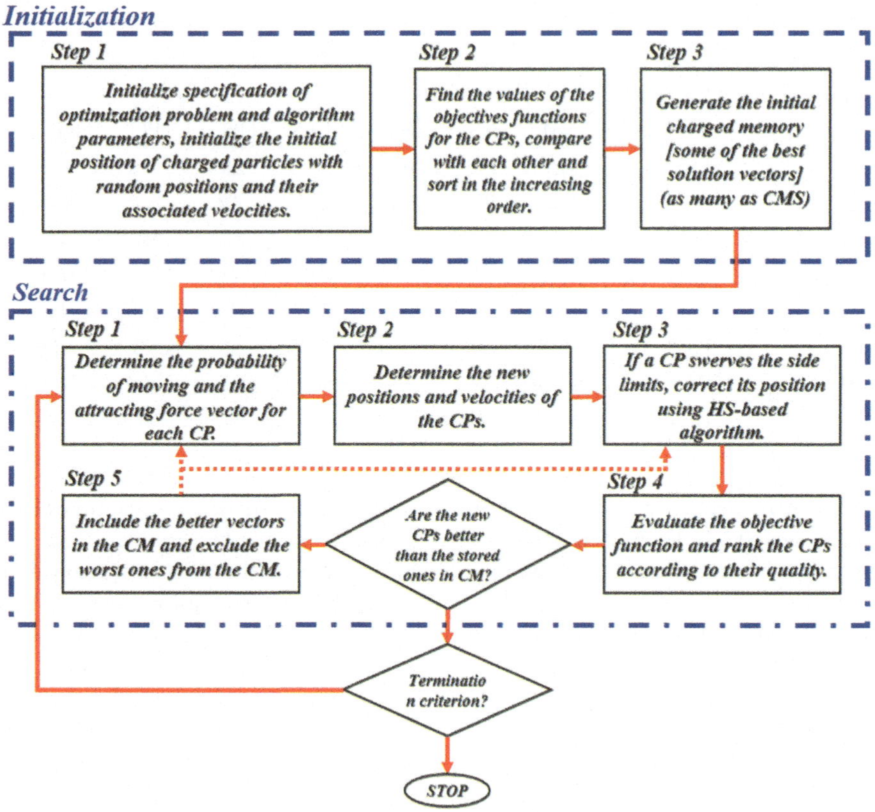

Fig. 9 The flowchart of the CSS algorithm [39]

tribe-CSS is partitioned into three different phases: the isolated phase, communing phase, and united phase. The isolated phase functions as the first phase of algorithm in which the CPs are randomly allocated to NCP isolated groups, called tribe. Search progression, which is limited to a predefined number of iterations, is the same in standard CSS at each tribe. It is notable that the CPs of a tribe are not allowed to share their individual experiences with the CPs assigned to other tribes. However, the communing phase as the second phase grants of the tribes are allowed to use CMs of each other and impart their best information. Although, this is the main dissimilarity between two first phases, the second phase finished by a prefixed number of iterations. Ultimately, all CPs comes to a unit tribe and the united phase is started and continues until the termination criterion is satisfied. Following steps explain the algorithm more:

Step 1 (Initialization)

Sub-step 1–1. CPs are initialized randomly as follows:

$$CP_i = x_{\min} + rand \cdot (x_{\max} - x_{\min}) \quad i = 1, 2, \ldots, N_{CP} \tag{1}$$

where CP_i represents the initial value of the ith CP; x_{min} and x_{max} are the minimum and the maximum permissible values for variables; rand is a random number in the range of [0,1] and N_{CP} represents the total number of existent CPs.

Sub-step 1–2. CPs are divided into N_{tribe} tribes haphazardly:

$$Tribe_k = \left\{ CP_i | i = 1, 2, \ldots, \frac{N_{CP}}{N_{tribe}} \right\} \quad k = 1, 2, \ldots, N_{tribe} \tag{2}$$

Sub-step 1-3. The charge for each CP is being determined at as:

$$q_i = \frac{fit(i) - fitworst}{fitbest_t - fitworst_t} \quad i = 1, 2, \ldots, \frac{N_{CP}}{N_{tribe}} \tag{3}$$

where *fitbestt* and *fitworstt* are the best fitness and the worst fitness values of all the CPs of the related tribe; fit(i) represents the fitness of the agent i, respectively. Also the separation distance r_{ij} between two charged particles is defined as follows:

$$r_{ij} = \frac{X_i - X_j}{(X_i - X_j)/2 - X_{best} + \varepsilon} \tag{4}$$

where X_i and X_j are the positions of the ith and jth CPs, respectively. X_{best} is the position of the best so far CP, and ε is a positive small number for singularity avoidance.

Sub-step 1-4. The initial value of velocity for each CP is set to zero:

$$v_{i,j}^{(0)} = 0 \quad i = 1, 2, \ldots, \frac{N_{CP}}{N_{tribe}} \tag{5}$$

Sub-step 1-5. A charged memory (CM) is initiated for each tribe to save the position as well as value of the fitness function for the tribe's best CP:

$$CM^{(0)} = Tribe^{(0)} \tag{6}$$

Step 2 (Searching Process).

Sub-step 2-1. Phase 1 (Isolated phase). With a limited iterations (e.g. for $\text{Iter}_{\max}/3$ times).

Sub-phase 1-1. Search process is started separately for each tribe.

Sub-phase 1-2. The resultant force for *j*th CP is evaluated as follows:

$$F_j = q_j \sum_{i,i \neq j}^{N_{CP}/N_{tribe}} \left(\frac{q_i}{a^3} r_{ij}.i_1 + \frac{q_i}{r_{ij}^2}.i_2 \right) ar_{ij} p_{ij} \left(X_i - X_j \right),$$
$$\begin{cases} j = 1, 2, \ldots, N \\ i = 1, 2, \ldots, \frac{N_{CP}}{N_{tribe}} \\ i_1 = 1, i_2 = 0 \Leftrightarrow r_{ij} < a, \\ i_1 = 1, i_2 = 0 \Leftrightarrow r_{ij} \geq a, \end{cases} \tag{7}$$

in which ar_{ij} and p_{ij} are the kind of forces (i.e. attraction or repulsion) and the probability of moving each CP in a tribe toward the others in the same tribe, respectively; these two parameters are determined as follows:

$$p_{ij} = \begin{cases} 1 \ \frac{fit(i) - fitbest}{fit(j) - fit(i)} > rand \lor fit(j) > fit(i), \\ 0 \ otherwise. \end{cases} \tag{8}$$

$$ar_{ij} = \begin{cases} +1, \ k_t < rand_{ij}, \\ -1, \ k_t < rand_{ij}, \end{cases} \tag{9}$$

where -1 represents repulsive force and $+1$ denotes the attractive force, and k_t is a parameter for controlling the effects of the kind of the force.

Sub-phase 1–3. The new position of the each CP at each tribe and its new velocity is being determined as follows:

$$X_{j,new} = rand_{j1} \cdot k_a \cdot \frac{F_j}{m_j} \cdot \Delta t^2 + rand_{j2} \cdot k_v \cdot V_{j,old} \cdot \Delta t^2 + X_{j,old}$$
$$V_{j,new} = \frac{X_{j,new} - X_{j,old}}{\Delta t}, \tag{10}$$

where k_a and k_v are the acceleration and velocity coefficients, respectively, and rand_{j1} and rand_{j2} are two uniformly distributed random numbers in the range of (0, 1).

Sub-step 2-2. Phase 2 (Communing phase).With a limited iterations (e.g. for $\text{Iter}_{\max}/3$ times).

Sub-phase 2-1. Search process is started separately for each tribe and its CM is shared with others.

Sub-phase 2-2. The resultant force for *j*th CP is evaluated as follows:

$$F_j = q_j \sum_{i,i \neq j}^{N_{CP}/N_{tribe}} \left(\frac{q_i}{a^3} r_{ij}.i_1 + \frac{q_i}{r_{ij}^2}.i_2 \right) ar_{ij} p_{ij} (X_i - X_j),$$

$$\begin{cases} j = 1, 2, \ldots, N \\ i = \left\{ 1, 2, \ldots, \frac{N_{CP}}{N_{tribe}} \right\} \cup B \\ i_1 = 1, i_2 = 0 \Leftrightarrow r_{ij} < a, \\ i_1 = 1, i_2 = 0 \Leftrightarrow r_{ij} \geq a, \end{cases} \tag{11}$$

where B represents the best CPs of the other tribes.

Sub-phase 2-3. The new position of the each CP at each tribe and its new velocity is being determined at each tribe as same in the Sub-phase 1-3.

Sub-step 2-3. Phase 3 (United phase). With a limited iterations (e.g. for $\text{Iter}_{\text{max}}/3$ times).

Sub-phase 3-1. All tribes are united to form a unit tribe.

Sub-phase 3-2. The resultant force for jth CP is evaluated at united tribe as in phase 1.

Sub-phase 3-3. The new position of CPs at united tribe and its new velocity is being determined as same in the Sub-phase 1-3.

Sub-phase 3-4. The terminating criterion is checked and the algorithm is repeated from Sub-phase 3-2 to Sub-phase 3-3 until this criterion is satisfied.

5 Design Example

5.1 Structural Details

In this paper, a 20-story steel building is considered as the structural model for optimization problem. This building is 80.77 m in elevation and 36.58 m by 30.48 m in plan. In the North-South (N-S) and East-West (E-W) directions, there are five and six bays respectively while in both directions, the bays are 6.10 m on center. This building has two underground levels called basement levels so that the first level below the ground level is the first basement (B-1) and the level underneath B-1 is the second basement (B-2). For the two basement levels and ground level, the floor-to-floor heights are 3.65 m, 3.65 m and 5.49 m respectively while the typical

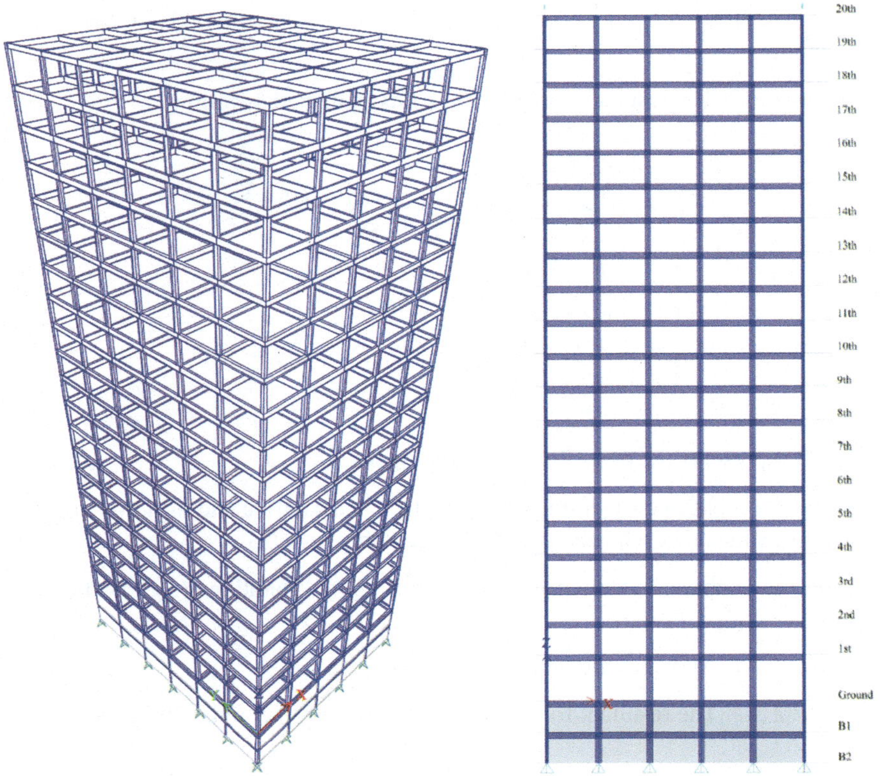

Fig. 10 Twenty-story steel building (Left) and the N-S perimeter frame (Right)

floor-to-floor heights are 3.96 m. The schematic view of the building along with one of the N-S perimeter frames are demonstrated in Fig. 10.

The load-resisting lateral system for this building is composed of high strength steel Moment Resisting Frames (MRFs) that are positioned in the perimeter of the plan. The structure's interior bays are comprised of simple framing alongside the composite floors. For beams, 248 MPa steel wide-flange (W) sections that are acting compositely with the floor slab are considered for the flooring system. For interior columns of the MRFs, 345 MPa steel W sections are selected while steel Box (B) columns are selected for the corner columns. The seismic mass of the structure and the design sections of the structural members are described in Table 5.

5.2 *Nonlinear Model*

When large ground motions such as the selected earthquakes based on the EBGMSM method occur, structural members can yield due to the nonlinear behavior in structural

Table 5 Seismic mass and design sections of the building

Summary	Level	Properties
Columns section (interior)	B-2–4th	W24 × 335
	4th–10th	W24 × 229
	10th–13th	W24 × 192
	13th–16th	W24 × 131
	16th–18th	W24 × 117
	18th–20th	W24 × 84
Columns section (corner)	B-2–1st	B38 × 5.08 cm
	1st–4th	B38 × 3.18 cm
	4th–13th	B38 × 2.54 cm
	13th–18th	B38 × 1.91 cm
	18th–20th	B38 × 1.27 cm
Beams section	B-2–4th	W30 × 99
	5th–10th	W30 × 108
	11th–16th	W30 × 99
	17th–18th	W27 × 84
	19th	W24 × 62
	20th	W21 × 50
Seismic mass	Ground level	5.32×10^5 kg
	1st level	5.63×10^5 kg
	2nd–9th level	5.52×10^5 kg
	20th level	5.84×10^5 kg
	Entire structure	1.11×10^7 kg

sections and materials. In this situation, the nonlinear behavior of the structure may be substantially different from a linear approximation. In order to describe this behavior, a bilinear hysteresis model, as shown in Fig. 11, is utilized to model the points of yielding (plastic hinges) in structural members.

The nonlinear parameters of the structural steel material (displayed in Fig. 11) is presented in Table 6. The nonlinear dynamic analysis is performed based on the Newmark-β method [46] that is presented in detail by Subbaraj and Dokainish [47] and implemented in MATLAB by Ohtori and Spencer [48].

5.3 *FLC Implementation*

An active control system with FLC as the control algorithm is implemented in the 20-story steel structure. The FLC for this building is based on acceleration feedback and the accelerations of the neighboring stories are utilized as input values for the

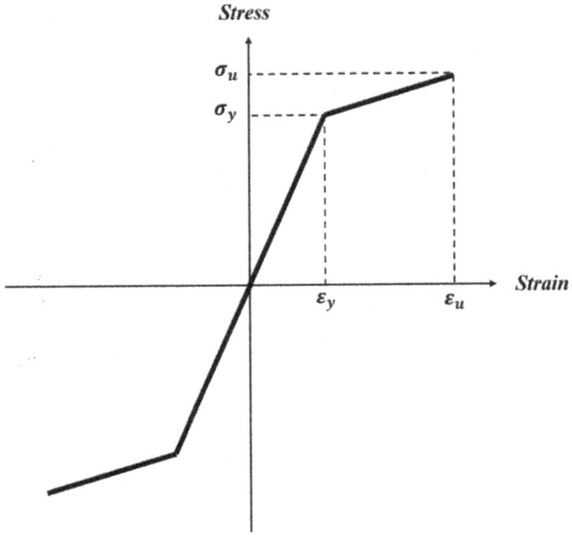

Fig. 11 Bilinear model for structural members

Table 6 Nonlinear parameters of the structural steel material

Properties	Value	Units
Modulus of Elasticity (E)	200,000	Mpa
Yield Strength (σ_y)	345	Mpa
Tensile Strength (σ_u)	450	Mpa
Yield Strain (ε_y)	0.001725	mm/mm
Tensile Stain (ε_u)	0.18	mm/mm

fuzzy controller. Five acceleration measurements on levels 4, 8, 12, 16 and 20 are selected for feedback on the building. Control actuators are located through the above ground stories of the building, connecting neighboring levels and these actuators are assumed to prepare the maximum control forces of 1000 KN. The simulators for the nonlinear evaluation model of the 20-story building with FLC is illustrated in Fig. 12.

5.4 Performance Criteria

The performance of the implemented fuzzy control strategy is checked according to the Performance Criteria (PC) specified for the selected building. These criteria are considered as a ratio of the controlled responses based on the FLC and the uncontrolled responses in most cases. A summary of the PC is presented in Table 7. The performance criteria in this research are divided into three classifications

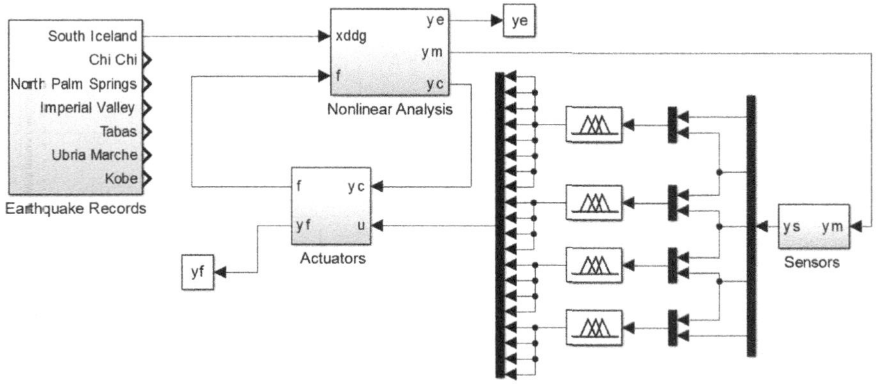

Fig. 12 Simulation of the nonlinear evaluation model for 20-Story building

as building responses, building damage and control devices requirements. The first classification is formulated based on the building responses and includes three criteria as the peak interstory drift ratio (PC1), peak story acceleration (PC2) and peak base shear (PC3). The second classification of the performance criteria discusses the building damage and includes three criteria. The ductility factor (PC4) refers to the maximum curvature at the ends of the structural members during the earthquakes. Dissipated energy at the ends of the structural members (PC5) and the ratio of the plastic hinges sustained by the structure (PC6) are the two other criteria in this classification. The third classification is based on the control devices requirements and includes three criteria. The maximum control force (PC7), control device stroke (PC8), and the power used for control (PC9) are the three criteria in this classification.

6 Statement of the Optimization Problem

Finding the best solution amongst all feasible solutions is called the optimization problem. The standard form of optimization problems is summarized in the following steps:

A function $f : B \to R$ from some set B to the real numbers.

An element $x_0 \in B$ such that $f(x_0) \leq f(x)$ for all $x \in B$ (minimization problem) or such that $f(x_0) \geq f(x)$ for all $x \in B$ (maximization problem).

In general, B is subset of the Euclidean space, usually determined by some set of equalities, inequalities or constraints that the components of B have to satisfy. The domain B of f is called the search space, while the components of B are called possible solutions or candidate solutions. f is a function that is called "objective function". One possible solution that minimizes (or maximizes, if that is the aim) the objective function is called an optimum solution. In structural control design, optimization problems are generally presented in terms of minimization.

Table 7 Summary of the performance criteria

Drift Ratio	Story Acceleration	Base Shear
$PC_1 = \max\limits_{7EQs}\left\{\frac{\max\limits_{t,i}\frac{\lvert d_i(t)\rvert}{h_i}}{\delta^{\max}}\right\}$	$PC_2 = \max\limits_{7EQs}\left\{\frac{\max\limits_{t,i}\lvert\ddot{x}_{ai}(t)\rvert}{\ddot{x}_a^{\max}}\right\}$	$PC_3 = \max\limits_{7EQs}\left\{\frac{\max\limits_{t,i}\left\lvert\sum\limits_i m_i\ddot{x}_{ai}(t)\right\rvert}{F_b^{\max}}\right\}$
Ductility	Dissipated Energy	Plastic Hinges
$PC_4 = \max\limits_{7EQs}\left\{\frac{\max\limits_{t,j}\frac{\lvert\varphi_j(t)\rvert}{\varphi_{yj}}}{\varphi_{\max}}\right\}$	$PC_5 = \max\limits_{7EQs}\left\{\frac{\max\limits_{t,j}\frac{\int dE_j}{F_{yj}\cdot\varphi_{yj}}}{E^{\max}}\right\}$	$PC_6 = \max\limits_{7EQs}\left\{\frac{N_d^c}{N_d}\right\}$
Control Force	Control Device Stroke	Control Power
$PC_7 = \max\limits_{7EQs}\left\{\frac{\max\limits_{t,l}\lvert f_l(t)\rvert}{W}\right\}$	$PC_8 = \max\limits_{7EQs}\left\{\frac{\max\limits_{t,l}\lvert y_l^a(t)\rvert}{x^{\max}}\right\}$	$PC_9 = \max\limits_{7EQs}\left\{\frac{\max\limits_{t}\lvert\sum P_l(t)\rvert}{\dot{x}^{\max}W}\right\}$

d_i—Controlled interstory drift for structural levels (m)
h_i— Height of the i-th level (m)
$\delta^{\max}$—Uncontrolled interstory drift ratio for structural levels
$\ddot{x}_{ai}$—Controlled acceleration of the ith level (m/s^2)
$\ddot{x}_a^{\max}$—Uncontrolled acceleration of the roof (m/s^2)
m_i—Seismic mass of the i-th level (kg)
$F_b^{\max}$—Maximum uncontrolled base shear (N)
ϕ_j—Controlled curvature for the end points of the j-th element
ϕ_{yj}—Yield curvature for the end points of the j-th element
$\phi^{\max}$—Maximum uncontrolled curvature in the structure
$\int dE_j$—Dissipated energy for the end points of the j-th element
F_{yj}—Yield moment for the end points of the j-th element
$E^{\max}$—Maximum uncontrolled dissipated energy of the structure
N_d—Maximum number of damaged joints for the uncontrolled structure
N_d^C—Maximum number of damaged joints for the controlled structure
f_l—Control force of the l-th control instrument (N)
W—Seismic weight of the structure (N)
y_l^a—Translation of the l-th control instrument (m)
$x^{\max}$—Uncontrolled displacement for structural levels (m)
P_l—Required power for the actuators
$\dot{x}^{\max}$—Uncontrolled velocity for structural levels (m/s)

In this paper, the optimization problem is formulated based on the optimum tuning of the FLC parameters. Selection of variables in the Tribe-CSS is based on the configuration of the membership functions and the rule base. According to the first and second inputs of the FLC, optimization variables are $a_1, a_2, \ldots, a_{11}$ as illustrated in Fig. 13a. In addition, for the output of the FLC, optimization variables are $b_1, b_2, \ldots, b_{15}$ as presented in Fig. 13b. It should be noted that for inputs and outputs of the FLC, the membership functions are assumed to be symmetric. For the rule base of the FLC, variables are $c_1, c_2, \ldots, c_{64}$ as presented in Table 8. Note that the

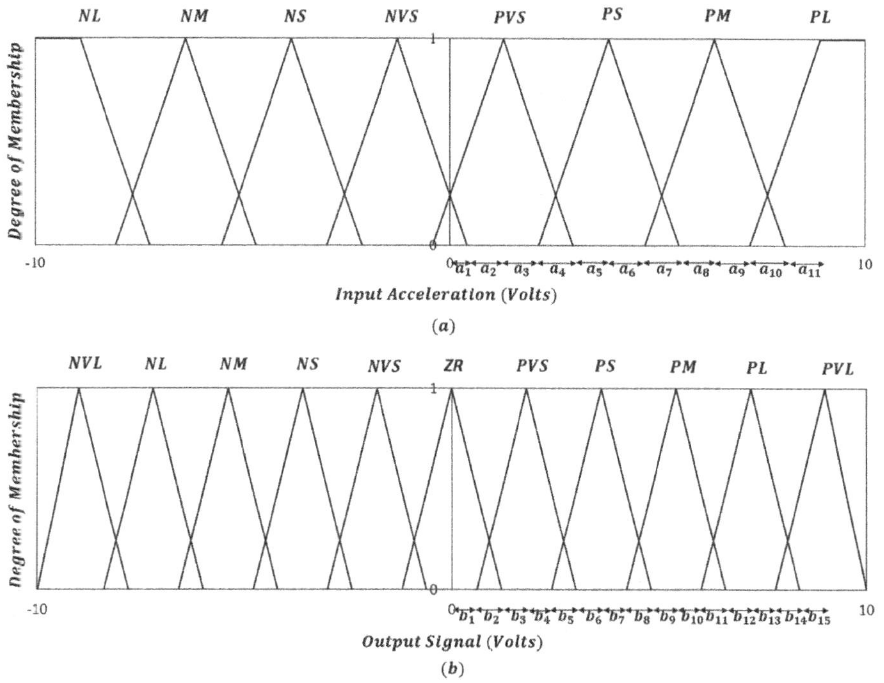

Fig. 13 Optimization variables of membership functions for fuzzy inputs (**a**) and output (**b**)

Table 8 Variables for the fuzzy rule base

Control Force								
	First Input							
Second Input	NL	NM	NS	NVS	PVS	PS	PM	PL
NL	PVL/c1	PL/c9	PM/c17	PS/c25	PVS/c33	ZR/c41	NVS/c49	NS/c57
NM	PL/c2	PM/c10	PS/c18	PS/c26	PVS/c34	ZR/c42	NVS/c50	NS/c58
NS	PM/c3	PS/c11	PS/c19	PVS/c27	PVS/c35	ZR/c43	NVS/c51	NS/c59
NVS	PM/c4	PS/c12	PVS/c20	PVS/c28	ZR/c36	NVS/c44	NS/c52	NM/c60
PVS	PM/c5	PS/c13	PVS/c21	ZR/c29	NVS/c37	NVS/c45	NS/c53	NM/c61
PS	PS/c6	PVS/c14	ZR/c22	NVS/c30	NVS/c38	NS/c46	NS/c54	NM/c62
PM	PS/c7	PVS/c15	ZR/c23	NVS/c31	NS/c39	NS/c47	NM/c55	NL/c63
PL	PS/c8	PVS/c16	ZR/c24	NVS/c32	NS/c40	NM/c48	NL/c56	NVL/c64

fractures in rule base refer to the weight of each rule that should be optimized by the Tribe-CSS.

In this paper, the optimization problem is considered as a single-objective problem while the objective function is formulated based on the nonlinear responses of the structure. The general state of the objective function is as following:

$$Obj = \frac{\sum_{i=1}^{n} P_i \frac{CR_i}{UR_i}}{\sum_{i=1}^{n} P_i} \tag{12}$$

where P_i is the weighing coefficient of the objective function that is considered as the PGAs of the selected earthquakes and the summation is utilized for the selected earthquake records. UR_i and CR_i are the uncontrolled and controlled responses of the building. The responses can be chosen based on the performance criteria discussed in Sec. 5.4 while in this paper, the objective function is formulated based on the ductility factor (PC4). It should be noted that all of the PC can be considered as the building major response in formulation of the objective function however in this paper the ductility factor is selected based on the great importance of this ratio in all seismic design codes and practices. The mentioned objective function is considered as follows:

$$Obj = \frac{\begin{array}{l} 0.93 \times (PC_4)_{SouthIceland} + 0.88 \times (PC_4)_{ChiChi} \\ +0.84 \times (PC_4)_{N.P.Springs} + 0.68 \times (PC_4)_{\text{Imp}.Valley} \\ +0.58 \times (PC_4)_{Tabas} + 0.94 \times (PC_4)_{U.Marche} \\ +0.56 \times (PC_4)_{Kobe} \end{array}}{(0.93 + 0.88 + 0.84 + 0.68 + 0.58 + 0.94 + 0.56)} \tag{13}$$

In this study the initial population is consists of 30 search agents and the termination criteria is considered based on the determined maximum number of iterations. The maximum number of iterations is determined as 100 iterations while this number of iterations is reachable according to the time cost issues in FLC tuning process.

7 Numerical Results

In this section, the convergence history for the best results of the Tribe-CSS and standard CSS is depicted based on the selected objective function. In addition, the performance criteria based on the best results of the optimization methods are calculated. The variation of some design variables for Tribe-CSS are also calculated for performing a valid judgment in performance evaluation of the upgraded algorithm. At the end, the performance of the Tribe-CSS is compared with various classical and advanced optimization algorithms.

7.1 Optimized FLC by Tribe-CSS and CSS

The best results of the Tribe-CSS and the standard CSS based on the convergence history of Obj is depicted in Fig. 14. The Tribe-CSS and CSS are capable of reducing the objective function by up to 0.9018 and 0.9134 respectively.

When the optimization process is finished, the best results of this process is utilized for evaluating the performance criteria and are mentioned as optimized PC. In Table 9, the optimized PC are displayed based on the best results of the Tribe-CSS. The results show the ability of the optimization algorithm in decreasing the PC for selected earthquake records. The inter-story drift ratio (PC1) for all of the earthquake records are decreased while the maximum reduction for PC1 is up to 19% (Kobe earthquake)

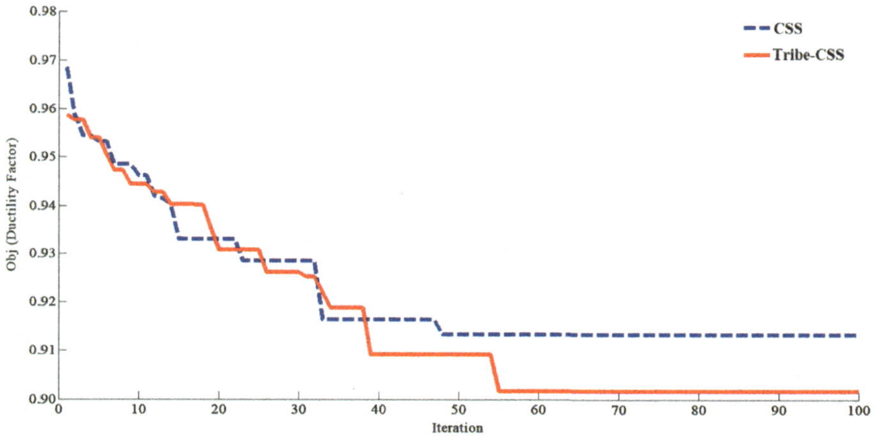

Fig. 14 The convergence history of Obj for the 20-story building by Tribe-CSS and CSS

Table 9 Optimized performance criteria for the 20-story building utilizing Tribe-CSS

Earthquakes							
PC	South Iceland	Chi Chi	N. P. Springs	Imperial Valley	Tabas	Umbria Marche	Kobe
PC1	0.9839	0.9857	0.9393	0.9867	0.8897	0.9815	0.8125
PC 2	0.9892	0.9514	0.9569	0.9933	0.7885	0.8911	0.9269
PC 3	0.9221	0.9026	0.8632	0.9867	0.8941	0.9067	0.8286
PC 4	0.9511	0.9544	0.8169	0.9455	0.8918	0.9340	0.7684
PC 5	1.0786	0.9834	0.7332	1.1843	0.7526	0.5667	0.5486
PC 6	1.0000	0.9583	0.8750	1.0000	0.8295	0.7500	0.7609
PC 7	0.0059	0.0039	0.0047	0.0068	0.0070	0.0069	0.0070
PC 8	0.1345	0.1330	0.1844	0.1219	0.1001	0.1295	0.2281
PC 9	0.0115	0.0074	0.0134	0.0107	0.0065	0.0098	0.0066

and the minimum reduction is for Imperial Valley earthquake that is up to 2%. The response of PC2 is also decreased for all of the earthquake records while the maximum reduction is for Tabas earthquakes that is by up to 22% while the minimum reduction is for Imperial Valley earthquake that is up to 1%. For PC3, the optimization process has been able to decrease the response due all of the earthquake records while the maximum and minimum reduction is for Kobe and Imperial Valley earthquakes that are about 18% and 2% respectively.

With respect to the building damage criteria (PC4 –PC6), the nonlinear behavior of the structure is concerned and the ability of optimization process is checked through the nonlinear responses of the structure. The maximum and minimum reductions for PC4 is up to 24% (Kobe) and 5% (Chi Chi) respectively. The dissipated energy (PC5) in the 20-story steel structure is decreased for five of the earthquake records by up to 46% while for other two (South Iceland and Imperial Valley), it is increased by up to 18%. The number of plastic hinges (PC6) as one of the important parameters in nonlinear evaluation of the structure is decreased by up to 5% (Chi Chi), 13% (North Palm Springs), 18% (Tabas), 25% (Umbria Marche), 24% (Kobe) while for South Iceland and Imperial Valley earthquakes, this ratio is unchanged according to the uncontrolled structural responses.

In Table 10, the optimized PC based on the best results of the CSS are displayed. In most cases, the optimized PC obtained by Tribe-CSS are lower than the values obtained by CSS and it brings out that using the upgrading process in CSS for FLC tuning makes the controlled responses of the structure smaller. It can be concluded that the responses of the structure obtained by the CSS is larger than the Tribe-CSS especially when a larger control force is considered. Note that the maximum amount of PC7 for optimized FLC based on the Tribe-CSS is calculated as 0.0070 while for CSS, it is calculated as 0.0076. It should be noted that the device stroke (PC8) and the control power (PC9) for the Tribe-CSS based optimized FLC have smaller values than the CSS based optimized FLC.

Table 10 Optimized performance criteria for the 20-story building utilizing CSS

Earthquakes							
PC	South Iceland	Chi Chi	N. P. Springs	Imperial Valley	Tabas	Umbria Marche	Kobe
PC1	1.0091	0.9879	0.9704	1.0669	0.8949	0.9894	0.8259
PC 2	0.9959	0.9548	0.9882	1.0084	0.8154	0.9527	0.9524
PC 3	0.9285	0.9295	0.8521	0.9818	0.9222	0.9286	0.8396
PC 4	0.9737	0.9668	0.8311	0.9508	0.8941	0.9378	0.7863
PC 5	1.0914	0.9911	0.8157	1.2195	0.7996	0.6272	0.5945
PC 6	1.0000	1.0000	0.8333	1.0600	0.8295	1.0000	0.7609
PC 7	0.0065	0.0045	0.0057	0.0066	0.0076	0.0071	0.0074
PC 8	0.1350	0.1333	0.1905	0.1280	0.1007	0.1298	0.2299
PC 9	0.0119	0.0075	0.0139	0.0121	0.0069	0.0098	0.0065

In Fig. 15, the ratio of the maximum curvature at the ends of the structural members to the yield curvature (ϕ_j/ϕ_{yj}) are summarized for the uncontrolled structure alongside the structure with CSS and Tribe-CSS based optimized FLC. This ratio (called Curvature Ratio) represents the nonlinear behavior of the structure and the control systems are supposed to reduce this ratio in order to prevent damages in structural members. For all of the seven earthquake records, the amount of this ratio in structure with Tribe-CSS based FLC is lower than the uncontrolled structure and the structure with CSS based FLC. The maximum difference between the structures with Tribe-CSS based optimized FLC and the uncontrolled structure is for Kobe earthquake that is about 23% while the maximum difference between the structures with CSS based FLC and the uncontrolled structure is about 21% that is for the same earthquake. The maximum difference between the structures with CSS and Tribe-CSS based FLC is for South Iceland earthquake that is about 3%. It can be concluded that the structure with Tribe-CSS based optimized fuzzy controller is capable of withstanding the severe earthquakes than the structure with the CSS based fuzzy controller.

Energy dissipation describes the irreversible processes occurring in a material subjected to loading cycles. Energy dissipation is also represents the nonlinear behavior of the structure. In Fig. 16, the ratio of the maximum dissipated energy ($\int dE_j$) to the yield curvature (ϕ_{yj}) and yield moment (F_{yj}) at the ends of the structural members are summarized for the uncontrolled structure alongside the structure with CSS and Tribe-CSS based optimized FLC. This ratio (called Energy Dissipation Ratio) represents the performance of the control system utilized in structure for controlling the earthquake-induced vibration of the structure. In structures with control systems, the control devices and algorithms are supposed to reduce this ratio in order to absorb the input energy of the earthquakes and reduce the energy dissipation in structural members. For all of the seven earthquake records, the amount of this ratio in structure with Tribe-CSS based FLC is lower than the uncontrolled structure and the structure with CSS based FLC. The maximum difference between the structures with Tribe-CSS based optimized FLC and the uncontrolled structure is for Umbria Marche earthquake that is about 43% while the difference between the structure with CSS based FLC and the uncontrolled structure is about 40% that is for

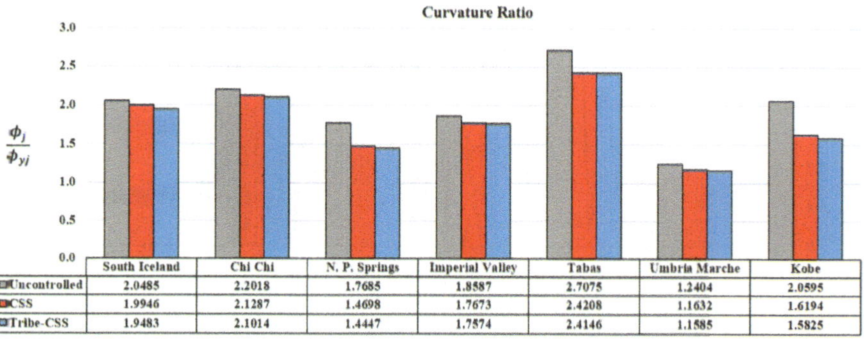

	South Iceland	Chi Chi	N. P. Springs	Imperial Valley	Tabas	Umbria Marche	Kobe
Uncontrolled	2.0485	2.2018	1.7685	1.8587	2.7075	1.2404	2.0595
CSS	1.9946	2.1287	1.4698	1.7673	2.4208	1.1632	1.6194
Tribe-CSS	1.9483	2.1014	1.4447	1.7574	2.4146	1.1585	1.5825

Fig. 15 Comparison of the curvature ratio for different earthquakes

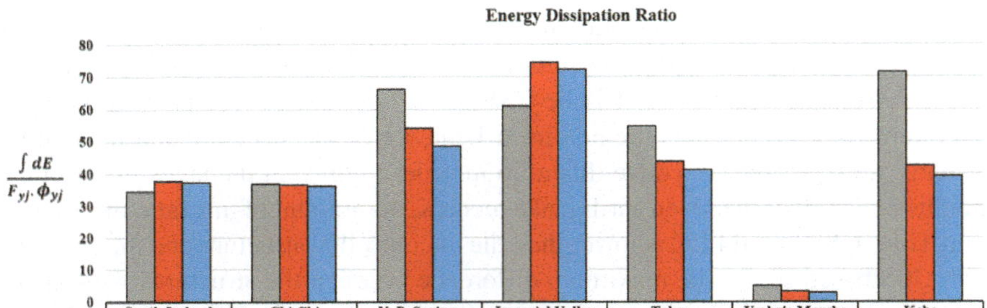

	South Iceland	Chi Chi	N. P. Springs	Imperial Valley	Tabas	Umbria Marche	Kobe
Uncontrolled	34.3801	36.7455	66.1896	61.0238	54.6725	4.9544	71.3812
CSS	37.5223	36.4184	53.9900	74.4185	43.7160	3.1073	42.4361
Tribe-CSS	37.0811	36.1351	48.5277	72.2701	41.1442	2.8077	39.1593

Fig. 16 Comparison of the energy dissipation ratio for different earthquakes

the Kobe earthquake. The maximum difference between the structures with CSS and Tribe-CSS based FLC is for North Palm Springs earthquake that is about 10%. It can be concluded that the structure with Tribe-CSS based optimized fuzzy controller is capable of reducing energy dissipation in structural members than the structure with the CSS based fuzzy controller in dealing with severe earthquakes.

7.2 *Variation of Design Variables*

Figure 17 shows the variation of some design variables for the optimization run corresponding to the best design overall. Whilst in the first iterations, the values of

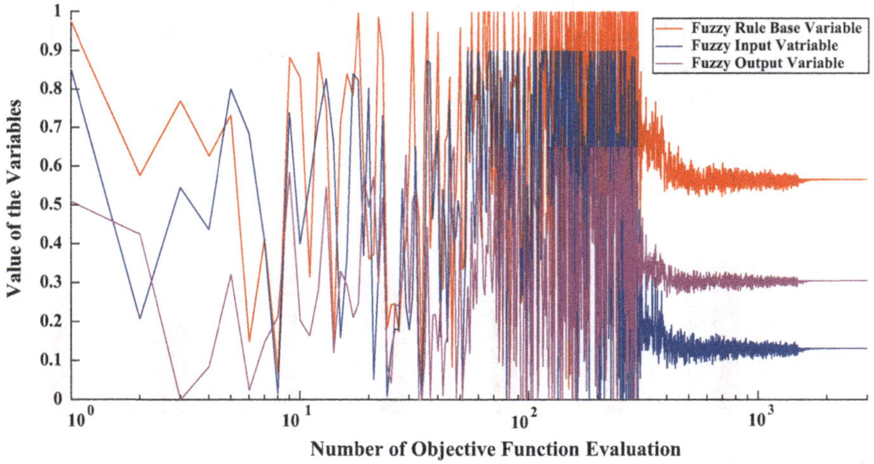

Fig. 17 Convergence history for the fuzzy input, output and rule base variables corresponding to the best design obtained by Tribe-CSS for the 20-story building

selected variables change considerably because of the high exploration power of the Tribe-CSS method; oscillations reduce as the optimization process progresses and then become marginal in the final iterations. This indicates that a local search is performed towards the end of the optimization process.

7.3 Comparing to Other Metaheuristics

With the purpose of comparing the performance of the Tribe-CSS with some other optimization techniques, seven different optimization process using the GA [49], PSO [50], ICA [51], ACOR [52], GWO [53], DA [54], and Jaya algorithm [55] are formulated. In the implementation of these algorithms for an FLC tuning process, the latest and developed versions of these methods are utilized. The convergence history for the best results of the selected algorithms in 100 iteration for 20-story building is depicted in Fig. 18. Among these methods, the Tribe-CSS can find 0.9018 for objective function which is the best one while the ICA with 0.9075, ACOR with 0.9084, PSO with 0.9125, CSS with 0.9134, Jaya with 0.9197, GA with 0.9242, GWO with 0.9267 and DA with 0.9282 have second to ninth places.

In Table 11, the maximum PC of seven earthquake records obtained by different metaheuristic algorithms are presented. The lowest amount of each PC among all optimization methods are underlined. It should be noted that for six of PC, the Tribe-CSS have the lowest possible values while for three of them, the PSO, Jaya and GWO have some privileges.

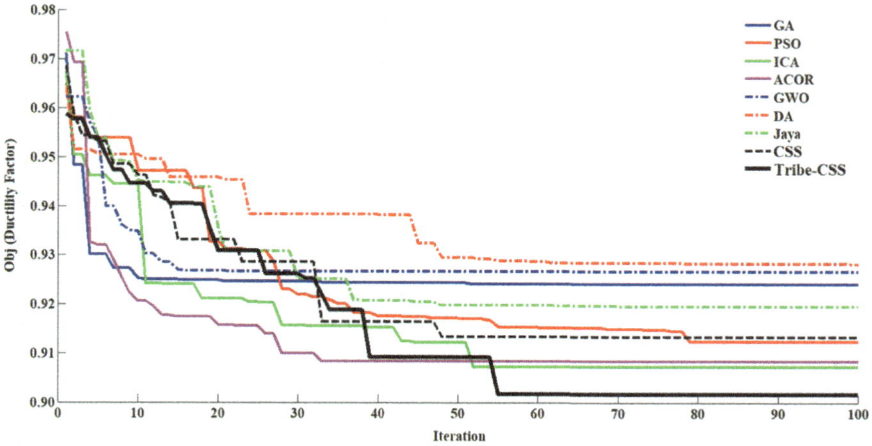

Fig. 18 The convergence history of *Obj* for the 20-story building obtained by different metaheuristics

Table 11 Maximum PC of seven earthquake records obtained by different metaheuristic

PC	Metaheuristics								
	GA	PSO	ICA	ACOR	GWO	DA	Jaya	CSS	Tribe-CSS
PC1	1.0101	1.0118	1.0654	1.0133	1.0360	1.0214	1.0797	1.0669	0.9867
PC 2	1.0405	1.0349	1.0687	1.0173	1.0155	1.0555	0.9959	1.0084	0.9933
PC 3	1.0242	1.0396	1.0153	1.0039	1.0218	1.0224	0.9813	0.9818	0.9867
PC 4	0.9875	0.9851	1.0108	0.9990	1.0028	0.9890	1.1053	0.9737	0.9544
PC 5	1.2235	1.2389	1.3845	1.1933	1.2536	1.2640	1.1988	1.2195	1.1843
PC 6	1.0222	1.0444	1.1000	1.0222	1.0222	1.0444	1.0600	1.0600	1.0000
PC 7	0.0070	0.0067	0.0071	0.0073	0.0070	0.0076	0.0075	0.0076	0.0070
PC 8	0.2381	0.2377	0.2310	0.2327	0.2349	0.2463	0.2290	0.2299	0.2281
PC 9	0.0109	0.0102	0.0152	0.0136	0.0101	0.0117	0.0161	0.0139	0.0134

8 Conclusions

This paper investigates the efficiency of the optimized fuzzy controller in reducing the responses of a 20-story steel building with nonlinear behavior under severe earthquake effects. The FLC is one of the recently proposed control algorithms that calculates the control force in structures with active control systems. Using human knowledge and experience in configuration of the membership functions and the rule base does not yield optimal control responses for a given structure so optimization of the FLC is concerned in this paper. The optimization problem consists of 11 variables for each fuzzy inputs, 15 variables for fuzzy output and 64 variables for the rule base. The Charged System Search (CSS) algorithm was considered as the optimization technique while an improved version of this algorithm was proposed as Tribe-CSS in order to improve the performance of the standard CSS. The objective function was formulated based on the ductility factor that concerns the maximum curvature at the ends of the structural members. The performance of the optimized controllers with standard and upgraded CSS was investigated through the nine performance criteria and the best results of these methods were compared to each other. The seismic inputs were considered based on the PSHA for city of Tabriz in Iran while the earthquake records were selected based on the EBGMSM method. For Tribe-CSS based optimized FLC, the performance criteria according to the building responses (PC1–PC3) are decreased by up to 19%, 22% and 18% respectively while for CSS based optimized FLC, these criteria are decreased by up to 17%, 18% and 16% respectively.

It could be concluded that the optimization of FLC based on Tribe-CSS results in smaller values for building responses. For performance criteria based on the building damages (PC4–PC6), the Tribe-CSS was able to reduce these criteria by up to 24%, 46% and 25% respectively. Comparing these values with corresponding values in CSS based FLC shows that the upgrading process makes the standard CSS very powerful in preventing damages of structural members. In terms of control devices requirement criteria (PC7–PC9), It should be noted that the Tribe-CSS is capable of reducing the maximum demanded control force while the performance of the controlled structure is much better. The comparing results of the maximum curvature and energy dissipation at the ends of structural members prove that the structure with optimized FLC based on Tribe-CSS is properly capable of withstanding the severe earthquakes than the structure with CSS based FLC. By comparing the results of the optimized controller by Tribe-CSS with other optimization techniques such as GA, PSO, ICA, ACOR, GWO, DA, and Jaya, it could be concluded that the Tribe-CSS is more effective in reducing the building responses and damages. Comparing the responses of the structure with optimized FLC based on different metaheuristics displays that when the selection of ground motions are considered based on the PSHA for a specific site with high seismic activity (such as Tabriz), an optimization process should be implemented to maintain an efficient control strategy in order to prevent damages in structural members. It should be concluded that most of the practical applications of the fuzzy logic controller were formulated based on the human knowledge or expert that could not provide optimal structural responses so the results obtained by Tribe-CSS would improve the performance of these controllers in practice.

References

1. Milne, W.G., Davenport, A.G.: Distribution of earthquake risk in Canada. Bull. Seismol. Soc. Am. **59**(2), 729–754 (1969)
2. McGuire, R.K., Veneziano, D., Van Dyck, J., Toro, G., O'Hara, T., Drake, L., Patwardhan A., Kulkarni R., Keeney R., Winkler R.: Seismic Hazard Methodology for the Central and Eastern United States: Volume 1: Part 2, Methodology (Revision 1), Electric Power Research Inst. (USA). (1988)
3. Cornell, C.A.: Engineering seismic risk analysis. Bull. Seismol. Soc. Am. **58**(5), 1583–1606 (1968)
4. Budnitz, R.J., Apostolakis, G., Boore, D.M.: Recommendations for Probabilistic Seismic Hazard Analysis: Guidance on Uncertainty and Use of Experts. Nuclear Regulatory Commission, Washington, DC (United States) (1997)
5. Mahsuli, M., Rahimi, H., Bakhshi, A.: Probabilistic seismic hazard analysis of Iran using reliability methods. Bull. Earthquake Eng., 1–27 (2018)
6. Kowsari, M., Eftekhari N., Kijko A., Dadras E.Y., Ghazi H., Shabani E., Quantifying seismicity parameter uncertainties and their effects on probabilistic seismic hazard analysis: a Case study of Iran. Pure Appl. Geophys., 1–6 (2018)
7. Keshavarz, A., Moghaddam, B.M.: Probabilistic seismic hazard analysis and determination of uniform hazard spectrum of Bushehr province assuming linear source model. J. Struct. Constr. Eng. **5**(1), 127–142 (2018)

8. Yaghmaei-Sabegh, S., Ebrahimi-Aghabagher, M.: Near-field probabilistic seismic hazard analysis with characteristic earthquake effects. Nat. Hazards **87**(3), 1607–1633 (2017)
9. Yazdani, A., Nicknam, A., Dadras, E.Y., Eftekhari, S.N.: Near-field probabilistic seismic hazard analysis of metropolitan Tehran using region-specific directivity models. Pure Appl. Geophys. **174**(1), 117–132 (2017)
10. Baghbani, M., Gholami, E., Barani, H.R., Birjand, I.: Seismic hazard analysis of Dam Siyaho in South Khorasan province (Eastern Iran). J. Tethys: **4**(3), 180–199 (2016)
11. Vafaie, J., Taghikhany, T., Tehranizade, M.: Near field uniform seismic hazard spectra of Tabriz zone. In: The 14th World Conference on Earthquake Engineering, Beijing, China, (2008)
12. Vafaie, J., Taghikhany, T., Tehranizadeh, M.: Near field effect on horizontal equal-hazard spectrum of Tabriz city in north-west of Iran. Int. J. Civil Eng. **9**(1), 49–56 (2011)
13. Razeghi, H.R., Amiri, G.G., Amrei, S.R., Rahimi, M.A.: Probabilistic seismic hazard analyses and obtaining uniform hazard spectra of Tabriz, Iran. KSCE J. Civil Eng. **20**(5), 1813–1823 (2016)
14. Moghaddam, H., Fanaie, N., Hamzehloo, H.: Uniform hazard response spectra and ground motions for Tabriz. J. Scientia Iranica **16**(3), 238–248 (2009)
15. Hessami, K., Pantosti, D., Tabassi, H., Shabanian, E., Abbassi, M.R., Feghhi, K., Solaymani, S.: Paleoearthquakes and slip rates of the North Tabriz Fault, NW Iran: preliminary results. Ann. Geophys. (2003)
16. Al-Dawod, M., Samali, B., Kwok, K., Naghdy, F.: Fuzzy controller for seismically excited nonlinear buildings. J. Eng. Mech. **130**(4), 407–415 (2004)
17. Edrees, T.: Structural control and identification of civil engineering structures (Doctoral dissertation). Luleå tekniska universitet (2015)
18. Kaveh, A.: Advances in Metaheuristic Algorithms for Optimal Design of Structures. Springer, Switzerland (2014)
19. Kaveh, A.: Applications of Metaheuristic Optimization Algorithms in Civil Engineering. Springer, Switzerland (2017)
20. Kaveh, A., Ghazaan, M.I.: Meta-heuristic Algorithms for Optimal Design of Real-Size Structures. Springer International Publishing (2018)
21. Fu, J., Lai, J., Liao, G., Yu, M., Bai, J.: Genetic algorithm based nonlinear self-tuning fuzzy control for time-varying sinusoidal vibration of a magnetorheological elastomer vibration isolation system. Smart Mater. Struct. (2018)
22. Soltani, M., Chaouech, L., Chaari, A.: Fuzzy sliding mode controller design based on euclidean particle swarm optimization. In: Real-Time Modelling and Processing for Communication Systems, pp. 95–122. Springer, Cham (2018)
23. Chen, Z., Yuan, X., Yuan, Y., Lei, X., Zhang, B.: Parameter estimation of fuzzy sliding mode controller for hydraulic turbine regulating system based on HICA algorithm. Renew. Energy **133**, 551–565 (2018)
24. Olivas, F., Valdez, F., Castillo, O., Gonzalez, C.I., Martinez, G., Melin, P.: Ant colony optimization with dynamic parameter adaptation based on interval type-2 fuzzy logic systems. Appl. Soft Comput. **53**, 74–87 (2017)
25. Amador-Angulo, L., Castillo, O.: A new fuzzy bee colony optimization with dynamic adaptation of parameters using interval type-2 fuzzy logic for tuning fuzzy controllers. Soft. Comput. **22**(2), 571–594 (2018)
26. Chamorro, H.R., Riaño, I., Gerndt, R., Zelinka, I., Gonzalez-Longatt, F., Sood, V.K.: Synthetic inertia control based on fuzzy adaptive differential evolution. Int. J. Electr. Power Energy Syst. **105**, 803–813 (2019)
27. Ray, P.K., Paital, S.R., Mohanty, A., Eddy, F.Y., Krishnan, A., Gooi, H.B., Amaratunga, G.A.: Firefly algorithm scaled fractional order fuzzy PID based PSS for transient stability improvement. In: 2018 19th International Carpathian Control Conference (ICCC). IEEE (2018)
28. Chrouta, J., Chakchouk, W., Zaafouri, A., Jemli, M.: Modelling and control of an irrigation station process using heterogeneous cuckoo search algorithm and fuzzy logic controller. IEEE Trans. Ind. Appl. (2018)

29. Giri, S., Bera, P.: Design of fuzzy-PI controller for hybrid distributed generation system using grey wolf optimization algorithm. In: Methodologies and Application Issues of Contemporary Computing Framework. Springer, Singapore (2018)
30. Chao, F., Zhou, D., Lin, C.M., Zhou, C., Shi, M., Lin, D.: Fuzzy cerebellar model articulation controller network optimization via self-adaptive global best harmony search algorithm. Soft. Comput. **22**(10), 3141–3153 (2018)
31. Li, C., Mao, Y., Zhou, J., Zhang, N., An, X.: Design of a fuzzy-PID controller for a nonlinear hydraulic turbine governing system by using a novel gravitational search algorithm based on Cauchy mutation and mass weighting. Appl. Soft Comput. **52**, 290–305 (2017)
32. Azizi, M., Ejlali, R.G., Ghasemi, S.A., Talatahari, S.: Upgraded Whale Optimization Algorithm for fuzzy logic based vibration control of nonlinear steel structure. Eng. Struct. **192**, 53–70 (2019)
33. Azizi, M., Ghasemi, S.A., Ejlali, R.G., Talatahari, S.: Optimum design of fuzzy controller using hybrid ant lion optimizer and Jaya algorithm. Artif. Intell. Rev., 1–32 (2019)
34. Azizi M., Ghasemi S.A., Ejlali R.G., Talatahari S., Optimal tuning of fuzzy parameters for structural motion control using multiverse optimizer. Struct. Des. Tall Special Build., e1652 (2019)
35. Talatahari, S., Motamedi, P., Farahmand, A.B., Azizi, M.: Tribe-charged system search for parameter configuration of non-linear systems with large search domains. Eng. Optim. (2019). https://doi.org/10.1080/0305215X.2019.1696786
36. Yildiz, A.R.: A comparative study of population-based optimization algorithms for turning operations. Inf. Sci. **210**, 81–88 (2012)
37. Kaveh, A., Pirgholizadeh, S., Khadem, H.O.: Semi-active tuned mass damper performance with optimized fuzzy controller using CSS algorithm. Asian J. Civil Eng. (BHRC) **16**(5), 587–606 (2015)
38. Abubaker, S., Nagan, S., Nasar, T.: Particle swarm optimized fuzzy control of structure with tuned liquid column damper. Global J. Pure Appl. Math. **12**(1), 875–886 (2016)
39. Kaveh, A., Talatahari, S.: A novel heuristic optimization method: charged system search. Acta Mech. **213**(3–4), 267–289 (2010)
40. Marasco, S., Cimellaro, G.P.: A new energy-based ground motion selection and modification method limiting the dynamic response dispersion and preserving the median demand. Bull. Earthquake Eng. **16**(2), 561–581 (2018)
41. Karakhanian, A.S., Trifonov, V.G., Philip, H., Avagyan, A., Hessami, K., Jamali, F., Bayraktutan, M.S., Bagdassarian, H., Arakelian, S., Davtian, V., Adilkhanyan, A.: Active faulting and natural hazards in Armenia, eastern Turkey and northwestern Iran. Tectonophysics **380**(3–4), 189–219 (2004)
42. Berberian, M., Yeats, R.S.: Patterns of historical earthquake rupture in the Iranian Plateau. Bull. Seismol. Soc. Am. **89**(1), 120–139 (1999)
43. Masson, F., Djamour, Y., Van Gorp, S., Chéry, J., Tatar, M., Tavakoli, F., Nankali, H., Vernant, P.: Extension in NW Iran driven by the motion of the South Caspian Basin. Earth Planet. Sci. Lett. **252**(1–2), 180–188 (2006)
44. Cimellaro, G.P., Marasco, S.: OPENSIGNAL: a software framework for earthquake record processing and selection. In: Second European conference on earthquake engineering and seismology (2ECSEES), Istanbul, Turkey Google Scholar (2014)
45. Cimellaro, G.P., Marasco, S.: A computer-based environment for processing and selection of seismic ground motion records: OPENSIGNAL. Front. Built. Environ. **1**, 17 (2015)
46. Newmark, N.M.: A method of computation for structural dynamics. J. Eng. Mech. Div. **85**(3), 67–94 (1959)
47. Subbaraj, K., Dokainish, M.A.: A survey of direct time-integration methods in computational structural dynamics—II. Implicit Methods. Comput. Struct. **32**(6), 1387–1401 (1989)
48. Ohtori, Y., Spencer, Jr, B.F.: A MATLAB-based tool for nonlinear structural analysis. In: Proceedings of the 13th Engineering Mechanics Conference (1999)
49. Holland, J.H.: Genetic algorithms. Sci. Am. **267**(1), 66–73 (1992)

50. Eberhart, R., Kennedy, J.: A new optimizer using particle swarm theory. In: Micro Machine and Human Science. IEEE (1995)
51. Atashpaz-Gargari E., Lucas C.: Imperialist competitive algorithm: an algorithm for optimization inspired by imperialistic competition. In: Evolutionary Computation. IEEE (2007)
52. Socha, K., Dorigo, M.: Ant colony optimization for continuous domains. Eur. J. Oper. Res. **185**(3), 1155–1173 (2008)
53. Mirjalili, S., Mirjalili, S.M., Lewis, A.: Grey wolf optimizer. Adv. Eng. Softw. **69**, 46–61 (2014)
54. Mirjalili, S.: Dragonfly algorithm: a new meta-heuristic optimization technique for solving single-objective, discrete, and multi-objective problems. Neural Comput. Appl. **27**(4), 1053–1073 (2016)
55. Rao, R.: Jaya: a simple and new optimization algorithm for solving constrained and unconstrained optimization problems. Int. J. Ind. Eng. Comput. **7**(1), 19–34 (2016)

Current Trends in the Optimization Approaches for Optimal Structural Control

Maziar Fahimi Farzam, Himan Hojat Jalali, Seyyed Ali Mousavi Gavgani, Aylin Ece Kayabekir, and Gebrail Bekdaş

Abstract This chapter describes some of the recent studies on the optimal control of structures. Initially, different passive, active and semi-active control devices are introduced and the importance of optimization in each case is discussed. A comprehensive literature review on the optimal design of various well-known control devices of each category i.e., Tuned Mass Damper (Active, Passive and Semi-Active modifications), Fluid Viscous Damper (Passive and Semi-Active), Viscoelastic Damper and Base Isolation (Passive and Semi-Active) and Active Tendon with special attention to the first studies and studies performed in the last decades, is presented. To suppress tall building vibrations subjected to wind and seismic loads, Tuned Mass Dampers (TMDs) modification have recently gained attention due to their simplicity, stability and reliability. Therefore, in the last section, the results of some recent studies on the optimization of Active and passive Tuned Mass Dampers and a recently proposed promising modification of TMD (i.e. Tuned Mass Damper Inerter) are comprehensively discussed.

Keywords Structural control · Optimization · Tuned mass damper

1 Introduction

The structural engineering community has always been interested in mitigating structural vibrations caused by dynamic lateral loads (such as earthquake, wind, wave, etc.) to provide a safe and uninterrupted performance of the structure. Therefore, researchers have attempted to address this challenge using different methods, among

M. F. Farzam · S. A. M. Gavgani
University of Maragheh, Maragheh, Iran

H. H. Jalali
University of Texas at Arlington, Arlington, TX, USA

A. E. Kayabekir · G. Bekdaş (✉)
Istanbul University-Cerrahpaşa, Istanbul, Turkey
e-mail: bekdas@istanbul.edu.tr

S. M. Nigdeli et al. (eds.), *Advances in Structural Engineering—Optimization*, Studies in Systems, Decision and Control 326,
https://doi.org/10.1007/978-3-030-61848-3_5

which structural control is one of the most popular and well-known approaches. In structural control, the dynamic structural parameters (e.g. the stiffness and damping) is altered by adding external devices and equipment to efficiently reduce the effects of the dynamic applied loads on the structural elements. As a result, the natural frequency, mode shapes and corresponding damping values of the structure will change such that the forces caused by the dynamic loads are reduced. In other words, structural control refers to the use of appropriate tools in a structure with the goal of improving its behavior. In general, structural control is subdivided into three different categories: passive, active and semi-active control; these devices have been widely used on a variety of structures, including buildings, bridges, power plants, offshore platforms, hospitals, schools, factories, etc.

It is important to note that in order to balance the associated expenses that is imposed on the projects due to the added control devices, they should be designed and located optimally in the structure to increase system efficiency and reduce the associated costs. Therefore, often the use of control devices is often intertwined with optimization methods. To this end, a significant number of studies have been carried out on optimal control of structures, which will be elaborated in the following sections of this chapter.

Initially, different passive, active and semi-active control devices and their features are introduced and followed up by the importance of optimization in each case. A comprehensive literature review on the optimal design of each category, with special attention to the first studies and studies performed in the last decades, is presented. Finally, results of some experimental studies on the optimization of various control devices are comprehensively discussed.

2 Passive Control

The most well-known and common type of control is passive-control, which dates back to the about a century ago. This method uses devices that do not require an external power source to impose the control force and works by dissipating the external imposed energy through the added devices. This in turn reduces the plastic deformation demand of the structural elements. A variety of passive control devices include but are not limited to tuned mass damper, viscous damper, viscoelastic damper, base isolation, etc. Many studies have pointed out the limited performance of these devices. However, researchers have been able to enhance their efficiency by integrating optimization methods, which has resulted in the extensive application of passive control devices [1, 2].

2.1 Tuned Mass Damper

Tuned mass damper (TMD) was first introduced by Frahm in 1909 to control the rolling motion and vibration of ships [3]. Den Hartog and Ormondroyd presented a comprehensive theory on the performance and limited efficiency of TMD [4], while Den Hartog optimized the design parameters (frequency and damping ratios) of TMDs in 1934 [5]. The main shortcoming of the formulation provided by Den Hartog was neglecting structural damping, which led [6] to incorporate damping and modify the aforementioned equations. Thereupon, a number of researchers presented experimental relationships on TMDs [7, 8] and some provided simplified design tables [9, 10]. In general, since 1934 numerous studies have been performed on the optimization of TMDs. Recently, with the advancement of computer science and optimization methods, the use of metaheuristic methods have become more common and popular [11, 12]. Hadi and Arfiadi for the first time found the optimal parameters of the tuned mass damper using Genetic Algorithm (GA) [13]. Similarly, researchers have used various optimization algorithms to find the optimal parameters of TMD, including but not limited to Particle Swarm [14], Harmony Search [15], Ant Colony [16], Artificial Bee Colony [17], Pollinated Flowers [18], Charged System Search [19], Colliding Bodies [20], etc. Furthermore, researchers have determined the optimal parameters of TMD using more realistic and enhanced structural models. For instance, Bekdaş et al. focused on determining the optimal parameters of TMD, considering soil-structure interaction; they considered minimizing the maximum roof acceleration transfer function amplitude as objective function [21]. Due to the large mass of the damper and the added costs, the majority of studies assume the mass as a predetermined parameter. Some studies have also compared and improved the optimization criteria for the optimal parameters of TMD (For instance, Kamgar and Khatibinia using whale optimization algorithm [22]). Additional and more detailed information on TMDs can be found in [1, 23, 24].

2.2 Fluid Viscous Damper

Fluid viscous dampers (FVD) are one the most widely used passive control devices in buildings, which dates back to over a century ago, where they were first employed in large-caliber artillery in the mid 1800s [25]. De Silva developed a gradient algorithm for the optimal design of dampers, with the focus on finding the optimal parameters of the damper, size of the damper and its location in the structure [26]. Gürgöze and Müller performed an analytical study on the optimal location of an FVD based on energy criteria [27]. Tsuji and Nakamura calculated the optimal damping coefficients of viscous dampers and provided a method to optimally distribute them in shear buildings for a specific set of earthquake records [28]. Takewaki presented the steepest direction search algorithm for finding the optimal location of a viscous dampers when combined with a TMD [29]. The algorithm is similar to the steepest descent method, but ensures the successive satisfaction of the optimality criteria. Assuming a constant floor height, Garcia introduced a simplified sequential search algorithm (SSSA) for determining the optimal location and damper parameters for multi-degree-of-freedom (MDOF) structures [30]. Singh and Moreschi determined the optimal size and location of different passive dampers, with special attention to viscous dampers, using genetic algorithm [31]. While GA is a powerful optimization tool, the main drawback is the high computational cost associated with the long duration of optimization. Martinez-Rodrigo and Romero performed a parametric strudy on 32 combinations of nonlinear viscous dampers and proposed a simple strategy for the optimum retrofit of a 6-story moment resisting frame using viscous dampers for a specific earthquake record [32]. Furthermore, Christopoulos and Filiatrault also proposed a new method for calculating the damping coefficients of viscous dampers [33]. Takewaki determined the optimal location of passive dampers, including FVD, based on a gradient-based approach [34].

In the past decade, researchers have also used novel optimization approaches and investigated special cases. For instance, Aydin et al. found the optimal distribution of viscous dampers in structures with soft stories using the steepest direction search method by taking the base shear force as objective function [35]. Similarly, the optimal distribution of viscous dampers in steel frames was investigated by Estekanchi and Basim using GA by minimizig the total damping coefficients in the structure [36]. The optimal distribution was selected based on their seismic performance using the Endurance Time method. Parcianello et al. focused on the optimal design of nonlinear viscous dampers to enhance the seismic behavior of structures [37]. Altieri et al. explored the optimal design of nonlinear viscous dampers using a reliabilty-based approach [38]. Akehashi and Takewaki studied the effect of the distribution of viscous dampers on the response of elastic-plastic structures and proposed a method for optimal placement of these dampers [39].

Another application of viscous damper is reducing the damage caused by vibration of two adjacent structures. Xu et al. used viscous dampers to control the vibration of two adjacent structures with different heights [40]. Some studies, such as Kandemir-Mazanoglu and Mazanoglu, evaluated the optimal capacity and location of viscous dampers between two adjacent structures [41]. Other researchers have focused on the optimization of viscous dampers when they are used in combination with other dampers, with the goal to overcome the weaknesses of each damper. Liu et al. investigated the optimal parameters of viscous dampers for base-isolated structures [42]. There are many other researches performed on determining the optimal parameters, size and location of viscous dampers, including [2, 34, 43].

2.3 *Viscoelastic Damper*

The third category of passive dampers that is presented in this chapter is viscoelastic (VE) dampers. The first application of these dampers, in the 1950s, was to control fatigue caused by vibration of airplane hulls [44]. Viscoelastic dampers were first introduced to civil engineering structures in 1969 in the twin towers of World Trade Center (WTC) in New York City, where approximately ten thousand VE dampers were installed to control the wind-induced vibrations. Similarly, in 1982 about 260 VE dampers were mounted in Columbia Center located in Seattle, Washington [45]. In general, VE dampers are used in structures, where shear deformation is expected. Zhang and Soong presented an optimization approach to minimize the number of VE dampers by finding the optimal location [46]. They concluded that the proposed method would yield a more economic and viable use of viscoelastic dampers. Hahn and Sathiavageeswaran focused on determining the optimal damping coefficient in VE dampers [47]. They further investigated the effect of distribution of these dampers along the height of the building and concluded that for uniform story stiffness, the dampers would be more efficient if placed in the lower floors. Based on intuitive criteria Wu et al. studied the optimal position of VE dampers in a three-dimensional assymetric frame [48]. Shukla and Datta determined the optimal location of VE dampers in MDOF structures using the root mean square (RMS) of story relative displacements (drift) as control criteria [49]. They found that the optimal number and locations of VE dampers depend on the nature of excitation force and the modeling of VEDs. Kim and Bang proposed a method to find the optimal distribution of VE dampers to minimize the torsional response of the structure [50]. Xu et al. carried out a synthethic optimization analysis for VE-equipped structures and found the optimal location and parameters of VE dampers [51]. They also performed a shake-table test of a 3 story building using a 1/5 scaled model for validation purposes. Furthermore, Park et al. used a gradient-based approach to find the optimal parameters of these dampers [52]. Similarly Fujita et al. and Pawlak and Lewandowski among other researchers determined the optimal location of VE dampers in structures [53, 54]. Zhu et al. investigated the optimum connecting dampers between two parallel structures using fluid viscous and viscoelastic dampers and compared their performance through the reduction in seismic response in adjacent structures [55]. They found that the performances of FVD and VE dampers on vibration mitigation in neighboring structures are similar. Lagaros et al. used Particle Swarm Optimization (PSO) to find the optimal location and parameters of VE dampers [43].

2.4 Base Isolation

Last, but not least, the optimization of base isolation (BI) passive devices is reviewed in this chapter. Its first use in the modern world dates back towards the end of the nineteenth century by John Milne, a British mining engineer [44]. The first use of a modern base isolation system was in 1969 that was installed to protect a three-story concrete school in Skopje, Macedonia from seismic events [56]. Different types of base isolators include but are not limited to elastomeric systems (natural rubber), lead rubber bearing (LRB), friction pendulum sliding (FPS) bearing, resilient-friction base isolator (R-FBI), high damping rubber bearing (HDRB), and some hybrid isolators

[57]. The isolators work more efficiently when they reduce the lateral stiffness of the system and therefore increase its natural period of vibration. As a result, many researches have attempted to enhance their performance by optimizing the numbers, size and their location within a structure. To this end, many researchers have tried to address this issue and to use different numerical and analytical methods to calculate the optimal properties of base isolators including but not limited to optimally criteria (OC) algorithm [58], constrained optimization procedure (COP) [59], artificial neural network [60], data mining [61], sequential quadratic programming (SQP) algorithm [62], etc. Many researchers have utilized different analytical, numerical and metaheuristic optimization methods to find the optimal parameters of base isolation devices. Pourzeynali and Zarif performed a mulit-objective optimization study on base-isolated high rise buildings using GA algorithm, considering both linear and nonlinear behavior for the base isolation device [63]. They found the proposed method to be efficient for obtaining the optimal parameters of the base isolation system based on minimizing the roof and base-isolation level displacements simultaneously. Nigdeli et al. determined the optimal parameters of the base isolation system for structures subjected to both near- and far-field records using the harmony search algorithm [64]; while Quaranta et al. carried out the same study using particle swarm optimization method [65]. Çerçevik et al. used three different metaheuristic optimization algorithms, i.e. crow search algorithm (CSA), whale optimization algorithm (WOA) and grey wolf optimizer (GWO) to find the optimal parameters of the isolator in base-isolated shear buildings [66]. They found that high damping ratio of the isolator does not necessarily warrant an optimal parameter for the base isolation. Zou et al. proposed an optimally criteria (OC) algorithm for the optimal design of base-isolated concrete structures [67]; while Mishra et al. performed a similar study using pattern search algorithm (PSA) on the design of seismic base isolators taking into account the uncertainty of the parameters with the goal of reducing earthquake effects [68]. Zhang and Shu proposed a performance-based optimization method to reduce structural, non-structural and isolation losses due to damages caused by earthquakes in base-isolated structures by performing a parametric study [69]. Mousazadeh et al. used multi-objective GA to present a practical method for the optimal design of LRB based on initial and life cycle cost of structures and used endurance time to evaluate the performance of the base-isolated building structures [70]. The optimal design of base isolators to control structural vibration is often perfromed by minimization of the RMS of structural responses and the variance of the responses is usually neglected; this could result in the sensitivty of the whole system to the input parameters. Therefore, Roy and Chakraborty used a standard gradiant-based MATLAB optimization routine to find an optimal and robust design for the base-isolation system [71].

In this section four passive control devices were introduced and a brief literature review on each was presented. It should be noted that there are other passive control devices that have been studied extensively to improve their performance using optimization methods in the literature, including but not limited to tuned liquid damper [72, 73], metallic yield damper [74, 75], friction damper [76, 77], U-shaped steel damper [78, 79].

3 Active Control

The somewhat low efficiency and disadvantages of passive control devices in reducing seismic effects resulted in the evolution of a new control concept, known as active control. The concept of active control was first adopted by Zuk and Clark (1970) for kinetic structures [80]. Active control has different components, including but not limited to sensors, control algorithm to calculate the control force, and actuator for applying the control force. Therefore, the optimization process not only includes the damper parameters and their locations, but also selecting the parameters of the control algorithm, location and number of sensors, power of the actuator, etc. Active tuned mass damper (ATMD) and active tendons are examples of active control devices [81].

3.1 Active Tuned Mass Damper

The ATMD is one of the most well-known active control devices that was first introduced by Lund [82]. The first optimization study on active control using TMD was performed by Chang and Soong [83]. Nishimura provided a closed-form solution of the optimum feedback gain and optimum parameters of the ATMD in the frequency domain under a harmonic excitation [84]. Nishimura et al. expanded their study to obtain and provide a closed-form solution for the optimum parameters of the ATMD under different types of excitations, i.e. harmonic, stationary white noise and non-stationary random vibrations such as earthquakes [85]. Xu performed a parametric study and proposed a method for selecting the design parameters of ATMD for tall buildings under wind excitation [86]; Results show that using acceleration sensors, instead of velocity and displacement, could reduce structural responses effectively. Furthermore, Wu and Yang employed ATMD to control the vibrations of a TV transmission tower in Nanjijng, China due to wind excitation using linear quadratic Gaussian (LQG), H_{∞} and continuous sliding mode control (CSMC) strategies based on acceleration feedbacks [87]. Li et al. used ATMD to reduce the translational and rotational responses of asymmetric structures subjected to earthquake excitations [88]. They obtained the optimal parameters of the ATMD by minimizing the translational and torsional displacement variances using the gradient search method (GSM).

Most of the control methods use optimization algorithms as a means to improving their performance. Therefore, a lot efforts have been placed on using metaheuristic approaches to optimize the control and actuator parameters. Jiang and Adeli proposed a novel neuro-genetic algorithm to find the optimum control force at each time step [89]. The proposed method, similar to the neural network optimization method, does not require training that results in the high-performance of the proposed method. Ahlawat and Ramaswamy proposed a method for the optimal design of multi-objective fuzzy logic controller and optimum parameters of ATMD based on genetic algorithm [90]. They selected the displacement and acceleration as objective functions, since these are representative of safety and comfort of residents, respectively. Soleymani and Khodadadi used a multi-objective adaptive genetic-fuzzy logic controller for the active control of a 76-story benchmark building using ATMD subjected to wind and seismic excitations [91]. Results showed an efficient performance in terms of suppressing base shear and interstory drifts. Kayabekir et al. used the modified harmony search to find the optimum parameters of ATMD using proposed proportional–integral–derivative (PID) type controllers [92]. They considered all the mechanical properties of the damper, i.e. mass, stiffness and damping coefficient and the PID controller parameters as design variables. Amini et al. used a new method to determine the optimal control force of a ATMD for a 10-story structure under near-fault ground motions using three different algorithms, i.e. Particle Swarm Optimization (PSO), Discrete Wavelet Transform (DWT) and Linear Quadratic Rregulator (LQR) [93]. Shariatmadar and Razavi investigated the optimal parameters of the fuzzy logic controller using the PSO method [94]. Furthermore, Amini and Bagheri used the colonial competitive algorithm to calculate the optimal control force [95]. Pourzeynali et al. used a combination of fuzzy logic and genetic algorithm for the active control of building structures with the goal of response reduction [96]. They determined the mass, frequency and damping ratios of the ATMD using the GA, while finding the overlap of parameters in the membership functions using the fuzzy logic control algorithm. Ozer et al. determined the optimal parameters of the controller of ATMD for a three story shear building using GA considering both single objective (response of the roof) and multi-objective (control energy and response of the roof) [97]. In general, some of the different control algorithms that are used for active control include but are not limited to Linear-Quadratic-Regulator (LQR) [98], Linear-Quadratic-Gaussian (LQG) [99], instantaneous optimal control [100], sliding mode control [101], Pole assignment [102], H_∞ [103], H_2 [104], Fuzzy logic [105], Proportional-Integral-Derivative (PID) [106, 107], Neural-Network [108], etc. Some more advanced algorithms include combination of different algorithms that employ different optimization methods, e.g. the LQR-PID optimized using the include combining different algorithm of the combined algorithms are optimized based on a cuckoo search (CS) algorithm [109], or the PID-FUZZY optimized using a salp swarm algorithm [110], etc.

3.2 Active Tendon

One of the other active control devices that is briefly presented in this chapter is active tendon system (ATS). The ATS was introduced by Eugene Freyssinet in 1960 [111]. This control system includes cable, hydraulic servomechanism system and a stiff steel frame that connects the cable and the actuator for applying the control force to the cable [112]. Suhardjo et al. proposed a method for the optimal active control of structures under wind excitation with the goal of reducing the floor accelerations [113]. The study was performed on a 60-story building using three different control systems, including the ATS. Chang and Lin introduced an optimized control algorithm for ATS, taking into account the time delay in the system [114]. Li et al. determined the optimal location of the actuator in ATS using GA, considering the complexity, discreteness and non-linearity of the optimal design problem [115]. The performance of the proposed method was assessed for a 16 story shear building under Tianjin earthquake record. Issa et al. studied the performance of ATS for vibration mitigation of frames and proposed a method for finding the optimal location of the tendons [116]. Rao and Sivasubramanian used the multiple start guided neighborhood search (MSGNS) algorithm to find the optimal location of the active tendons [117]. In this optimization method four important design criteria in active control are implemented, i.e. controlled peak displacement, controlled peak drift ratio, controlled peak absolute acceleration and average control force. The performance of the MSGNS algorithm is assessed for a 10-story, three-bay shear building under 14 earthquake records and results of the proposed algorithm is compared to three popular metaheuristic approaches: simulated annealing (SA), tabu search (TS) and GA.

Many optimization studies have also been performed on other active control devices including, but not limited to active brace [118], active viscous damper [119], active aerodynamical devices [120], active isolators [121], active tuned liquid damper [122].

4 Semi-active Control

Active control has also not been well accepted among engineers due to some drawbacks. One of the biggest shortcomings of these systems are its dependence of an external energy source, high energy consumption, problems associated with power outages during earthquake and also possibility of structural instability due to the applied external force [81]. Combining the different features of passive and active devices, the concept of semi-active dampers was first introduced in 1970 for car suspension system [123]. Semi-active (SA) control systems utilize different devices such as semi-active tuned liquid damper, semi-active friction damper, semi-active variable stiffness damper, Electrorheological (ER) fluid damper, Magnetorheological (MR) fluid damper, etc.

4.1 Magnetorheological Fluid Damper

Magnetorheological fluid damper is one of the most widely used semi-active control devices due to benefits associated with its properties. The first studies on the use of semi-active dampers in civil engineering was performed by Dyke and Spencer [124–126]. The first building that used MR dampers for vibration control was the Tokyo National Museum of Emerging Science and Innovation [127]. Research on optimization of MR dampers initiated with finding the optimal modeling parameters of the damper. Liu et al. used GA to determine the optimal parameters of the Bouc-Wen model in MR fluid dampers [128]. Azar et al. combined two of the most popular metaheuristic algorithms, charged system search (CSS), grey wolf (GW) and proposed the hybrid CSS-GW algorithm to obtain the optimal modeling [129]. Razman et al. and Yang et al. performed a similar analysis using the PSO algorithm [130, 131]. Xiao et al. used the ant colony optimization algorithm (ACO) to determine the optimal parameters for the Bingham and Bouc-Wen models of MR dampers [132]. Zhu et al. proposed a new method to obtain the optimal parameters of MR dampers and showed the superior performance of the developed model to that of GA and PSO [133]. Talatahari et al. used the adaptive charged system search optimization algorithm to identify the optimal parameters of the Bouc-Wen model for MR dampers [134]. Zaman and Sikder determined the optimal parameters of the Bouc-Wen model using the modified firefly algorithm and compared their approach to the GA and differential evolution (DE) in terms of accuracy and convergence rate [135]. It should be noted that the Bouc-Wen model is one of the most widely used hysteresis models that has many applications in other fields such as, Physics, Chemistry, Engineering, Science and Economics. Therefore, a significant number of research has been performed on the optimization of its parameters, specially using metaheuristic algorithms, e.g. GA [136, 137], PSO [138, 139], Jaya [140], artificial bee colony [141], etc. Spencer Jr et al. focused on finding the optimal parameters of a 20 tons MR damper that was modeled by the modified Bouc-Wen model based on voltage using a least-squares optimization method [142]. Yang et al. used a current driver to effectively reduce the response time of the MR damper by connecting the dampers coils in parallel [143]. Giuclea et al. found the optimal parameters of MR dampers using the inverse method, in which they utilized the modified Bouc-Wen model based on GA [144, 145]. The parameters of the model were calibrated based on experimental measurements, including current, displacement, velocity and force time histories at different flow values. Shu and Li used the optimized genetic algorithm to find the optimal parameters of the modified Bouc-Wen model [146].

Some researchers have tried to introduce improved models for the MR damper and to optimize the new modified model using different approaches. Kwok et al. used the Kwok model for simulating the behavior of MR dampers and used PSO for the determination of damper parameters [147]. Furthermore, Kwok et al. proposed an asymmetric Bouc-Wen model for modeling the behavior of MR dampers that unlike the conventional model was able to capture asymmetric hysteresis behavior [148]. They used GA for determination of the optimal parameters of the Bouc-Wen model. One of the issues that has drawn attention in optimization of MR dampers, is enhancing the performance of the controller (applied voltage to the damper). For instance, the behavior improvement in the fuzzy logic controller is due to the improvement of membership functions, control rules, scale coefficients, etc. Yan and Zhou proposed a method based on GA for designing structures controlled by MR damper with fuzzy logic controller [149]. The objective functions in this study were reduction of displacement and acceleration of the structure. They concluded that the proposed method has great flexibility and high reliability. In fact, the merit of this research was to provide a fuzzy correlation between the controller (structural response) and the output (command voltage) based on proper rules using GA. Askari and Davaie-Markazi investigated the rules, membership functions and scale factor of the fuzzy logic controller to determine and manage the optimal required voltage for the MR damper in nonlinear structures using a recursive optimization algorithm through the Nondominated Sorting Genetic Algorithm (NSGAII) [150]. The objective functions used in this study were the nondimensional peak floor acceleration and ductility indices. Moreover, Shook et al. used GA to overcome the complexities associated with the input of the fuzzy logic controller and to optimize the controller in buildings equipped with MR dampers [151]. To this end, they used 4 objective functions, i.e. reducing the peak drift, peak acceleration and root mean squared (RMS) of the inter-story drift and acceleration of the first two modes. Furthermore, Shook et al. employed MR dampers and fuzzy logic controller to reduce torsional response in buildings under earthquake excitations [152]. They used GA for improving the fuzzy logic controller and showed that it results in a significant reduction in responses. Ali and Ramaswamy used the microgenetic algorithm (μ-GA) and PSO to optimize the fuzzy logic control parameters of the MR damper [153]. They investigated two scenarios in their study: optimizing the fuzzy logic control parameters with fixed fuzzy rules, and optimizing both the fuzzy logic control parameters and the fuzzy rules. In general, optimization of the parameters of the membership functions and fuzzy logic rules were performed using 10 variables and finally the performance of the controllers were assessed in SDOF and MDOF structures. Huang et al. used the Bouc-Wen model and fuzzy logic controller for the semi-active control of buildings equipped with MR dampers and used GA for the optimization of fuzzy logic rules [154]. Results showed an enhanced performance of the controller to mitigate seismic vibrations. Bitaraf et al. used the multi-objective genetic-based fuzzy logic control approach to optimize the parameters of the fuzzy logic controller in MR dampers [155]. Bozorgvar and Zahrai proposed a semi-active control approach using MR dampers through a neuro-fuzzy logic controller that consists of an adaptive neuro-fuzzy inference system (ANFIS) to determine the input voltage of the MR damper [156]. The fuzzy membership and the ANFIS output functions were tuned simultaneously using GA. The seismic performance of the control system was assessed for a 3-story shear building and compared to the neural network predictive control (NNPC) algorithm, linear quadratic Gaussian (LQG) and clipped optimal control (COC) systems. Uz and Hadi used a binary-coded GA for optimization of MR dampers for vibration control of adjacent structures [157]. The controller and damper were idealized using fuzzy logic and modified Bouc-Wen model, respectively. Results of their study were compared to those using LQR and H2/LQG optimization algorithms, with the goal to reduce general damper-associated costs. Raeesi et al. proposed an optimized alternative to the inverse Takagi-Sugeno-Kang (TSK) fuzzy inference system for a MR damper, in which they determined the optimal parameters of the TSK using the improved grasshopper optimization algorithm (IGOA) [158].

A part of the literature is devoted to the application of neural network and neuro-fuzzy algorithm to improve the performance of the fuzzy logic controller, in which the input and output of the fuzzy logic controller are the acceleration and applied voltage to the MR damper, respectively [159]. Hiramoto et al. used GA to find the Lyapunov functions for optimal control of structures [160]. Hashemi et al. employed a controller based on neural network and wavelet to determine the optimal control force of the MR damper [161]. In addition, they used a modified and localized GA to optimize the network weight, biases and wavelet functions. The proposed modification offers fast convergence in addition to the global search provided by the conventional GA.

Another application of optimization algorithms in MR dampers is the optimal location and numbers of sensors and actuators. Shi et al. employed several optimization techniques for finding the optimal placement of MR dampers in high-rise buildings [162]. Ok et al. used GA and stochastic linearization method, and showed that a limited number of MR dampers could reduce the structural responses effectively [163]. Bao et al. proposed a genetic-gradient based algorithm to determine the optimal location of the MR damper and the optimal feedback gain [164]. They used the clipped-optimal-control (COC) algorithm to find the required external voltage. Li et al. proposed a two-phased approach based on GA to obtain the optimal placement of MR dampers in nonlinear buildings [165]. The first phase, includes determining the active control force using the GA algorithm, which is an important aspect in semi-active control. The optimal placement of the MR dampers is then determined in the second phase using the results of the first phase. Amini and Karami used GA to minimize the objective functions, including the coefficient matrices that define the optimal location of actuators and control force in semi-active control [166]. Elmeligy and Hassan studied the optimal number of MR dampers in a 3-story benchmark shear building and its effects on displacement and acceleration responses by performing a parametric study [167]. They found that placement of one damper resulted in an improved performance, while increasing the number of dampers did not enhance the performance of the structure. Bhaiya et al. developed an optimal and efficient design for semi-active control of structures using a limited number of sensors and MR dampers based on GA [168]. Zabihi-Samani and Ghanooni-Bagha proposed a fuzzy logic controller for MR dampers that was based on the wavelet-based cuckoo-search for vibration mitigation of structures under earthquake excitation to find the optimal location and number of MR dampers and sensors [169]. The proposed algorithm included the discrete wavelet transformation (DWT), fuzzy logic controller (FLC), modified Bouc-Wen model and geometrical nonlinearity. The discrete wavelet transformation was used to obtain the energy distribution of earthquake excitation over the frequency bands, which was transmitted to the fuzzy logic controller. The nonlinear behavior of MR damper was modeled using the modified Bouc-Wen model.

4.2 Semi-active TMD

The second semi-active control device that is discussed in this chapter and widely studied, is the semi-active TMD that was first introduced by Hrovat et al. to mitigate wind-induced vibrations in high-rise buildings [170]. Setareh et al. used a semi-active MR device inside a pendulum TMD and created the semi-active pendulum tuned mass damper (SAPTMD) to control large floor vibrations [171]. The equations of motion of the coupled SAPTMD–floor system were developed based on an equivalent SDOF of the SAPTMD device. The optimal parameters were found by an optimization software. The results of the analytical modeling were compared to an experimental full-scale test and its equivalent passive device. Kaveh et al. investigated the performance of an active TMD in parallel with an MR damper on the behavior of 10-story shear building under four earthquake records [172]. They used the optimal groundhook and fuzzy logic controller by means of the charged system search (CSS) algorithm to find the optimal required voltage. It in noteworthy that in this study the fuzzy rules were optimized.

4.3 Semi-active Base Isolation

One of the common semi-active devices used for structural control is the semi-active base isolator that uses different control system including variable orifice damper [173, 174], variable friction damper [175]. MR dampers, as mentioned earlier, is one of the other semi-active control devices that can be used separately [176–178] or in combination with base isolation systems [179–181]. Kim and Roschke used a combination of MR damper and base isolation for semi-active control of structures [182]. They determined the input voltage for the MR damper using fuzzy logic, and employed GA to find the appropriate fuzzy control rules and optimize the membership function parameters. Furthermore, Kim and Roschke used a combination of MR damper and friction pendulum sliding (FPS) bearing for semi-active control of buildings [183]. The dynamic behavior of the MR damper and FPS bearing was modeled based on neuro-fuzzy logic. A fuzzy logic controller was used to tune the MR damper, while the optimal parameters of the membership functions and finding the appropriate fuzzy logic rules was determined with the aid of multi-objective GA. Results showed that the proposed method was able to find the optimal membership functions and fuzzy logic rules and could enhance the semi-active control algorithm. Ozbulut et al. employed GA to find the optimal parameters of an adaptive fuzzy neural controller (AFNC) in semi-active base isolation systems [184]. They used variable friction dampers to develop the semi-active control device.

In general, the semi-active base isolation device is used for reducing the drift at the base isolation level. The drawback of this method is that in most applications, it increases the peak acceleration of the structure, which is unfavorable. Therefore, Mohebbi and Dadkhah investigated the optimal design of the semi-active base isolation consisting of linear base isolation with low damping and MR damper to control the building acceleration and base drift [185]. In addition, they performed a multi-objective optimization study using GA with a linear combination of peak acceleration and base drift as the objective functions. They used the H2/LQG and COC algorithms to determine the input voltage of the MR damper. Results showed that implementation of base isolation and MR damper can lead to a significant reduction in acceleration and base drift.

4.4 Other Semi-active Dampers

Another device that has been proposed for vibration mitigation and structural control is the semi-active viscous dampers, which were proposed to enhance the performance of the conventional viscous dampers [186]. Bakhshinezhad and Mohebbi proposed an efficient method for the design of semi-active fluid viscous dampers (SAFVD) for vibration mitigation of nonlinear structures [187]. This multi-objective function optimization study aimed to minimize the interstory drift and absolute acceleration as life safety and convenience criteria, respectively using non-dominated sorting genetic algorithm (NSGA-II). The damping coefficient of the damper and the control algorithm parameters were considered as design parameters. The uncertainty in the base excitation has been taken into account. Results showed that the proposed method is able to significantly reduce the structural responses and provide enhanced safety and convenience for the occupants.

Semi-active hydraulic dampers are another type of SA control devices. Kazemi Bidokhti et al. used semi-active hydraulic dampers along with fuzzy logic controller [188]. Furthermore, they used GA to find the optimal parameters of the fuzzy logic controller. The objective function in this study was minimizing the peak interstory drift and absolute acceleration.

5 Current Trends in Optimization of Structural Control Devices

The previous sections provided a brief review of the literature on optimization of control devices. In this section some examples of the state-of-the-art research is explained more in detail.

5.1 Particle Swarm Optimization and Active Control of Building Structures Using ATMD

Particle Swarm Optimization (PSO) was first introduced by Eberhart and Kennedy [189]. They initially intended to develop a computational intelligence that eliminates manual interference and can reach a converged solution using bird flocking and fish schooling models. They also showed that this method is computationally efficient and fast. Leung et al. employed the particle PSO algorithm to find the optimal parameters (mass ratio, damping ratio and frequency ratio) of a TMD under non-stationary base excitation [14]. The following describes the application of PSO for finding the optimal parameters of an ATMD that is used to control the seismic response of a linear 10-story shear building under 28 far-field (FF) and near-field ground motions using MATLAB. Complete details and results of the investigation can be found in [190] and limited results are presented herein. Fuzzy logic controller (FLC) and Mamdani Inference system are used in this study due to the random nature of seismic excitation to obtain the control force. PSO is used to optimize the performance of the controller by finding the optimal actuator power, taking into account the effect of the actuator saturation. The mass, stiffness and damping of all the stories were considered to be uniform as 360 ton, 650,000 kN/m and 6200 kNs/m, respectively [19], which leads to a fundamental period of vibration of 0.98 s. Figure 1 shows a sketch of the 10-story shear building, in which mi, ki and ci are the ith story mass, stiffness and damping, respectively; while m_ATMD, c_ATMD, k_ATMD and u_ATMD are the mass, damping, stiffness and the actuator control force of the ATMD, respectively. The mass of the ATMD is considered as 0.03 of the total mass of the structure. The stiffness and damping of the ATMD was considered as 3750 kN/m and 151.5 kNs/m based on [13]. The equations of motion, written in state space, for n degrees of freedom (DOF) are shown in Eq. (1), where A, B, H and Z are the system, control force, excitation matrices and state vector, respectively, and are defined in Eqs. (2) to (5).

$$\dot{Z}(t) = AZ(t) + Bu(t) + Hf(t) \tag{1}$$

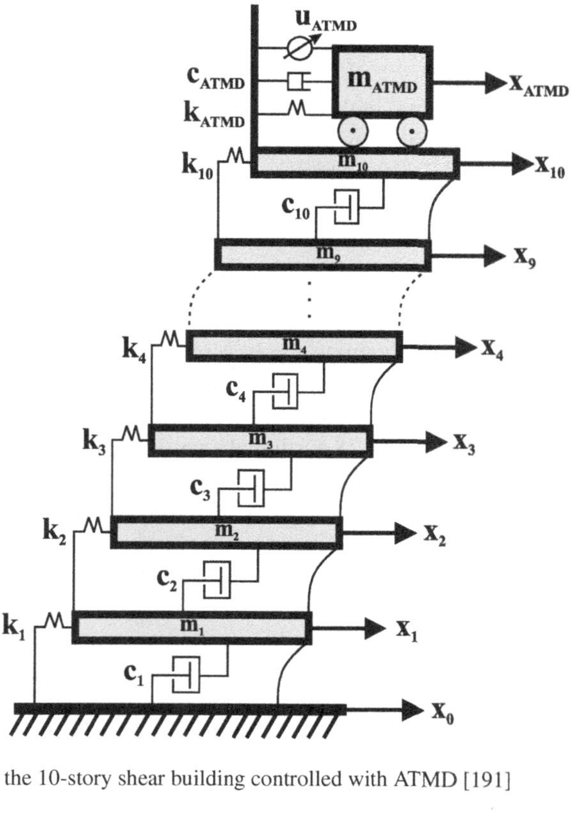

Fig. 1 Sketch of the 10-story shear building controlled with ATMD [191]

$$A = \begin{bmatrix} 0 & I \\ -M^{-1}K & -M^{-1}C \end{bmatrix}_{2n \times 2n} \tag{2}$$

$$B = \begin{bmatrix} 0 \\ M^{-1}D \end{bmatrix}_{2n \times m} \tag{3}$$

$$H = \begin{bmatrix} I \\ M^{-1}E \end{bmatrix}_{2n \times r} \tag{4}$$

$$Z = \begin{bmatrix} x(t) \\ \dot{x}(t) \end{bmatrix}_{2n \times 1} \tag{5}$$

In Eqs. 1–5 ***M***, ***K***, ***C***, ***u***(t) and ***f***(t) are the mass, stiffness, damping matrices, excitation and control force vectors, respectively.

As mentioned earlier, optimization of the actuator power is performed using the PSO algorithm. The objective function is taken as the roof displacement and the actuator saturation, which is a common practical issue, is considered as a constraint. The optimization ensures that the actuator power is selected such that the peak roof displacement is minimized. The different steps of employing the PSO algorithm is as follows:

- The initial location of the particles, in this case the actuator power, is randomly generated.
- The objective function value, in this case the roof displacement, is determined.
- The corresponding objective function values for each particle are compared to each other, and the minimum is selected as the best global cost. The particle that has generated the least value is considered as the best particle.
- The previous cycle is repeated for the following particles until the best particle is introduced as the result of the optimization. The new particles are selected based on the previous location, velocity, best global cost and best particle.

It should be noted that by obtaining the best particle, which corresponds to the optimum actuator power the control design is optimized for each earthquake record. The performance of the employed method is assessed using linear time history analysis of the 10-story building under 28 earthquake records with different characteristics (Table 1). All records are scaled to 0.3 g to eliminate the effect of peak ground acceleration (PGA) variability. In Table 1, the earthquake records are divided in 4 suites, each containing 7 earthquake records, i.e. far-fault, near-fault (fling step), near-fault (forward-rupture directivity) and near-fault (without pulse).

To better assess the performance of the ATMD control device, the results for peak roof displacement are compared to the uncontrolled structure and controlled with passive TMD. Figure 2 shows the comparison of roof displacement time histories for some of the records, where active control showed the biggest reduction in responses (records 4, 12, 17 and 27).

Table 1 Earthquake records features [190]

No.	Earthquake	Station	Comp	PGA (g)
(a) Near-fault records (fling-step)				
1	Kocaeli	Yarimca (YPT)	EW	0.23
2	Chi-Chi	TCU052	NS	0.44
3	Chi-Chi	TCU068	EW	0.5
4	Chi-Chi	TCU074	EW	0.59
5	Chi-Chi	TCU084	EW	0.98
6	Chi-Chi	TCU102	EW	0.29
7	Chi-Chi	TCU128	EW	0.14
(b) Near-fault records (forward-rupture directivity)				
8	Cape mendocino	Petrolia	90	0.66
9	Northridge	Olive view	360	0.84
10	Erzincan	Erzincan	EW	0.5
11	Parkfield	Fault zone 1	FN	0.5
12	Morgan hill	Anderson dam	340	0.29
13	Superstition hills	Parachute test site	315	0.45
14	Imperial-valley	Brawley airport	225	0.16
(c) Far-fault records				
15	Kern county	Taft	111	0.18
16	Imperial valley	Calexico	225	0.27
17	Loma Prieta	Presidio	0	0.1
18	Northridge	Century CCC	90	0.26
19	Northridge	Moorpark	180	0.29
20	Northridge	Montebello	206	0.18
21	San Fernando	Castaic	291	0.27
(d) Near-fault records (without pulse)				
22	Imperial valley-06	Bonds corner	140	0.59
23	Imperial valley-06	chihuahua	12	0.26
24	Northridge-01	Saticoy	90	0.34
25	Loma Prieta	Capitola	0	0.51
26	Parkfield	Array #8	50	0.24
27	Superstition hills	Superstition Mtn	45	0.58
28	Northridge-01	Rinaldi	228	0.87
28	Northridge-01	Rinaldi	228	0.87

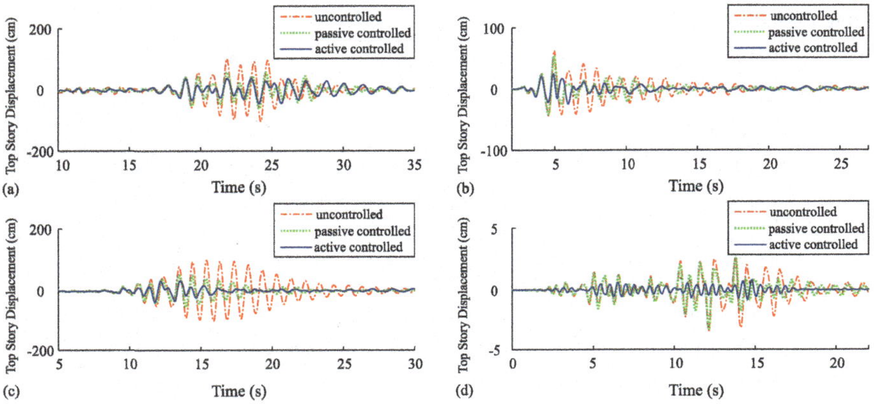

Fig. 2 Roof displacement time history for records no. **a** 4, **b** 12, **c** 17 and **d** 27

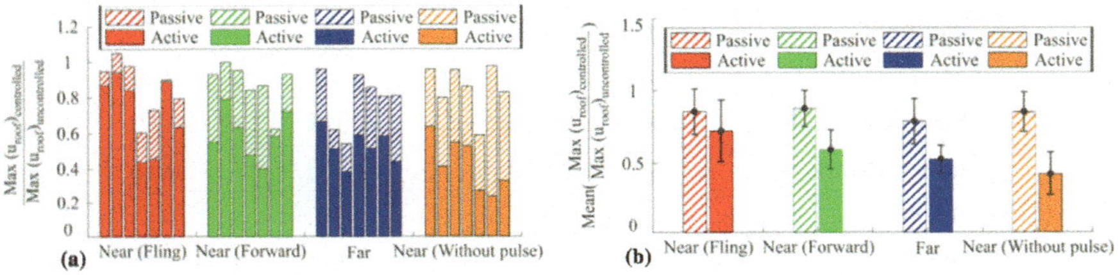

Fig. 3 **a** Normalized peak roof displacement, **b** Average normalized peak displacement with optimum actuator power

For better comparison, the peak roof displacement for each 4 suites of earthquake is shown in Fig. 3a. Results are normalized to the peak roof displacement of the uncontrolled structure under all 28 earthquake records. In Fig. 3, the active and passive control results are shown with solid and hatched bar charts, respectively. Figure 3a shows that in almost all cases the active control strategy has lead to a more efficient performance, with response reductions as high as about 80% for near-fault records without pulse. Since for each suite of earthquake excitation, 7 records are used, the average of peak roof displacement throughout each suite is also reported herein, as suggested by ASCE 7-10 [192]. Figure 3b clearly shows the superior performance of active control strategy even in case of near-fault records with fling step. The best and the worst performance of the optimum ATMD are realized under near-fault earthquake without pulse (58% average response reduction) and fling step (28% average response reduction), respectively.

5.2 *Harmony Search Algorithm (HSA) and Active Control of Jacket Platform Using ATMD*

One of the major applications of structural control is in offshore structures, which are subjected hydrodynamic and wave loads. Several methods exist for the analysis of offshore platform under wave loads. The American Petroleum Institute [193] recommends using response spectrum, push-over and time history analysis. The time history analysis is an appropriate analysis method due to the dynamic and nonlinear nature of waves; however, due to the long computational duration it has not been widely accepted by the industry. Therefore, researchers have tried to come up with different methods to address this issue. Vamvatsikos and Cornell proposed applying the incremental dynamic analysis (IDA) for the analysis of structures [194]. Golafshani et al. proposed the Incremental Wave Analysis (IWA) [195], which is similar to IDA; however, instead of a three hour analysis, it is based on the peak wave height and therefore reduces the computational efforts and is therefore, more efficient. Zeinoddini et al. proposed the Endurance Wave Analysis (EWA) for the analysis of jacket platforms [196], which is based on the endurance time analysis (ETA) [197]. In this method a fixed reduced time of 100 s, as opposed to standard three hours, is used for analysis. Mohajernasab and Dastan Diznab proposed the Modified Endurance Wave Analysis (MEWA) method and its application, which determines the optimum time of wave record based on the period of the peak of the spectrum (peak spectral period) [198, 199]. In other words, time duration for each wave load was proposed as a factor of peak spectral period instead of being a fixed value. This modification, called Time Duration Factor (TDF), considers the effects of significant frequencies of the power spectral density and reduces total time of the records. The following presents the dynamic behavior of a jacket platform, located in the Persian Gulf, which is controlled with an active tuned mass damper (ATMD). The analysis takes into account the effect of actuator saturation. The control force and the optimum actuator power are determined using fuzzy logic controller (FLC) and Harmony Search Algorithm (HSA) [200], respectively. Furthermore, fluid-structure interaction (FSI), the effect of added mass due to the accelerated motion of the fluid are considered. Complete details and results of the investigation can be found in [201] and limited results are presented herein.

A linear lumped mass model of the jacket platform is used in the current study, where the lumped mass characteristics (mass and stiffness) were determined based on the properties of the real structure such that the simplified model exhibited the same natural periods of the detailed model as reported by Mohajernasab et al. [198]. The equivalent seven degrees of freedom system is shown in Fig. 4a, where the platform deck is located at the 7th level. Damping in the structure is simulated using Rayleigh damping with 2% damping ratio for the first two modes. Random wave and constrained new-wave theories are utilized in generation of the wave records based on [199]. A wave with a 100-year return period was considered in this study with significant wave height, peak spectral period, and peak level of 5.83 m, 7.1 s,

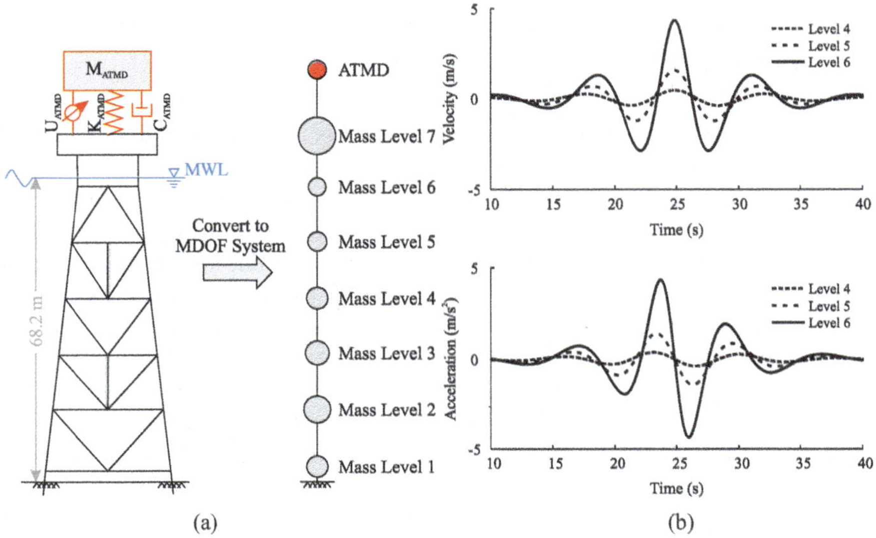

Fig. 4 Active controlled platform and its simplified lumped mass model [201]

and 4.77 m, respectively. Figure 4b presents an example of the wave velocity and acceleration used for the current analysis.

The equation of motion of the controlled jacket platform under wave loads is shown in Eq. (6), where $\boldsymbol{M}$, $\boldsymbol{K}$, $\boldsymbol{C}$, $\boldsymbol{X}(t)$ (t), $\dot{\boldsymbol{X}}(t)$ and $\ddot{\boldsymbol{X}}(t)$ are the mass, stiffness, damping matrices, displacement, velocity and acceleration vectors, respectively.

$$\boldsymbol{M}\ddot{\boldsymbol{X}}(t) + \boldsymbol{C}\dot{\boldsymbol{X}}(t) + \boldsymbol{K}\boldsymbol{X}(t) = \boldsymbol{F}_I + \boldsymbol{F}_D + \boldsymbol{F}_C \quad (6)$$

Similarly, $\boldsymbol{F}_I$, $\boldsymbol{F}_D$, and $\boldsymbol{F}_C$, are the inertial, drag and control forces, respectively. The inertial and drag forces are determined based on Eqs. (7) and (8), where the "·" is the element-wise product operator [199]; while the control force is determined based on fuzzy logic control algorithm. In Eqs. (7) and (8), $\boldsymbol{U}$ and $\dot{\boldsymbol{U}}$ are the wave velocity and acceleration and ρ is the mass density of water, which is considered as 1024 kg/m^3. In addition, C_m and C_d are the inertia and drag coefficients, which are taken as 1.2 and 1.05 according to API [193]; A and V are the cross-sectional area and volume vector.

$$\boldsymbol{F}_I = \rho \boldsymbol{C}_m \boldsymbol{V} \cdot \dot{\boldsymbol{U}} \quad (7)$$

$$\boldsymbol{F}_D = \rho \boldsymbol{C}_d \boldsymbol{A}\boldsymbol{V} \cdot \left(\boldsymbol{U} - \dot{\boldsymbol{X}}\right) \cdot \left|\boldsymbol{U} - \dot{\boldsymbol{X}}\right| \quad (8)$$

It should be noted that in Eq. (6), M is the combined mass matrix considering the effect of added mass due to fluid acceleration, which is shown in Eq. (9). In this equation $\boldsymbol{M}_0$ is the mass of the platform structure.

$$M = M_0 + \rho(C_m - 1)V \cdot \ddot{X} \tag{9}$$

As mentioned earlier, Harmony Search Algorithm (HSA) is employed for the optimization of the actuator power for the active control. To this end, minimizing the displacement of the deck (Level 7) is considered as the objective function. Figure 5 summarizes the optimization procedure for HSA used in this study.

Simulation of the controlled and uncontrolled structures are performed in MATLAB and SIMULINK software. Figure 6 depicts the outline of the model

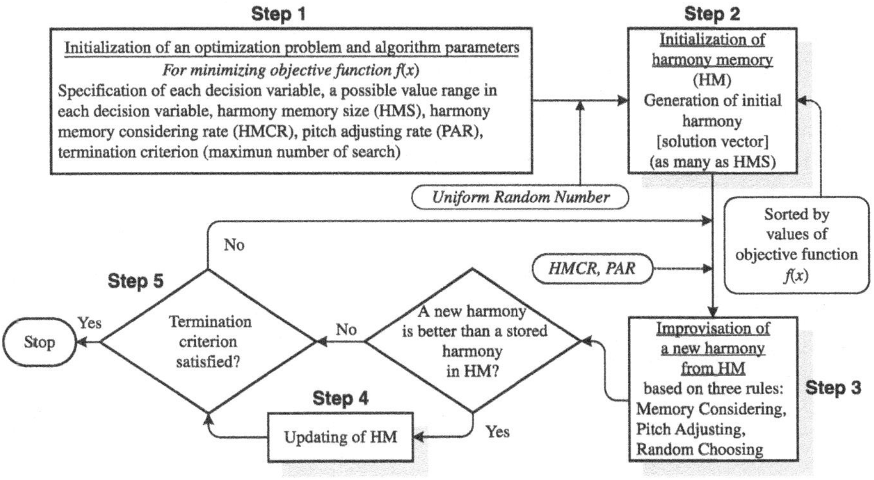

Fig. 5 Optimization procedure of the harmony search algorithm [200]

Fig. 6 SIMULINK model [201]

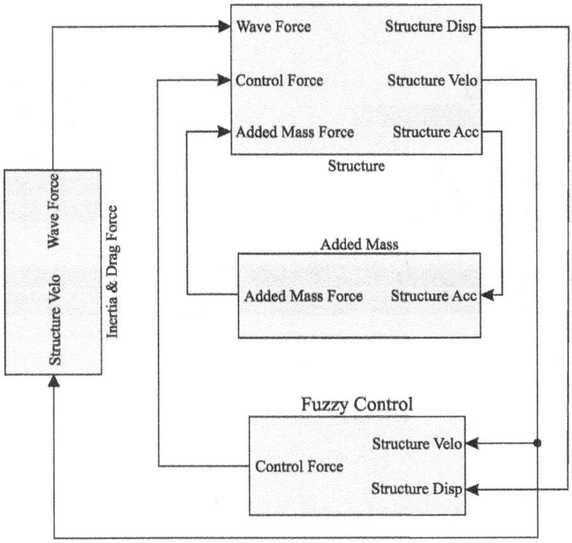

prepared in SIMULINK. As mentioned earlier, the fluid-structure interaction, the effect of added mass due to accelerated motion of the body in the fluid, and the effect of actuator saturation are considered.

In order to better assess the proposed approach to control the vibration of the platform, performance of both passive TMD and ATMD are assessed under the defined wave loads. The circular frequency, structural damping and mass ratios of the TMD was considered as 2.68 rad/s, 2% and 3%, with an optimal frequency and damping ratio of 0.964 and 0.106, respectively.

Figure 7 compares the time history responses for the displacement and acceleration at the deck level for the uncontrolled, passive and active TMD control approaches. Figure 7 clearly demonstrates the superior performance of the ATMD to control the displacement and acceleration responses during the duration of the wave load at the deck level.

Peak deck displacement, velocity and acceleration, as well as the root mean square (RMS) of the responses over time are considered as the performance criteria and are shown in Fig. 8. Results are presented as the normalized performance criteria

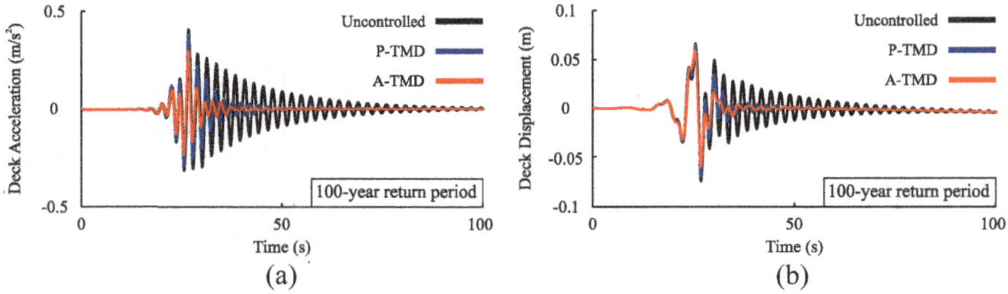

Fig. 7 Time history responses at the deck level for **a** displacement and **b** acceleration for the uncontrolled, passive and active TMD cases [201]

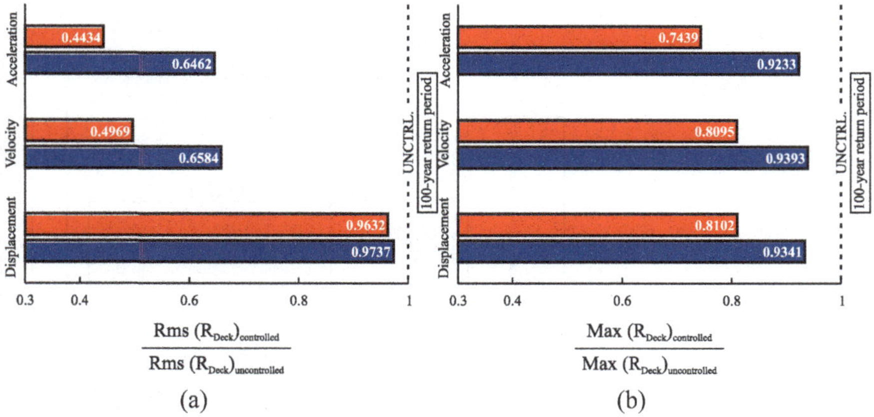

Fig. 8 Comparison of **a** Peak, **b** RMS of responses for the active and passive TMDs [201]

(response of controlled to uncontrolled structure) for better comparison. In Fig. 8, blue and red colors represent the normalized structural response of passive and active controlled platform, respectively.

Figure 8 clearly indicates that the active controlled platform, both in terms of peak and RMS responses, outperforms the passive controlled platform. The effect is more significant when comparing the accelerations. In fact, the ATMD has resulted in a peak acceleration reduction of 26%, while the passive TMD shows only an 8% reduction in acceleration. The RMS of acceleration shows a 36% and 56% reduction for the passive TMD and ATMD controlled platform, respectively.

5.3 Soil-Structure Interaction (SSI) and TMD Optimization Using Metaheuristic Methods

Soil-structure-interaction (SSI) is one of the topics that has drawn the attention of many researchers due to its complex effects on the control of building structures. Different metaheuristic algorithms such as ant colony [16], artificial bee colony [17], shuffled complex evaluation [202], HSA [203], bat algorithm [203] have been utilized for control of structures with optimal TMD parameters considering soil–structure interaction (SSI) (Fig. 9). The following presents an approach to determine the optimum TMD parameters taking into account the effect of SSI. The optimization is performed through different metaheuristic algorithms, including HSA [204], teaching-learning based optimization (TLBO) [205], flower pollination (FPA) [206], and Jaya Algorithm (JA) [207] and two of its different forms, i.e. Jaya algorithm with Lévy flight (JALF) and two-phase Jaya algorithm (2PJA). Complete details and results of the investigation can be found in [21] and limited results are presented herein.

The proposed methods for optimization are employed for the optimal control of a 40-story benchmark building [208], with the objective function to minimize the acceleration transfer function. Three different soil conditions, i.e. soft, medium and dense soil, are considered for investigating the SSI effects (Table 2). The story parameters of the 40-story benchmark building can be found in [21].

As mentioned earlier, the objective function for the optimization is minimizing the transfer function of the peak acceleration of the top story. Equations (10) and (11) show the objective functions (in dB) in case of fixed-base structure and when considering SSI, respectively.

$$f(x) = 20\mathbf{Log}_{10}|\mathbf{max}(\boldsymbol{TF}_N(\omega))| \tag{10}$$

$$f(x) = 20\mathbf{Log}_{10}|\mathbf{max}(\boldsymbol{TF}_N(\omega) + \boldsymbol{TF}_0(\omega) + \boldsymbol{TF}_\theta(\omega)z_N)| \tag{11}$$

Table 3 shows the optimum TMD parameters based on different optimization algorithms. The constraint of the current optimization problem are considered as the

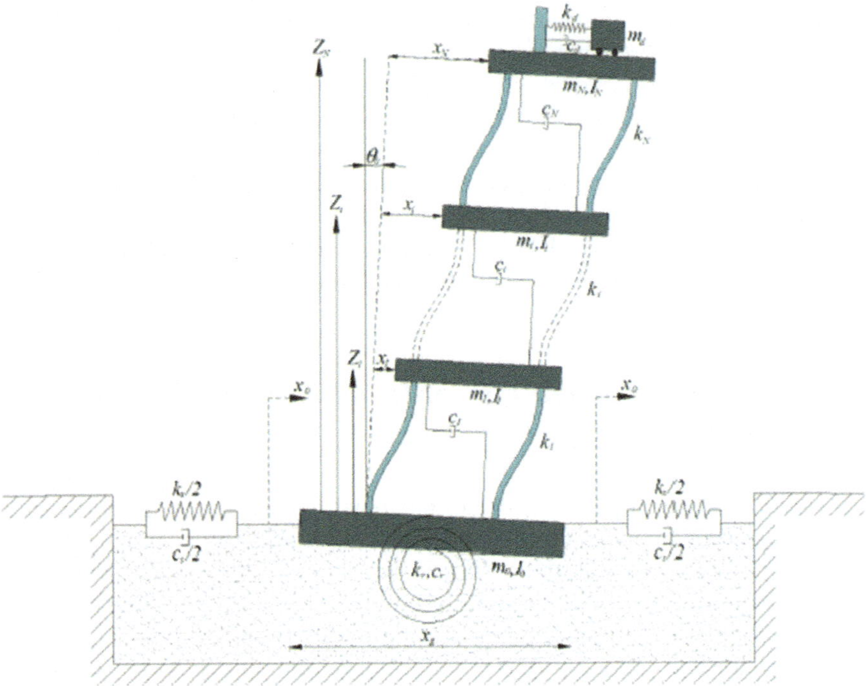

Fig. 9 Sketch of the structural control considering SSI [21]

Table 2 Soil properties used [21]

Soil type	Swaying damping c_s (Ns/m)	Rocking damping c_r (Ns/m)	Swaying stiffness k_s (N/m)	Rocking stiffness k_r (N/m)
Soft	2.19×10^8	2.26×10^{10}	1.91×10^9	7.53×10^{11}
Medium	6.9×10^8	7.02×10^{10}	1.8×10^{10}	7.02×10^{12}
Dense	1.32×10^9	1.15×10^{11}	5.75×10^{10}	1.91×10^{13}

mass of TMD between 1 and 2% of the total mass of the superstructure. For each optimization method 20 iterations are used. f_{best} and f_{ave} present the best and average of the 20 iterations. The standard deviation (σ) of the 20 iteration are also presented in Table 3. The numbers of objective function evaluations to reach the best results is shown as C_{best}; therefore, it can act as a means for the computational efforts to complete the optimization.

Table 3 shows that the optimum TMD parameters are definitely a function of the soil type. It also shows that almost all algorithms show the same optimum parameters, with a small standard deviation. In terms of computational effort, the original Jaya algorithm outperforms all the other algorithms including JALF and 2PJA and is therefore, more robust.

Table 3 Optimum TMD parameters [21]

		HS	FPA	TLBO	JA	JA	JALF
Fixed based	m_d (t)	784	784	784	784	784	784
	T_d (s)	3.9489	3.9496	3.9496	3.9496	3.9496	3.9496
	ξ_d	0.1313	0.1317	0.1317	0.1317	0.1317	0.1317
	f_{best}	17.3344	17.3285	17.3285	17.3285	17.3285	17.3285
	f_{ave}	17.3357	17.3285	17.3285	17.3285	17.3285	17.3285
	σ	0.00294	3.65E–15	3.65E–15	3.65E–15	3.65E–15	3.65E–15
	C_{best}	4318	1450	4800	1300	3525	1750
Dense soil	m_d (t)	784	784	784	784	784	784
	T_d (s)	3.9977	3.9969	3.9969	3.9969	3.9969	6.0324
	ξ_d	0.1100	0.1102	0.1102	0.1102	0.1102	0.1458
	f_{best}	17.0762	17.0669	17.0669	17.0669	17.0669	17.3247
	f_{ave}	17.0925	17.0669	17.0669	17.0669	17.0669	17.3247
	σ	0.00800	7.27E−13	7.25E−07	1.83E−14	1.58E−05	2.10E−07
	C_{best}	4346	1525	4650	1100	3050	2050
Medium soil	m_d (t)	784	784	784	784	784	784
	T_d (s)	4.2196	4.2196	4.2196	4.2196	4.2196	4.2196
	ξ_d	0.1269	0.1267	0.1267	0.1267	0.1267	0.1267
	f_{best}	17.4443	17.4394	17.4394	17.4394	17.4394	17.4394
	f_{ave}	17.4577	17.4394	17.4394	17.4394	17.4394	17.4394
	σ	0.00813	1.21E−12	4.61E−07	9.50E−15	3.87E−06	4.67E−11
	C_{best}	3713	1800	8600	1475	4800	2000
Soft soil	m_d (t)	784	784	784	784	784	784
	T_d (s)	6.0313	6.0324	6.0324	6.0324	6.0323	6.0324
	ξ_d	0.1448	0.1458	0.1458	0.1458	0.1458	0.1458
	f_{best}	17.3384	17.3247	17.3247	17.3247	17.3248	17.3247
	f_{ave}	17.3409	17.3247	17.3247	17.3247	17.3249	17.3247
	σ	0.00497	7.52E−12	6.12E−06	8.23E−11	0.00013	1.08E−07
	C_{best}	2395	1700	6850	1650	4550	3050

The performance of the TMD is assessed both in the frequency and time domains. The acceleration transfer function for the roof and the top story displacement are selected as the performance criteria in the frequency and time domains, respectively. Figure 10 shows both the controlled and uncontrolled roof displacements and roof acceleration transfer functions for the fixed base structure and the structure on soft soil. The time domain analysis results in Fig. 10 is presented for the structure under the Chi-Chi Taiwan Earthquake (FN component of CHY101 record) based on FEMA P-695 [209]. Both time and frequency domain results show the efficiency of the TMD in response reduction, especially considering SSI for the structure supported on soft

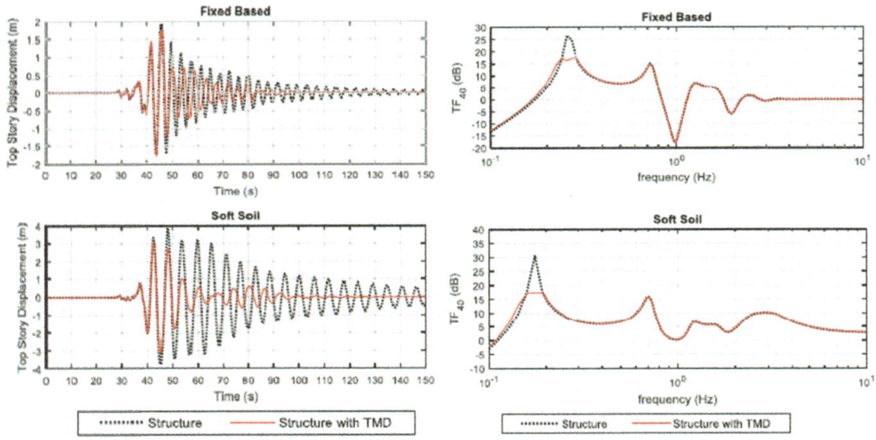

Fig. 10 Comparison of (a) roof displacement and (b) roof acceleration transfer function for the uncontrolled structure and the TMD-equipped structure for fixed base case and SSI case with soft soil [21]

soil. Considering SSI also enhances the rapid damping behavior as the responses are reduced much faster, as compared to the fixed-base structure.

5.4 Modified Harmony Search Algorithm (MHSA) for Optimization of ATMD

The following presents a modification to the previously discussed HSA for optimizing ATMDs based on [92]. The modified approach is used for the optimum design of a proportional–integral–derivative (PID) type controller, as well as the mechanical properties of the ATMD (mass, stiffness and damping ratios). The modification included in this study considers the best solution with a defined probability, as well as updating the harmony memory, pitch adjusting and considering rates. Complete details and results of the investigation can be found in [92] and limited results are presented herein.

The proposed method is used for the control of a 10-story shear building with uniform properties for each floor including mass, stiffness, and damping equal to 360 tons, 650 MN/m and 6.2 MNs/m, respectively [3]. The mass of the TMD is considered in the range of 1% to 5% of the total mass of the structure; while the period of vibration of the ATMD is selected at 0.5–1.5 times the fundamental period of the uncontrolled structure. The optimization for the ATMD is performed using a Matlab Simulink Block Diagram and a velocity feedback control is employed for PID controller optimization. To this end, roof displacement is selected as the objective function with the constraint of limiting the ATMD stroke. Different stroke limitations are considered for comparison purposes. The optimization process flowchart is shown in Fig. 11. Details of the optimization process can be found in [92].

The robustness of the proposed method is assessed through 22 far-field records according to FEMA P-695 [209]. It should be noted that the optimum parameters for the ATMD will be different based on different ATMD strokes. The roof story

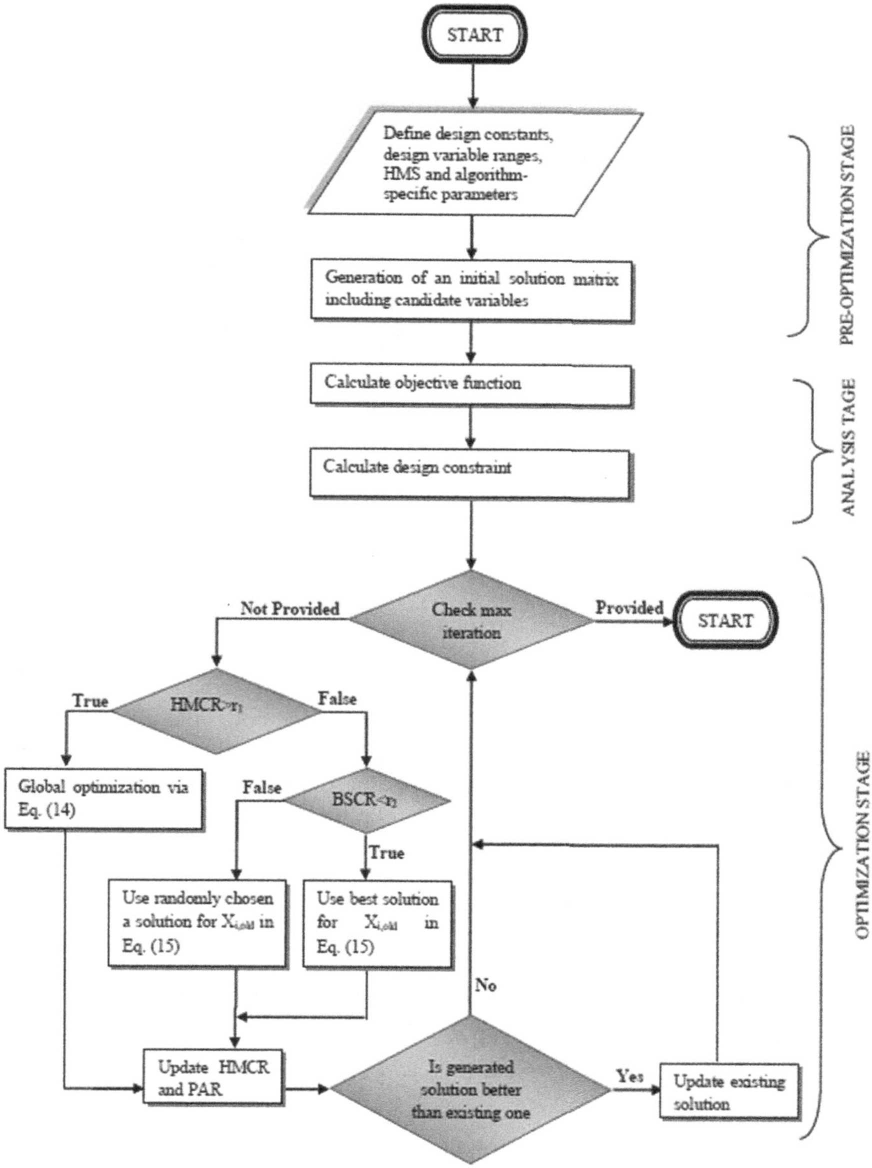

Fig. 11 Optimization flowchart [92]

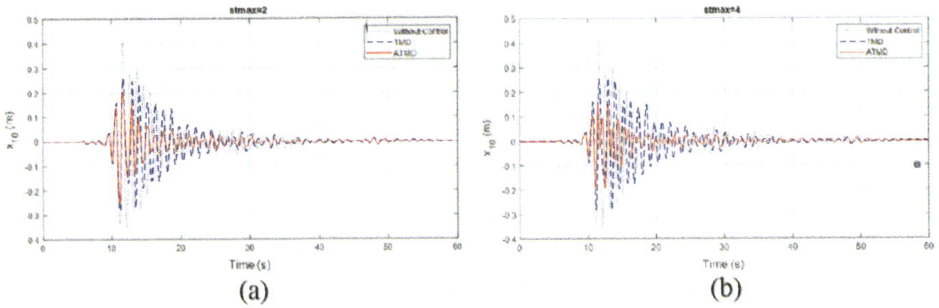

Fig. 12 Comparison of the roof displacement for different ATMD strokes: **a** st_max = 2 and **b** st_max = 4 for the uncontrolled, passive TMD and ATMD [92]

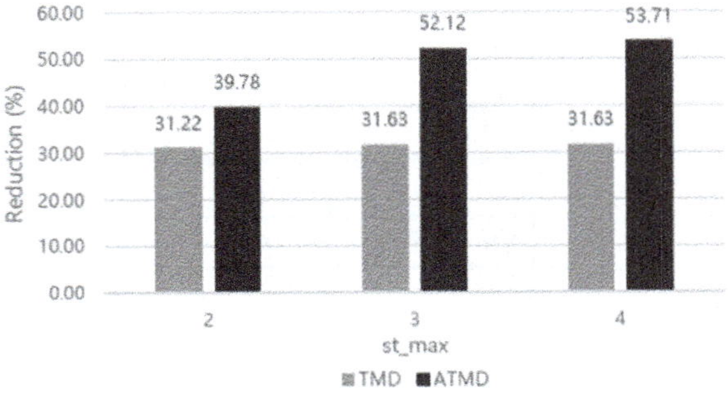

Fig. 13 Reduction percentage of roof displacement for different TMD and ATMD strokes [92]

displacement under the critical record for a mass ratio equal to 5% is shown in Fig. 12 for different ATMD strokes. In Fig. 12, st_max is a user-defined normalized value for the ATMD stroke limitation. From Fig. 12, the superior performance of the ATMD controlled using the proposed MHSA and PID controller, both in terms of reducing peak response and reaching the steady state, is evident. The efficiency is even more with high ATMD stroke. Figure 13 compares the reduction of roof displacement for both TMD and ATMD, with different stroke limitations. The ATMD shows a maximum of 54% reduction in roof displacement, while TMD can only mitigate up to 31%.

5.5 H_2 and H_∞ Optimization Algorithms for Robust Optimization of TMDs Under Near-Fault and Far-Fault Records

The following presents a robust optimum design of TMD under a suite of Near-Fault (FF) (with forward directivity and fling step characteristics), and Far-Fault

(FF) records. One of the objective functions is to minimize the peak amplitude of the roof displacement response function in the frequency domain (called H_∞ norm). The other objective function is to reduce the total vibrational energy of the system for all frequencies (called H_2 norm); in other words, the area under the frequency response curve of the system is minimized [210]. The optimum parameters are independent of external excitation frequency content, and the control methods are therefore robust. Complete details and results of the investigation can be found in [211] and limited results are presented herein.

The mathematical expression for H_∞ and H_2 norms is shown in Eqs. (12) and (13), respectively. In these equations, $\bar{\sigma}$ and $\sup_\omega$ represent the biggest singular value of the transfer function matrix (**H**) and the smallest upper bound of $\bar{\sigma}$ for all frequencies, respectively.

$$f(x) = 20\mathbf{Log}_{10}|\mathbf{max}(\boldsymbol{TF}_N(\omega))| \tag{12}$$

$$f(x) = 20\mathbf{Log}_{10}|\mathbf{max}(\boldsymbol{TF}_N(\omega))| \tag{13}$$

General steps for the calculation of the TMD optimum parameters for this equivalent structure are:

1. Frequency analysis is performed and natural frequencies and modal shapes are realized.
2. Dynamical properties of the first mode (mass, stiffness, and damping) are determined, and an equivalent SDOF structure is constructed.
3. CBO algorithm is employed to find the TMD optimum parameters as a function of mass ratio (μ) and different inherent structural damping ratio ξ_s In this procedure, minimizing the H_∞ and H_2 norm (Eqs. (12) and (13), respectively) of the equivalent structure roof displacement are selected as the objective function separately. Additionally, the frequency and damping ratios of the TMD are selected as the design variables. In Matlab vector notation, the ranges of these variables are considered as follows: f = [0.55:0.01:1.2], ξ = [0:0.005:0. 5].
4. The optimum design of TMD (i.e., optimum frequency and damping ratios) as a function of mass ratio and inherent damping ratio is computed. Design graphs are developed in which optimum parameters are calculated based on the proposed method and compared with the closed-form formulae presented in the literature.

The Colliding Bodies Optimization (CBO) [212]:
Set initial position for 2 N-CBs randomly

Repeat

For each CB the objective function is calculated.

The mass of CBs is assigned proportioned inversely to its fitness value.

The CBs are lined up in ascending order based on their mass.
The organized CBs are divided into two parts.

The CBs in the second part move toward their relevant CBs in the first part.

CBs are colliding with each other and their velocity after collision is evaluated.

The new positions of CBs are calculated in terms of their after-collision velocities.

Until the termination criteria is fulfilled.

Output: founded best solution.

For the current study, the number of CBs and steps that are used in the optimization process are 20 and 30, respectively.

The described method and algorithm is employed for finding the optimal TMD parameters for a benchmark 10-story shear building presented by Sadek et al. [8]. The structural properties are reported in Table 4. The TMD is added at the roof level to control the structural responses under external excitations. Optimum free parameters at different values of mass ratio (μ) and four different inherent structural damping ratios (ξ_s= 0, 2, 5, and 10%.) are calculated, and optimum values of the free parameters are plotted as a function of μ in Figs. 14 and 15.

In order to assess the performance of the optimal TMD in vibration mitigation of the 10-story benchmark building, normalized peak and normalized RMS of roof displacement and acceleration under different FF and NF earthquake records are selected as performance criteria. As an example the normalized controlled to uncontrolled maximum and RMS of the roof acceleration under FF and NF earthquake records with forward directivity or fling step characteristics with 5% inherent structural damping is shown in Fig. 16.

The resulting statistical assessment shows that the H_∞ objective function is superior to H_2 objective function for optimum design of TMDs under NF and FF earthquake excitations. Furthermore, the proposed approaches reduce the structural response under all three different suites of earthquake vibrations and are therefore robust. Three characteristics of the proposed procedures are their simple implementation, robustness, and less computational cost.

Table 4 Structural parameters of the 10-story benchmark building [8]

Story	Mass (× 10^3 kg)	Stiffness (×10^6 N/m)	Circular Freq.(rad/s)	Story	Mass (×10^3 kg)	Stiffness (×10^6 N/m)	Circular Freq.(rad/s)
1	179	62.47	0.50037	6	134	46.79	4.29197
2	170	59.26	1.32631	7	125	43.67	4.83577
3	161	56.14	2.15121	8	116	40.55	5.27169
4	152	53.02	2.93387	9	107	37.43	5.59050
5	143	49.91	3.65320	10	98	34.31	5.78653

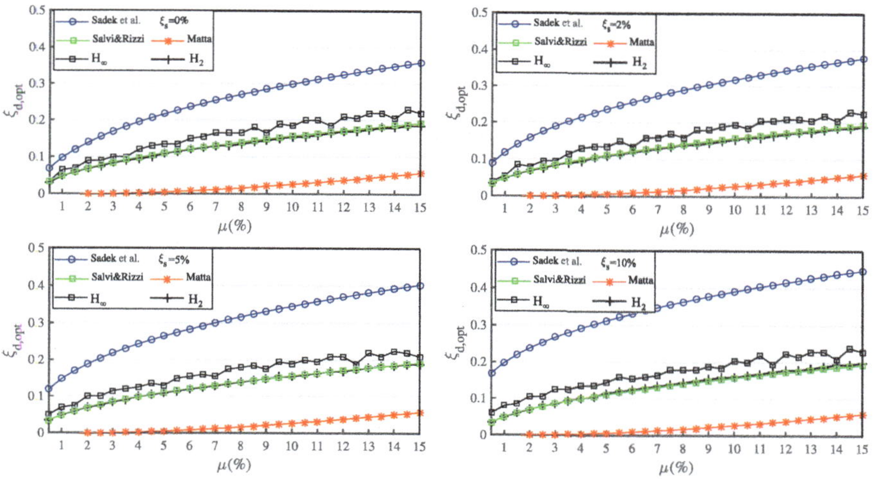

Fig. 14 Optimum damping ratio (ξ_d-opt) as a function of mass ratio (μ) and four different inherent damping ratios [211]

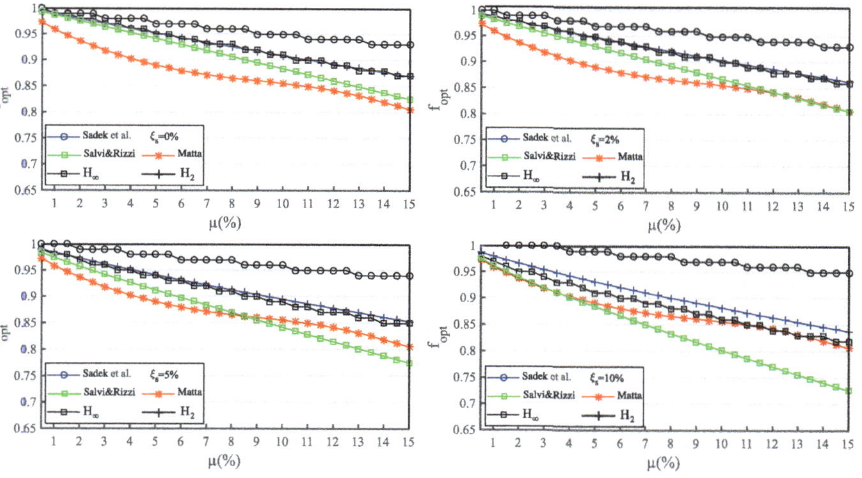

Fig. 15 Optimum frequency ratio (f_{opt}) as a function of mass ratio (μ) and four different inherent damping ratios [211]

5.6 Robust Optimal Control of Tuned Mass Damper Inerter (TMDI)

In classical TMD tuning, due to architectural and constructional limitations, the mass ratio (μ) is usually preselected and under 5%. However, for seismic applications, due to the transient and non-stationary characteristics of earthquake-induced excitations, a larger TMD mass will improve the control performance significantly. In this

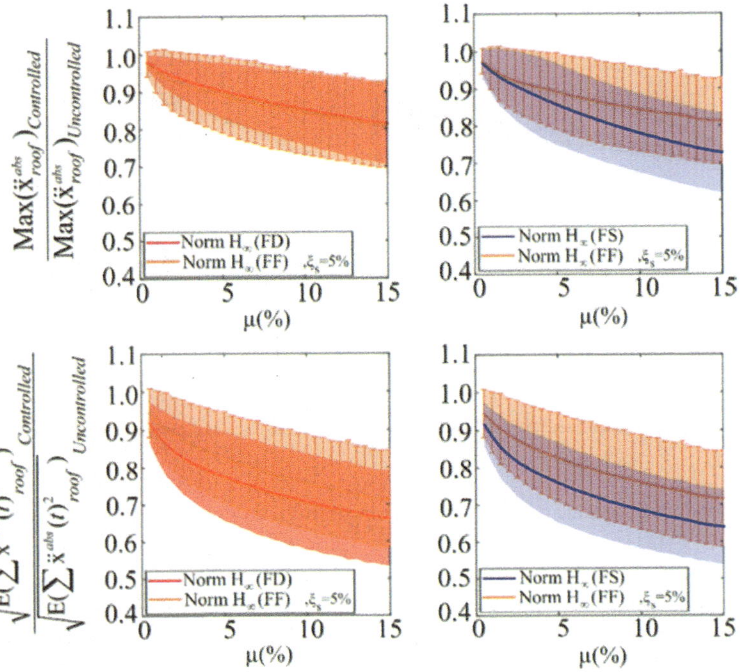

Fig. 16 normalized controlled to uncontrolled maximum and RMS of the roof acceleration under FF and NF earthquake records with forward directivity or fling step characteristics with 5% inherent structural damping [211]

context, various researchers investigated the application of unconventional TMDs with large masses, even by treating the top floor(s) of buildings as the "attached" TMD mass connected to the bottom floors via isolators. Consequently, the TMD mass in such "unorthodox" configurations may reach up to 50 percent or more of the whole building weight [213]; however, the cost of TMD implementation, gravity loads of the primary structure and the uncertainty and complexity of the optimum TMD tuning increases accordingly. In 2002 Smith [214] introduced a mechanical device, the "inerter", that acts as an artificial mass which can be of two orders higher magnitude than the attached physical mass [215]. The inerter consists of a linear two-terminal mechanical device which produces resisting forces proportional to the relative acceleration of its ends (Fig. 17b), and the constant of proportionality b (i.e. inertance) with units of mass [216]. Lately, application of this element in structural vibration control has been studied [217]. Marian and Giaralis [218] used the mass-amplification effect of the inerter and proposed a novel device, called the tuned mass damper inerter (TMDI). The main concept of a controlled SDOF with a recently introduced TMDI is depicted in Fig. 17c. For a preselected mass ratio (μ) with the same definitions of free vibration parameters for optimal tuning of TMDs, the TMDI free vibration parameters for optimum design are defined by Eq. (14), where β is inertance ratio and could be preselected similar to the mass ratio.

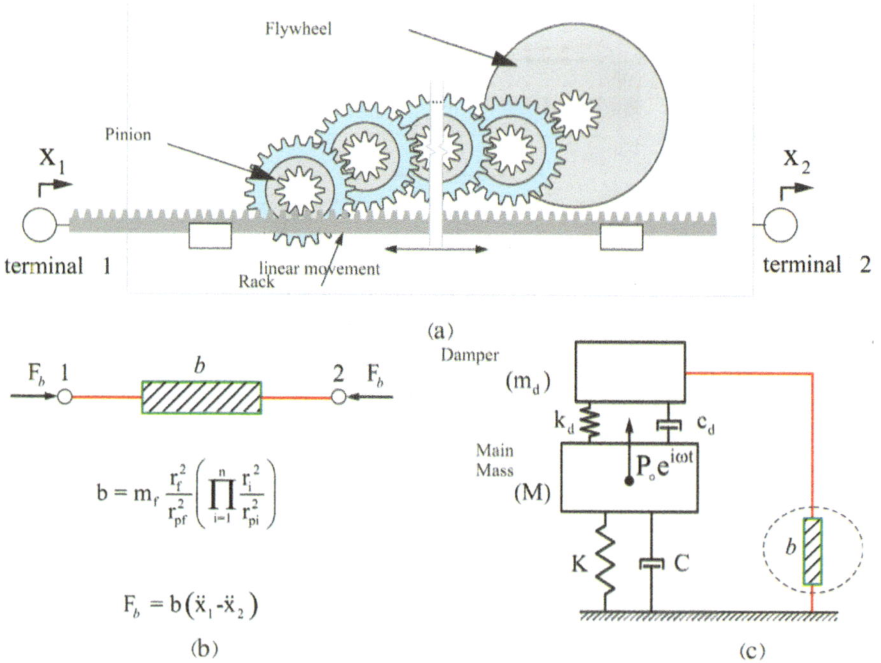

Fig. 17 **a** Rack and pinion inerter device [5], **b** ideal inerter, inertance constant and its related force [220], **c** main concept of classical TMDI [221]

$$\beta = \frac{b}{M};\ f = \frac{\sqrt{\frac{k_d}{m_d+b}}}{\omega_s};\ \zeta_d = \frac{c_d}{2(m_d + b)\omega_d} \tag{14}$$

The following describes the optimum design of a tuned mass damper inerter (TMDI) to control a benchmark 10-story base-excited linear shear building under seismic excitations. Mass and inertance ratios are preselected, and optimum free vibration parameters of the TMDI (i.e., natural frequency and damping ratios) are calculated for single degree-of-freedom (SDOF) and multi degree-of-freedom (MDOF) models with different configurations of single and double inerter TMDIs at different locations using Colliding Bodies Optimization (CBO) technique [212]. Four different inherent damping values are considered for each analysis. The objective function herein is considered as minimizing H_∞ norm of roof displacement transfer. Additionally, the performance of a TMDI-equipped building with optimal free parameters and its robustness is assessed in both the frequency and time domains. For sake of brevity only limited results are presented herein. Additional information can be found in [219].

The optimum parameters (frequency and damping ratio) for a single attached inerter (TMD-SI) with terminals connected to the TMD and 8th story (TMD-SI8) has been calculated and results are presented in Figs. 18 for 4 level of structural

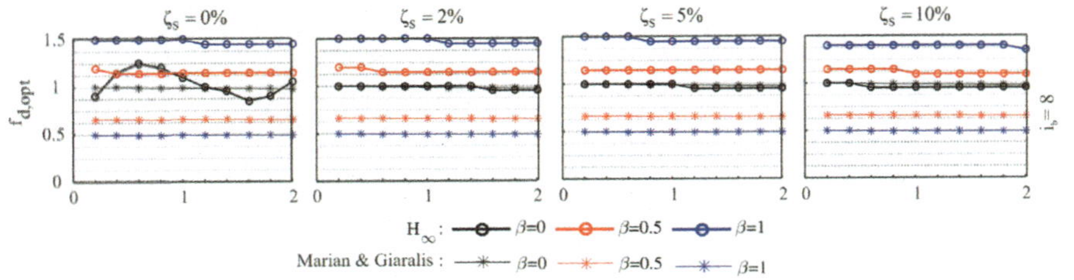

Fig. 18 Optimum damping ratio ($\zeta_{d\text{-opt}}$) and frequency ratio ($f_{d\text{-opt}}$) vs. mass ratio (μ) for different inherent damping ratios with $i_b = 8$ (TMD-SI8) [219]

damping. It should be noted that attaching the second terminal of the inerter to lower stories than the 8th story is impractical and has not been investigated as a result. The closed form solution by Marian and Giaralis [222], while developed for the equivalent SDOF, is also presented for comparison purposes. Figure 18 shows that for the range of mass ratio of this study, the optimal damping and frequency ratios of the damped structure remain almost constant for different values of mass ratio. It can be shown that as the second terminal of the inerter is attached to a higher story, its effect on the optimum parameters is reduced and its response will get closer to the TMD-only case. This raises a concern that when the optimal parameters of the TMDI are calculated with the structure modelled as an equivalent SDOF system, the effectiveness of the TMDI might be compromised, depending on where the second terminal of the inerter is connected.

The performance of the 10-story building equipped with the optimized TMDI (parameters obtained from the MDOF model) is evaluated. Figure 19 shows the normalized FR for roof displacement for different values of inertance ratio when mass ratio (μ) is equal to 0.4%. According to Fig. 19, not only the inerter is showing a better performance in the first mode for TMD-SI8, but it also exhibits significant contribution in controlling higher modes. For most cases the inertance demonstrates a better performance in the first mode than the TMD only-equipped structure. The greatest reduction in response in this case (TMD-SI8) is when $\beta = 1$.

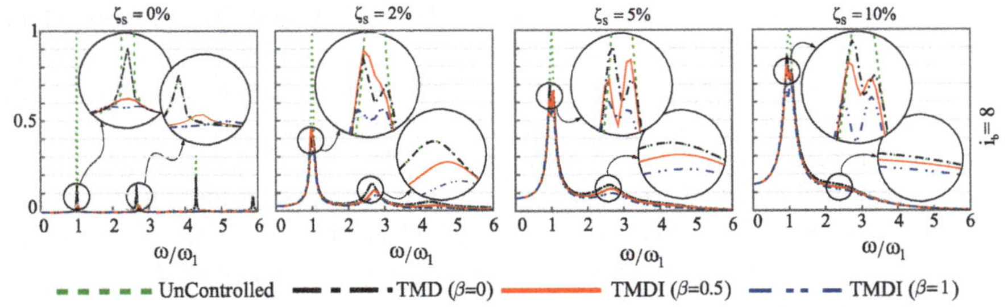

Fig. 19 Normalized frequency response of roof displacement for the 10-story shear building optimally tuned with ib = 8 (TMD-SI8) for $\mu = 0.4\%$ [219]

According to [219] the SDOF-based optimized TMDI approach should be used with caution and it is recommended to optimize and employ the TMDI using the MDOF model. Results indicated the superior performance of the proposed well-tuned damper with proper configuration in comparison to a same weight/mass classical TMD.

Since the performance of the TMDI depends on the frequency content of the earthquake, a statistical performance assessment was carried out by [223], for which the results are briefly reported herein. The H_2 and H_∞ of three different objective functions (i.e., minimizing roof displacement, minimizing 7th-story displacement, and maximizing the TMD stroke) are selected as objective functions for two different single and double inerter configurations in a 10-story benchmark shear building. Optimum free vibration parameters of the TMDI (i.e., natural frequency and damping ratios) are calculated using a metaheuristic technique (CBO, explained in 5.5) for two different structural damping values, i.e. 2 and 5%. Additionally, the performance of the optimum designed damper and its robustness under 25 Far-Fault (FF) ground motions (Table 5) are assessed in the time domain with 12 different performance criteria (i.e., the mean and standard deviation of the maximum and norm of response histories for roof displacement, roof acceleration and the total kinetic energy of the building). It should be noted that norm of response stands for average of response over the time duration of earthquake, while mean represents the average of response under the 25 records.

Table 5 Far field ground motions used for performance assessment of TMDI [223]

No.	Record	Station	No.	Record	Station
1	Loma Prieta	Presidio	14	Northridge	Ranchos-Palos
2	Northridge	Laguna	15	Northridge	Terminal Island
3	Northridge	Century	16	Northridge	Buena Park
4	Northridge	Moorpark	17	Imperial Valley	Calexico
5	Landers	Baker	18	Whittier-Nar.	Tarzana
6	Northridge	Baldwin-P.	19	San Fernando	Castaic, Old Ridge
7	Parkfield	Parkfield	20	Big Bear	Desert Hot Spr.
8	Kern County	Taft	21	Kern County	Santa Barbara
9	Northridge	Santa FE	22	N. Palms Spr.	Temecula
10	Northridge	Montebello	23	N. Palms Spr.	Anza Tule
11	Northridge	La-Habra	24	Northridge	Saturn Street Sch.
12	Northridge	Lakewood	25	Whittier-Nar.	Glendora
13	Northridge	La Puente			

Figure 20 shows the performance of the TMD-SI8-equipped building with a mass ratio of 0.4% and inherent damping of 2% under the 25 far field earthquake records.

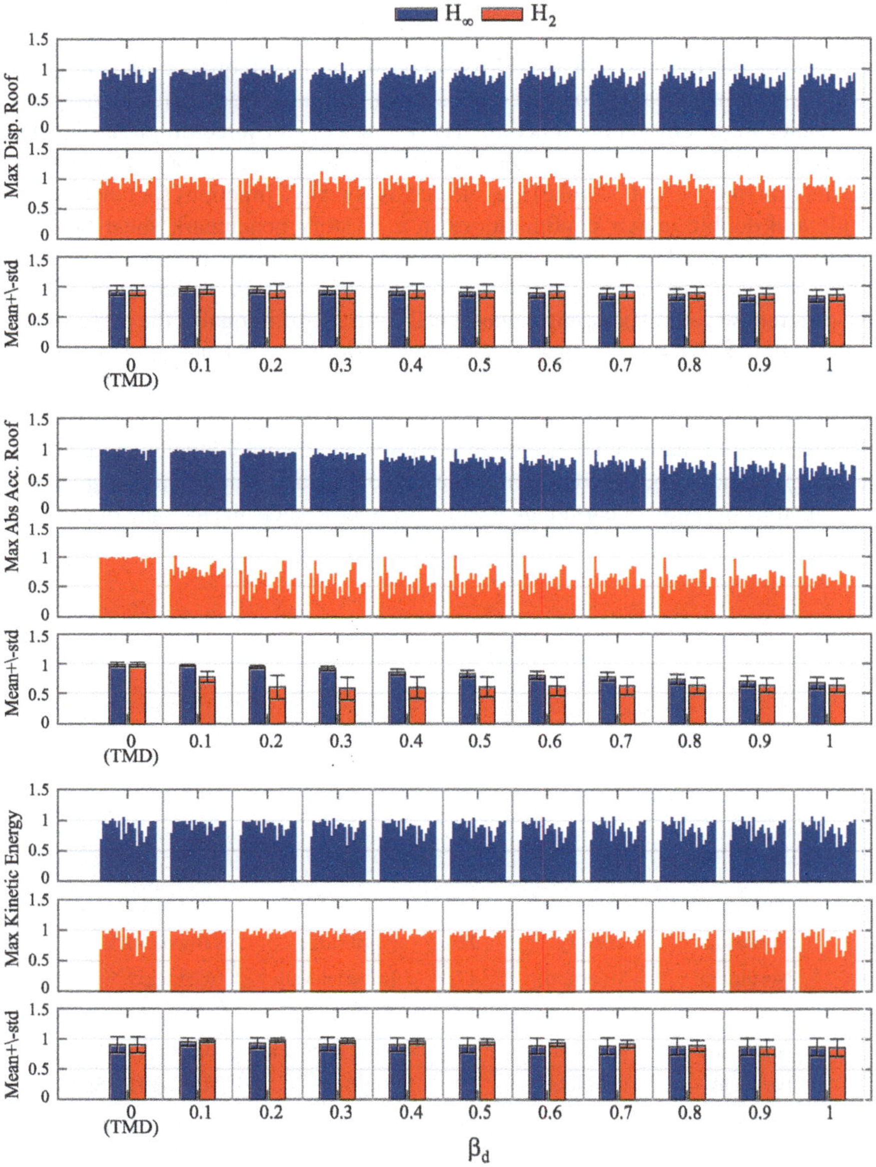

Fig. 20 Max and Mean(Max) ±std of different performance criteria for the 10-story shear building optimally tuned with $i_b = 8$ (TMD-SI8) under 25 benchmark earthquakes as a function of inertance ratio for $\mu = 0.4\%$ and $\zeta_s = 2\%$

The optimal free vibration parameters of the TMDI were obtained based on minimizing the H_∞ and H_2 norms of the roof displacement. Results have been presented for different inertance ratios (the horizontal axis) in the form of normalized response (response of controlled to uncontrolled structure). The results from the analyses using H_∞ and H_2 norms are shown in blue and red colors, respectively. Figure 20 presents the maximum values of the performance indices. The maximum response was selected to assess the performance of TMDI on reducing the maximum response for each earthquake record and evaluate its local effects. In Fig. 20, for each performance criterion, three row of results are presented: the first and second rows show the performance index for the 25 records by minimizing H_∞ and H_2 norms of the roof displacement, respectively; while the third row presents the mean and standard deviation of the performance index for each case for statistical evaluation and comparison purposes. Figure 20 shows that regardless of the control method (H_∞ or H_2), the biggest reduction in response (either norm/max for each record or mean over 25 records) is obtained when the roof acceleration is selected as the performance index. This can be beneficial for example for reducing the response of acceleration-sensitive equipment. According to Fig. 20 the response ratio is almost always less than one, which emphasizes the robustness of the current control method. The resulting statistical assessment shows reliable performance of the designed damper under different FF earthquake excitations, and the rather superior performance of the proposed TMDI in comparison to the classical TMD.

References

1. Elias, S., Matsagar, V.: Research developments in vibration control of structures using passive tuned mass dampers. Ann. Rev. Control **44**, 129–156 (2017)
2. De Domenico, D., Ricciardi, G., Takewaki, I.: Design strategies of viscous dampers for seismic protection of building structures: a review. Soil Dynam. Earthq. Eng. **118**, 144–165 (2019)
3. Singh, M.P., Singh, S., Moreschi, L.M.: Tuned mass dampers for response control of torsional buildings. Earthq. Eng. Struct. Dynam. **31**(4), 749–769 (2002)
4. Den Hartog, J., Ormondroyd, J.: Theory of the dynamic vibration absorber. ASME J. Appl. Mech. **50**(7), 11–22 (1928)
5. Den Hartog, J.P.: Mechanical Vibrations. Courier Corporation (1985)
6. Ioi, T., Ikeda, K.: On the dynamic vibration damped absorber of the vibration system. Bull. JSME **21**(151), 64–71 (1978)
7. Villaverde, R.: Reduction seismic response with heavily-damped vibration absorbers. Earthq. Eng. Struct. Dynam. **13**(1), 33–42 (1985)
8. Sadek, F., Mohraz, B., Taylor, A.W., Chung, R.M.: A method of estimating the parameters of tuned mass dampers for seismic applications. Earthq. Eng. Struct. Dynam. **26**(6), 617–635 (1997)
9. Warburton, G., Ayorinde, E.: Optimum absorber parameters for simple systems. Earthq. Eng. Struct. Dynam. **8**(3), 197–217 (1980)
10. Randall, S., Halsted III, D., Taylor, D.: Optimum vibration absorbers for linear damped systems (1981)
11. Kaveh, A., Ghazaan, M.I.: Meta-heuristic Algorithms for Optimal Design of Real-Size Structures. Springer (2018)

12. Kaveh, A.: Applications of Metaheuristic Optimization Algorithms in Civil Engineering. Springer (2017)
13. Hadi, M.N., Arfiadi, Y.: Optimum design of absorber for MDOF structures. J. Struct. Eng. **124**(11), 1272–1280 (1998)
14. Leung, A.Y., Zhang, H., Cheng, C., Lee, Y.: Particle swarm optimization of TMD by non-stationary base excitation during earthquake. Earthq. Eng. Struct. Dynam. **37**(9), 1223–1246 (2008)
15. Bekdaş, G., Nigdeli, S.M.: Estimating optimum parameters of tuned mass dampers using harmony search. Eng. Struct. **33**(9), 2716–2723 (2011)
16. Farshidianfar, A., Soheili, S.: Ant colony optimization of tuned mass dampers for earthquake oscillations of high-rise structures including soil–structure interaction. Soil Dynam. Earthq. Eng. **51**, 14–22 (2013)
17. Farshidianfar, A., Soheili, S.: ABC optimization of TMD parameters for tall buildings with soil structure interaction. Interact. Multiscale Mech. **6**(4), 339–356 (2013)
18. Nigdeli, S.M., Bekdas, G., Yang, X.: Optimum tuning of mass dampers for seismic structures using flower pollination algorithm. Int. J. Theor. Appl. Mech **1**, 264–268 (2016)
19. Kaveh, A., Mohammadi, S., Hosseini, O.K., Keyhani, A., Kalatjari, V.: Optimum parameters of tuned mass dampers for seismic applications using charged system search. Iranian J. Sci. Technol. Trans. Civil Eng. **39**(C1), 21 (2015)
20. Farzam, M.F., Kaveh, A.: Optimum design of tuned mass dampers using colliding bodies optimization in frequency domain. Iranian J. Sci. Technol. Trans. Civil Eng. 1–16 (2019)
21. Bekdaş, G., Kayabekir, A.E., Nigdeli, S.M., Toklu, Y.C.: Tranfer function amplitude minimization for structures with tuned mass dampers considering soil-structure interaction. Soil Dynam. Earthq. Eng. **116**, 552–562 (2019)
22. Kamgar, R., Khatibinia, M.: Optimization criteria for design of tuned mass dampers including soil–structure interaction effect. Iran Univer. Sci. Technol. **9**(2), 213–232 (2019)
23. Parulekar, Y., Reddy, G.: Passive response control systems for seismic response reduction: a state-of-the-art review. Int. J. Struct. Stab. Dyn. **9**(01), 151–177 (2009)
24. Saaed, T.E., Nikolakopoulos, G., Jonasson, J.-E., Hedlund, H.: A state-of-the-art review of structural control systems. J. Vib. Control **21**(5), 919–937 (2015)
25. Taylor, D.P.: Smart buildings and viscous dampers—a design engineer's perspective. Struct. Des. Tall and Spec. Build. **19**(4), 369–372 (2010)
26. De Silva, C.: An algorithm for the optimal design of passive vibration controllers for flexible systems. J. Sound Vib. **75**(4), 495–502 (1981)
27. Gürgöze, M., Müller, P.: Optimal positioning of dampers in multi-body systems. J. Sound Vib. **158**(3), 517–530 (1992)
28. Tsuji, M., Nakamura, T.: Optimum viscous dampers for stiffness design of shear buildings. Struct Des. Tall Build. **5**(3), 217–234 (1996)
29. Takewaki, I.: Soil–structure random response reduction via TMD-VD simultaneous use. Comput. Methods Appl. Mech. Eng. **190**(5–7), 677–690 (2000)
30. Garcia, D.L.: A simple method for the design of optimal damper configurations in MDOF structures. Earthq. Spectra. **17**(3), 387–398 (2001)
31. Singh, M.P., Moreschi, L.M.: Optimal placement of dampers for passive response control. Earthq. Eng. Struct. Dynam. **31**(4), 955–976 (2002)
32. Martinez-Rodrigo, M., Romero, M.: An optimum retrofit strategy for moment resisting frames with nonlinear viscous dampers for seismic applications. Eng. Struct. **25**(7), 913–925 (2003)
33. Christopoulos, C., Filiatrault, A.: Principles of Passive Supplemental Damping and Seismic Isolation. Luss Press (2006)
34. Takewaki, I.: Building Control with Passive Dampers: Optimal Performance-Based Design for Earthquakes. Wiley (2011)
35. Aydin, E., Boduroglu, M., Guney, D.: Optimal damper distribution for seismic rehabilitation of planar building structures. Eng. Struct. **29**(2), 176–185 (2007)
36. Estekanchi, H.E., Basim, M.C.: Optimal damper placement in steel frames by the Endurance time method. Struct Des Tall Spec Build **20**(5), 612–630 (2011)

37. Parcianello, E., Chisari, C., Amadio, C.: Optimal design of nonlinear viscous dampers for frame structures. Soil Dynam. Earthq. Eng. **100**, 257–260 (2017)
38. Altieri, D., Tubaldi, E., De Angelis, M., Patelli, E., Dall'Asta, A.: Reliability-based optimal design of nonlinear viscous dampers for the seismic protection of structural systems. Bullet. Earthq. Eng. 16(2):963–982 (2018)
39. Akehashi, H., Takewaki, I.: Comparative investigation on optimal viscous damper placement for elastic-plastic MDOF structures: transfer function amplitude or double impulse. Soil Dynam. Earthq. Eng. **130**, 105987 (2020)
40. Xu, Y., He, Q., Ko, J.: Dynamic response of damper-connected adjacent buildings under earthquake excitation. Eng. Struct. **21**(2), 135–148 (1999)
41. Kandemir-Mazanoglu, E.C., Mazanoglu, K.: An optimization study for viscous dampers between adjacent buildings. Mech. Syst. Signal Process. **89**, 88–96 (2017)
42. Liu, Y., Wu, J., Donà, M.: Effectiveness of fluid-viscous dampers for improved seismic performance of inter-storey isolated buildings. Eng. Struct. **169**, 276–292 (2018)
43. Lagaros, N.D., Plevris, V., and Mitropoulou, C.C.: Design optimization of active and passive structural control systems (2013)
44. Housner, G., Bergman, L.A., Caughey, T.K., Chassiakos, A.G., Claus, R.O., Masri, S.F., Skelton, R.E., Soong, T., Spencer, B., Yao, J.T.: Structural control: past, present, and future. J. Eng. Mech. **123**(9), 897–971 (1997)
45. Nielsen, E.J., Lai, M.-L., Soong, T., Kelly, J.M.: Viscoelastic damper overview for seismic and wind applications. In: Smart Structures and Materials 1996: Passive Damping and Isolation. 1996. International Society for Optics and Photonics
46. Zhang, R.-H., Soong, T.: Seismic design of viscoelastic dampers for structural applications. J. Struct. Eng. **118**(5), 1375–1392 (1992)
47. Hahn, G., Sathiavageeswaran, K.: Effects of added-damper distribution on the seismic response of buildings. Comput. Struct. **43**(5), 941–950 (1992)
48. Wu, B., Ou, J.-P., Soong, T.: Optimal placement of energy dissipation devices for three-dimensional structures. Eng. Struct. **19**(2), 113–125 (1997)
49. Shukla, A., Datta, T.: Optimal use of viscoelastic dampers in building frames for seismic force. J. Struct. Eng. **125**(4), 401–409 (1999)
50. Kim, J., Bang, S.: Optimum distribution of added viscoelastic dampers for mitigation of torsional responses of plan-wise asymmetric structures. Eng. Struct. **24**(10), 1257–1269 (2002)
51. Xu, Z.-D., Shen, Y.-P., Zhao, H.-T.: A synthetic optimization analysis method on structures with viscoelastic dampers. Soil Dynam. Earthq. Eng. **23**(8), 683–689 (2003)
52. Park, J.H., Kim, J., Min, K.W.: Optimal design of added viscoelastic dampers and supporting braces. Earthq. Eng. Struct. Dynam. **33**(4), 465–484 (2004)
53. Fujita, K., Moustafa, A., Takewaki, I.: Optimal placement of viscoelastic dampers and supporting members under variable critical excitations. Earthq. Struct. **1**(1), 43–67 (2010)
54. Pawlak, Z., Lewandowski, R.: Optimization of viscoelastic dampers as described by the fractional rheological model. In: Proceedings of the tenth international conference on computational structures technology, Valencia, Spain, Cd-Rom, Paper. (2010)
55. Zhu, H., Ge, D., Huang, X.: Optimum connecting dampers to reduce the seismic responses of parallel structures. J. Sound Vib. **330**(9), 1931–1949 (2011)
56. Morgan, T.A.: The use of innovative base isolation systems to achieve complex seismic performance objectives. University of California, Berkeley (2007)
57. Ganji, M., Kazem, H.: Comparing seismic performance of steel structures equipped with viscous dampers and lead rubber bearing base isolation under near-field earthquake. Civ. Eng. J. **3**(2), 124–136 (2017)
58. Zou, X.: Integrated design optimization of base-isolated concrete buildings under spectrum loading. Struct. Multi. Optim. **36**(5), 493 (2008)
59. Oliveto, G., Oliveto, N.D., Athanasiou, A.: Constrained optimization for 1-D dynamic and earthquake response analysis of hybrid base-isolation systems. Soil Dynam. Earthq. Eng. **67**, 44–53 (2014)

60. Yucel, M., Öncü-Davas, S., Nigdeli, S.M., Bekdas, G., Sevgen, S.: Estimating of analysis results for structures with linear base isolation systems using artificial neural network model. Int. J. Cont. Syst. Robot. 3 (2018)
61. Huang, P.C., Wan, S., Yen, J.Y.: A novel method of searching appropriate ranges of base isolation design parameters through entropy-based classification. Struct. Contr. Health Monit.: Offic. J. Int. Assoc. Struct. Contr. Monit. Eur. Assoc. Control Struct. **16**(4), 385–405 (2009)
62. Fan, J., Long, X., Zhang, Y.: Optimum design of lead-rubber bearing system with uncertainty parameters. Struct. Eng. Mech. **56**(6), 959–982 (2015)
63. Pourzeynali, S., Zarif, M.: Multi-objective optimization of seismically isolated high-rise building structures using genetic algorithms. J. Sound Vib. **311**(3–5), 1141–1160 (2008)
64. Nigdeli, S.M., Bekdaş, G., Alhan, C.: Optimization of seismic isolation systems via harmony search. Eng. Optim. **46**(11), 1553–1569 (2014)
65. Quaranta, G., Marano, G.C., Greco, R., Monti, G.: Parametric identification of seismic isolators using differential evolution and particle swarm optimization. Appl. Soft Comput. **22**, 458–464 (2014)
66. Çerçevik, A.E., Avşar, Ö., Hasançebi, O.: Optimum design of seismic isolation systems using metaheuristic search methods. Soil Dynam. Earthq. Eng. **131**, 106012 (2020)
67. Zou, X.-K., Wang, Q., Li, G., Chan, C.-M.: Integrated reliability-based seismic drift design optimization of base-isolated concrete buildings. J. Struct. Eng. **136**(10), 1282–1295 (2010)
68. Mishra, S.K., Roy, B.K., Chakraborty, S.: Reliability-based-design-optimization of base isolated buildings considering stochastic system parameters subjected to random earthquakes. Int. J. Mech. Sci. **75**, 123–133 (2013)
69. Zhang, J., Shu, Z.: Optimal design of isolation devices for mid-rise steel moment frames using performance based methodology. Bull. Earthq. Eng. **16**(9), 4315–4338 (2018)
70. Mousazadeh, M., Pourreza, F., Basim, M.C., Chenaghlou, M.: An efficient approach for LCC-based optimum design of lead-rubber base isolation system via FFD and analysis of variance (ANOVA). Bull. Earthq. Eng. **18**(4), 1805–1827 (2020)
71. Roy, B.K., Chakraborty, S.: Robust optimum design of base isolation system in seismic vibration control of structures under random system parameters. Struct. Saf. **55**, 49–59 (2015)
72. Hu, J. Xu, J.: Parameter optimization and control characteristics analysis of TLMD system based on phase deviation. J. Shanghai Jiaotong University (Sci), 1–12 (2019)
73. Park, S., Glade, M., Lackner, M.A.: Multi-objective optimization of orthogonal TLCDs for reducing fatigue and extreme loads of a floating offshore wind turbine. Eng. Struct. **209**, 110260 (2020)
74. Manchalwar, A., Bakre, S.: Optimization of metallic damper location for seismic response control. J. Vibr. Eng. Technol. **7**(3), 261–275 (2019)
75. Javanmardi, A., Ibrahim, Z., Ghaedi, K., Ghadim, H.B., Hanif, M.U.: State-of-the-art review of metallic dampers: testing, development and implementation. Arch. Comput. Meth. Eng. **27**(2), 455–478 (2020)
76. Jarrahi, H., Asadi, A., Khatibinia, M., Etedali, S.: Optimal design of rotational friction dampers for improving seismic performance of inelastic structures. J. Build Eng. **27**, 100960 (2020)
77. Zou, S., Heisha, W., Tan, P.: Performance-based dynamic optimal design for isolated structures with multiple-coupling friction dampers. Adv. Struct. Eng. 1369433220908110 (2020)
78. Deng, K., Pan, P., Su, Y., Xue, Y.: Shape optimization of U-shaped damper for improving its bi-directional performance under cyclic loading. Eng. Struct. **93**, 27–35 (2015)
79. Khatibinia, M., Jalaipour, M., Gharehbaghi, S.: Shape optimization of U-shaped steel dampers subjected to cyclic loading using an efficient hybrid approach. Eng. Struct. **197**, 108874 (2019)
80. Morales-Beltran, M., Teuffel, P.: Towards smart building structures: adaptive structures in earthquake and wind loading control response–a review. Intell. Build. Int. **5**(2), 83–100 (2013)
81. Casciati, F., Rodellar, J., Yildirim, U.: Active and semi-active control of structures–theory and applications: a review of recent advances. J. Intell. Mater. Syst. Struct. **23**(11), 1181–1195 (2012)
82. Lund, R.: Active Damping of Large Structures in Winds, Structural Control, vol. 459, p. 470. North-Holland Publishing Company (1980)

83. Chang, J.C., Soong, T.T.: Structural control using active tuned mass dampers. J. Eng. Mech. Div. **106**(6), 1091–1098 (1980)
84. Nishimura, I.: Acceleration feedback method applied to active tuned mass damper. In: First European Conference on Smart Structures and Materials. International Society for Optics and Photonics (1992)
85. Nishimura, I., Sakamoto, M., Yamada, T., Koshika, N., Kobori, T.: Acceleration feedback method applied to active-passive composite tuned mass damper. J. Struct. Control **1**(1–2), 103–116 (1994)
86. Xu, Y.: Parametric study of active mass dampers for wind-excited tall buildings. Eng. Struct. **18**(1), 64–76 (1996)
87. Wu, J., Yang, J.: Active control of transmission tower under stochastic wind. J. Struct. Eng. **124**(11), 1302–1312 (1998)
88. Li, C., Li, J., Qu, Y.: An optimum design methodology of active tuned mass damper for asymmetric structures. Mech. Syst. Sig. Process. **24**(3), 746–765 (2010)
89. Jiang, X., Adeli, H.: Neuro-genetic algorithm for non-linear active control of structures. Int. J. Numer. Meth. Eng. **75**(7), 770–786 (2008)
90. Ahlawat, A., Ramaswamy, A.: Multiobjective optimal fuzzy logic control system for response control of wind-excited tall buildings. J. Eng. Mech. **130**(4), 524–530 (2004)
91. Soleymani, M., Khodadadi, M.: Adaptive fuzzy controller for active tuned mass damper of a benchmark tall building subjected to seismic and wind loads. Struct. Des. Tall and Spec. Build. **23**(10), 781–800 (2014)
92. Kayabekir, A.E., Bekdaş, G., Nigdeli, S.M., Geem, Z.W.: Optimum design of PID controlled active tuned mass damper via modified harmony search. Appl. Sci. **10**(8), 2976 (2020)
93. Amini, F., Hazaveh, N.K., Rad, A.A.: Wavelet PSO-based LQR algorithm for optimal structural control using active tuned mass dampers. Comput.-Aided Civ. Infrastruct. Eng. **28**(7), 542–557 (2013)
94. Shariatmadar, H., Razavi, H.M.: Seismic control response of structures using an ATMD with fuzzy logic controller and PSO method. Struct. Eng. Mech. **51**(4), 547–564 (2014)
95. Amini, F., Bagheri, A.: Optimal control of structures under earthquake excitation based on the colonial competitive algorithm. Struct. Des. Tall and Spec. Build. **23**(7), 500–511 (2014)
96. Pourzeynali, S., Lavasani, H., Modarayi, A.: Active control of high rise building structures using fuzzy logic and genetic algorithms. Eng. Struct. **29**(3), 346–357 (2007)
97. Ozer, H.O., Sayin, A., Korkmaz, N., Yagız, N.: Sliding mode control optimized by genetic algorithm for building model. In: Proceedings of International Conference on Vibration Problems (2013)
98. Li, Q., Liu, D., Fang, J., Tam, C.: Multi-level optimal design of buildings with active control under winds using genetic algorithms. J. Wind Eng. Ind. Aerodyn. **86**(1), 65–86 (2000)
99. Ankireddi, S., Yang, H.T.: Multiple objective LQG control of wind-excited buildings. J. Struct. Eng. **123**(7), 943–951 (1997)
100. Wong, K.K., Yang, R.: Predictive instantaneous optimal control of elastic structures during earthquakes. Earthquak. Eng. Struct. Dynam. **32**(14), 2161–2177 (2003)
101. Khatibinia, M., Mahmoudi, M., Eliasi, H.: Optimal sliding mode control for seismic control of buildings equipped with atmd. Iran Univ. Sci. Technol. **10**(1), 1–15 (2020)
102. Chang, C.-C., Yu, L.-O.: A simple optimal pole location technique for structural control. Eng. Struct. **20**(9), 792–804 (1998)
103. Huo, L., Song, G., Li, H., Grigoriadis, K.: Robust control design of active structural vibration suppression using an active mass damper. Smart Mater. Struct. **17**(1), 015021 (2007)
104. Palazzo, B., Petti, L.: Optimal structural control in the frequency domain: control in norm H_2 and H_∞. J. Struct. Control **6**(2), 205–221 (1999)
105. Ahlawat, A., Ramaswamy, A.: Multiobjective optimal structural vibration control using fuzzy logic control system. J. Struct. Eng. **127**(11), 1330–1337 (2001)
106. Othman, S.M., Rahmat, M.F.a., Rozali, S.M., Salleh, S.: PID parameters optimization using PSO technique for nonlinear electro hydraulic actuator. J. Teknologi **77**(28), 67–72 (2015)

107. Arif Şen, M., Tinkir, M., Kalyoncu, M.: Optimisation of a PID controller for a two-floor structure under earthquake excitation based on the bees algorithm. J. Low Freq. Noise, Vib. Active Control **37**(1), 107–127 (2018)
108. Kim, J.-T., Jung, H.-J., Lee, I.-W.: Optimal structural control using neural networks. J. Eng. Mech. **126**(2), 201–205 (2000)
109. Heidari, A.H., Etedali, S., Javaheri-Tafti, M.R.: A hybrid LQR-PID control design for seismic control of buildings equipped with ATMD. Front. Struct. Civ. Eng. **12**(1), 44–57 (2018)
110. Baygi, S.M.H., Karsaz, A., Elahi, A.: A hybrid optimal PID-Fuzzy control design for seismic exited structural system against earthquake: a salp swarm algorithm. In: 2018 6th Iranian Joint Congress on Fuzzy and Intelligent Systems (CFIS). IEEE (2018)
111. Soong, T.: State-of-the-art review: active structural control in civil engineering. Eng. Struct. **10**(2), 74–84 (1988)
112. Soong, T., Spencer Jr., B.: Active structural control: theory and practice. J. Eng. Mech. **118**(6), 1282–1285 (1992)
113. Suhardjo, J., Spencer Jr., B., Kareem, A.: Frequency domain optimal control of wind-excited buildings. J. Eng. Mech. **118**(12), 2463–2481 (1992)
114. Chang, C.-C., Lin, C.-C.: H_{∞} drift control of time-delayed seismic structures. Earthquak. Eng. Eng. Vibr. **8**(4), 617–626 (2009)
115. Li, Q., Liu, D., Fang, J.: Optimum design of actively controlled structures using genetic algorithms. Adv. Struct. Eng. **2**(2), 109–118 (1999)
116. Issa, J., Mukherjee, R., Shaw, S.W.: Vibration suppression in structures using cable actuators. J. Vib. Acoust. 132(3) (2010)
117. Rao, A.R.M., Sivasubramanian, K.: Optimal placement of actuators for active vibration control of seismic excited tall buildings using a multiple start guided neighbourhood search (MSGNS) algorithm. J. Sound Vib. **311**(1–2), 133–159 (2008)
118. Nigdeli, S.M.: Active brace control of frame structures under earthquake excitation. Nat. Cataclysms Glob. Prob. Mod. Civilization **102** (2011)
119. Gluck, J., Ribakov, Y.: Active viscous damping system with amplifying braces for control of MDOF structures. Earthquak. Eng. Struct. Dynam. **31**(9), 1735–1751 (2002)
120. Gupta, H., Soong, T., Dargush, G.: Active aerodynamic bi-directional control of structures II: tall buildings. Eng. Struct. **22**(4), 389–399 (2000)
121. Scruggs, J., Taflanidis, A., Beck, J.: Reliability-based control optimization for active base isolation systems. Struct. Control Health Monit.: Offic. J. Int. Assoc. Struct. Control Monit. Eur. Association Control Struct. **13**(2–3), 705–723 (2006)
122. Balendra, T., Wang, C., Yan, N.: Control of wind-excited towers by active tuned liquid column damper. Eng. Struct. **23**(9), 1054–1067 (2001)
123. Xu, J., Yang, X., Li, W., Zheng, J., Wang, Y., Fan, M.: Research on semi-active vibration isolation system based on electromagnetic spring. Mech. Ind. **21**(1), 101 (2020)
124. Dyke, S., Spencer Jr., B., Sain, M., Carlson, J.: Modeling and control of magnetorheological dampers for seismic response reduction. Smart Mater. Struct. **5**(5), 565 (1996)
125. Dyke, S., Spencer Jr., B., Sain, M., Carlson, J.: Seismic response reduction using magnetorheological dampers. IFAC Proc. Volumes **29**(1), 5530–5535 (1996)
126. Spencer, B., Dyke, S., Sain, M.K.: Magnetorheological dampers: a new approach to seismic protection of structures. In: Proceedings of 35th IEEE Conference on Decision and Control. IEEE (1996)
127. Jung, H.-J., Lee, I.W., Spencer Jr, B.F. State-of-the-art of MR damper-based control systems in civil engineering applications. In: Proceedings of US-Korea Workshop on Smart Infra-Structural Systems (2002)
128. Liu, Y.-Q., Yang, S.-p., Liao, Y., Zhang, G.: Parameter identification of Bouc-Wen model for MR damper based on genetic algorithm. Zhendong yu Chongji (J. Vib. Shock) **30**(7), 261–265 (2011)
129. Azar, B.F., Veladi, H., Talatahari, S., Raeesi, F.: Optimal design of magnetorheological damper based on tuning Bouc-Wen model parameters using hybrid algorithms. KSCE J. Civil Eng. **24**(3), 867–878 (2020)

130. Razman, M.A., Priyandoko, G., Yusoff, A.R.: Bouc-wen model parameter identification for a MR fluid damper using particle swarm optimization. In: Advanced Materials Research. Trans Tech Publ (2014)
131. Yang, Y., Ding, Y., Zhu, S. parameter identification of MR damper model based on particle swarm optimization. In: Proceedings of the 11th International Conference on Modelling, Identification and Control (ICMIC2019). Springer (2020)
132. Xiao, Z., Zhang, Z., Quansah, A.: Combined identification of parameters in the mechanical model of magnetorheological damper (2019)
133. Zhu, H., Rui, X., Yang, F., Zhu, W., Wei, M.: An efficient parameters identification method of normalized Bouc-Wen model for MR damper. J. Sound Vib. **448**, 146–158 (2019)
134. Talatahari, S., Kaveh, A., Rahbari, N.M.: Parameter identification of Bouc-Wen model for MR fluid dampers using adaptive charged system search optimization. J. Mech. Sci. Technol. **26**(8), 2523–2534 (2012)
135. Zaman, M.A., Sikder, U.: Bouc-Wen hysteresis model identification using modified firefly algorithm. J. Magn. Magn. Mater. **395**, 229–233 (2015)
136. Giuclea, M., Sireteanu, T., and Mitu, A.: Use of genetic algorithms for fitting the Bouc-Wen model to experimental hysteretic curves. Rev. Roum. Sci. Techn.–Mec. Appl. **54**(1), 3–10 (2009)
137. Bartkowski, P., Zalewski, R., Chodkiewicz, P.: Parameter identification of Bouc-Wen model for vacuum packed particles based on genetic algorithm. Arch. Civ. Mech. Eng. **19**(2), 322–333 (2019)
138. Ye, M., Wang, X.: Parameter estimation of the Bouc-Wen hysteresis model using particle swarm optimization. Smart Mater. Struct. **16**(6), 2341 (2007)
139. Charalampakis, A., Dimou, C.: Identification of Bouc-Wen hysteretic systems using particle swarm optimization. Comput. Struct. **88**(21–22), 1197–1205 (2010)
140. Son, N.N., Van Kien, C., Anh, H.P.H.: Parameters identification of Bouc-Wen hysteresis model for piezoelectric actuators using hybrid adaptive differential evolution and Jaya algorithm. Eng. Appl. Artif. Intell. **87**, 103317 (2020)
141. Talatahari, S., Mohaggeg, H., Najafi, K., Manafzadeh, A.: Solving parameter identification of nonlinear problems by artificial bee colony algorithm. Math. Prob. Eng. **2014** (2014)
142. Spencer Jr., B., Dyke, S., Sain, M., Carlson, J.: Phenomenological model for magnetorheological dampers. J. Eng. Mech. **123**(3), 230–238 (1997)
143. Yang, G., Spencer Jr., B., Carlson, J., Sain, M.: Large-scale MR fluid dampers: modeling and dynamic performance considerations. Eng. Struct. **24**(3), 309–323 (2002)
144. Giuclea, M., Sireteanu, T., Stancioiu, D., Stammers, C.W.: Modeling of magneto rheological damper dynamic behavior by genetic algorithms based inverse method. Rom. Acad. **5**(1), 000–000 (2004)
145. Giuclea, M., Sireteanu, T., Stancioiu, D., Stammers, C.W.: Model parameter identification for vehicle vibration control with magnetorheological dampers using computational intelligence methods. Proc. Inst. Mech. Eng. Part I: J. Syst. Control Eng. **218**(7), 569–581 (2004)
146. Shu, G., Li, Z.: Parametric identification of the Bouc-Wen model by a modified genetic algorithm: application to evaluation of metallic dampers. Earthquak. Struct. **13**(4), 397–407 (2017)
147. Kwok, N., Ha, Q., Nguyen, T., Li, J., Samali, B.: A novel hysteretic model for magnetorheological fluid dampers and parameter identification using particle swarm optimization. Sens. Actuators, A **132**(2), 441–451 (2006)
148. Kwok, N., Ha, Q., Nguyen, M., Li, J., Samali, B.: Bouc-Wen model parameter identification for a MR fluid damper using computationally efficient GA. ISA Trans. **46**(2), 167–179 (2007)
149. Yan, G., Zhou, L.L.: Integrated fuzzy logic and genetic algorithms for multi-objective control of structures using MR dampers. J. Sound Vib. **296**(1–2), 368–382 (2006)
150. Askari, M., Davaie-Markazi, A.H.: Multi-objective optimal fuzzy logic controller for nonlinear building-MR damper system. In: 2008 5th International multi-conference on systems, signals and devices. IEEE (2008)

151. Shook, D.A., Roschke, P.N., Lin, P.-Y., Loh, C.-H.: GA-optimized fuzzy logic control of a large-scale building for seismic loads. Eng. Struct. **30**(2), 436–449 (2008)
152. Shook, D.A., Roschke, P.N., Lin, P.-Y., Loh, C.-H.: Semi-active control of a torsionally-responsive structure. Eng. Struct. **31**(1), 57–68 (2009)
153. Ali, S.F., Ramaswamy, A.: Optimal fuzzy logic control for MDOF structural systems using evolutionary algorithms. Eng. Appl. Artif. Intell. **22**(3), 407–419 (2009)
154. Huang, Z.-S., Wu, C., Hsu, D.-S.: Semi-active fuzzy control of mr damper on structures by genetic algorithm. J. Mech. **25**(1), N1–N6 (2009)
155. Bitaraf, M., Ozbulut, O.E., Hurlebaus, S., Barroso, L.: Application of semi-active control strategies for seismic protection of buildings with MR dampers. Eng. Struct. **32**(10), 3040–3047 (2010)
156. Bozorgvar, M., Zahrai, S.M.: Semi-active seismic control of buildings using MR damper and adaptive neural-fuzzy intelligent controller optimized with genetic algorithm. J. Vib. Control **25**(2), 273–285 (2019)
157. Uz, M.E., Hadi, M.N.: Optimal design of semi active control for adjacent buildings connected by MR damper based on integrated fuzzy logic and multi-objective genetic algorithm. Eng. Struct. **69**, 135–148 (2014)
158. Raeesi, F., Azar, B.F., Veladi, H., Talatahari, S.: An inverse TSK model of MR damper for vibration control of nonlinear structures using an improved grasshopper optimization algorithm. In: Structures. Elsevier (2020)
159. Schurter, K.C., Roschke, P.N.: Neuro-fuzzy control of structures using acceleration feedback. Smart Mater. Struct. **10**(4), 770 (2001)
160. Hiramoto, K., Matsuoka, T., Sunakoda, K.: Inverse lyapunov approach for semi-active control of civil structures. Struct. Control Health Monit. **18**(4), 382–403 (2011)
161. Hashemi, S.M.A., Haji Kazemi, H., Karamodin, A.: Localized genetically optimized wavelet neural network for semi-active control of buildings subjected to earthquake. Struct. Control Health Monit. **23**(8), 1074–1087 (2016)
162. Shi, Y., Xin, N., Ningwei, W.: Optimal placement of MR damper set in structures. Earthquak. Eng. Eng. Vib.-Chin. Ed. **24**(3), 175–178 (2004)
163. Ok, S.-Y., Song, J., Park, K.-S.: Optimal design of hysteretic dampers connecting adjacent structures using multi-objective genetic algorithm and stochastic linearization method. Eng. Struct. **30**(5), 1240–1249 (2008)
164. Bao, Y., Huang, C., Zhou, D., Zhao, Y.-J.: Semi-active direct velocity control method of dynamic response of spatial reticulated structures based on MR dampers. Adv. Struct. Eng. **12**(4), 547–558 (2009)
165. Li, L., Song, G., Ou, J.: A genetic algorithm-based two-phase design for optimal placement of semi-active dampers for nonlinear benchmark structure. J. Vib. Control **16**(9), 1379–1392 (2010)
166. Amini, F., Karami, K.: Capacity design by developed pole placement structural control. Struct. Eng. Mech. **39**(1), 147–168 (2011)
167. Elmeligy, O.M., Hassan, M.: Optimum allocation of MR dampers within semi-active control strategies of three-degree-of-freedom systems. Int. J. Recent Contrib. Eng. Sci. IT (IJES) **4**(4), 45–49 (2016)
168. Bhaiya, V., Bharti, S., Shrimali, M., Datta, T.: Genetic algorithm based optimum semi-active control of building frames using limited number of magneto-rheological dampers and sensors. J. Dyn. Syst. Measur. Control **140**(10) (2018)
169. Zabihi-Samani, M., Ghanooni-Bagha, M.: Optimal semi-active structural control with a wavelet-based cuckoo-search fuzzy logic controller. Iranian J. Sci. Technol. Trans. Civ. Eng. **43**(4), 619–634 (2019)
170. Hrovat, D., Barak, P., Rabins, M.: Semi-active versus passive or active tuned mass dampers for structural control. J. Eng. Mech. **109**(3), 691–705 (1983)
171. Setareh, M., Ritchey, J.K., Murray, T.M., Koo, J.-H., Ahmadian, M.: Semiactive tuned mass damper for floor vibration control. J. Struct. Eng. **133**(2), 242–250 (2007)

172. Kaveh, A., Pirgholizadeh, S., Khadem, H.O.: Semi-active tuned mass damper performance with optimized fuzzy controller using CSS algorithm (2015)
173. Madden, G.J., Symans, M.D., Wongprasert, N.: Experimental verification of seismic response of building frame with adaptive sliding base-isolation system. J. Struct. Eng. **128**(8), 1037–1045 (2002)
174. Wongprasert, N., Symans, M.: Experimental evaluation of adaptive elastomeric base-isolated structures using variable-orifice fluid dampers. J. Struct. Eng. **131**(6), 867–877 (2005)
175. Narasimhan, S., Nagarajaiah, S.: Smart base isolated buildings with variable friction systems: H_∞ controller and SAIVF device. Earthquak. Eng. Struct. Dynam. **35**(8), 921–942 (2006)
176. Choi, K.M., Lee, H.-J., Cho, S.-W., Lee, I.W.: Modified energy dissipation algorithm for seismic structures using magnetorheological damper. KSCE J. Civ. Eng. **11**(2), 121–126 (2007)
177. Moon, S.-J., Huh, Y.-C., Jung, H.-J., Jang, D.-D., Lee, H.-J.: Sub-optimal design procedure of valve-mode magnetorheological fluid dampers for structural control. KSCE J. Civ. Eng. **15**(5), 867 (2011)
178. Bharti, S., Dumne, S., Shrimali, M.: Earthquake response of asymmetric building with MR damper. Earthquak. Eng. Eng. Vib. **13**(2), 305–316 (2014)
179. Ramallo, J., Johnson, E., Spencer Jr., B.: "Smart" base isolation systems. J. Eng. Mech. **128**(10), 1088–1099 (2002)
180. Lee, H.J., Yang, G., Jung, H.J., Spencer, B.F., Lee, I.W.: Semi-active neurocontrol of a base-isolated benchmark structure. Struct. Control Health Monit.: Offic. J. Int. Assoc. Struct. Control Monit. Eur. Assoc. Control Struct. **13**(2–3), 682–692 (2006)
181. Ali, S.F., Ramaswamy, A.: Hybrid structural control using magnetorheological dampers for base isolated structures. Smart. Mater. Struct. **18**(5), 055011 (2009)
182. Kim, H.-S., Roschke, P.N.: Design of fuzzy logic controller for smart base isolation system using genetic algorithm. Eng. Struct. **28**(1), 84–96 (2006)
183. Kim, H.S., Roschke, P.N.: Fuzzy control of base-isolation system using multi-objective genetic algorithm. Comput.-Aid. Civ. Infrastruct. Eng. **21**(6), 436–449 (2006)
184. Ozbulut, O.E., Bitaraf, M., Hurlebaus, S.: Adaptive control of base-isolated structures against near-field earthquakes using variable friction dampers. Eng. Struct. **33**(12), 3143–3154 (2011)
185. Mohebbi, M., Dadkhah, H.: Multi-objective semi-active base isolation system. Iran Univ. Sci. Technol. **7**(3), 319–338 (2017)
186. Symans, M.D.: Development and experimental study of semi-active fluid damping devices for seismic protection of structures (1996)
187. Bakhshinezhad, S., Mohebbi, M.: Multi-objective optimal design of semi-active fluid viscous dampers for nonlinear structures using NSGA-II. Structures, Elsevier (2020)
188. Kazemi Bidokhti, K., Moharrami, H., Fayezi, A.: Semi-active fuzzy control for seismic response reduction of building frames using SHD dampers. Struct. Control Health Monit. **19**(3), 417–435 (2012)
189. Eberhart, R., Kennedy, J.: A new optimizer using particle swarm theory. In: MHS'95. Proceedings of the Sixth International Symposium on Micro Machine and Human Science. IEEE (1995)
190. Fahimi Farzam, M., Alinejad, B., Mousavi Gavgani, S.A.: Seismic control of 10-Storey Shear frame using active tuned mass dampers and optimization by particle swarm optimization algorithm. Amirkabir J. Civ. Eng. (2019)
191. Guclu, R., Yazici, H.: Vibration control of a structure with ATMD against earthquake using fuzzy logic controllers. J. Sound Vib. **318**(1–2), 36–49 (2008)
192. Committee, A.: Minimum Design Loads for Buildings and Other Structures (ASCE/SEI 7-10). American Society of Civil Engineering, Reston, Virginia, Structural Engineering Institute (2010)
193. RP2A-WSD, A.: American petroleum institute recommended practice for planning, designing and constructing fixed offshore platforms—working stress design. American Petroleum Institute, Washington (2007)

194. Vamvatsikos, D., Cornell, C.A.: Incremental dynamic analysis. Earthquake Eng. Struct. Dynam. **31**(3), 491–514 (2002)
195. Golafshani, A., Bagheri, V., Ebrahimian, H., Holmas, T.: Incremental wave analysis and its application to performance-based assessment of jacket platforms. J. Constr. Steel Res. **67**(10), 1649–1657 (2011)
196. Zeinoddini, M., Nikoo, H.M., Estekanchi, H.: Endurance wave analysis (EWA) and its application for assessment of offshore structures under extreme waves. Appl. Ocean Res. **37**, 98–110 (2012)
197. Estekanchi, H., Valamanesh, V., Vafai, A.: Application of endurance time method in linear seismic analysis. Eng. Struct. **29**(10), 2551–2562 (2007)
198. Mohajernasab, S., Dastan Diznab, M.A., Tabeshpour, M.R., Mehdigholi, H., Seif, M.S.: Application of new-wave theory in the endurance wave method to assess offshore structures under the persian gulf wave conditions. J. Mar. Eng. **9**(18), 71–82 (2014)
199. Dastan Diznab, M.A., Jahanmard, V., Tabeshpour, M.R., Seif, M.S., Mehdigholi, H.: Wave endurance time: a new concept for structural assessment under extreme waves. Proc. Instit. Mech. Eng. Part M: J. Eng. Marit. Environ. **230**(2), 364–377 (2016)
200. Lee, K.S., Geem, Z.W.: A new meta-heuristic algorithm for continuous engineering optimization: harmony search theory and practice. Comput. Methods Appl. Mech. Eng. **194**(36–38), 3902–3933 (2005)
201. Fahimi Farzam, M., Alinejad, B., Bekdaş, G., Mousavi Gavgani, S.A.: Control of a jacket platform under wave load using ATMD and optimization by HSA. In: 6th International Conference on Harmony Search, Soft Computing and Applications (ICHSA 2020) (2020)
202. Farshidianfar, A., Soheili, S.: Optimization of TMD parameters for earthquake vibrations of tall buildings including soil structure interaction. Iran Univ. Sci. Technol. **3**(3), 409–429 (2013)
203. Bekdaş, G., Nigdeli, S.M.: Metaheuristic based optimization of tuned mass dampers under earthquake excitation by considering soil-structure interaction. Soil Dynam. Earthquak. Eng. **92**, 443–461 (2017)
204. Geem, Z.W., Kim, J.H., Loganathan, G.V.: A new heuristic optimization algorithm: harmony search. Simulation **76**(2), 60–68 (2001)
205. Rao, R.V., Savsani, V.J., Vakharia, D.: Teaching–learning-based optimization: a novel method for constrained mechanical design optimization problems. Comput. Aided Des. **43**(3), 303–315 (2011)
206. Yang, X.-S.: Flower pollination algorithm for global optimization. In: International Conference on Unconventional Computing and Natural Computation. Springer (2012)
207. Rao, R.: Jaya: A simple and new optimization algorithm for solving constrained and unconstrained optimization problems. Int. J. Ind. Eng. Comput. **7**(1), 19–34 (2016)
208. Liu, M.-Y., Chiang, W.-L., Hwang, J.-H., Chu, C.-R.: Wind-induced vibration of high-rise building with tuned mass damper including soil–structure interaction. J. Wind Eng. Ind. Aerodyn. **96**(6–7), 1092–1102 (2008)
209. FEMA, P.: Quantification of Building Seismic Performance Factors. Washington, DC (2009)
210. Asami, T., Nishihara, O., Baz, A.M.: Analytical solutions to H_∞ and H_2 optimization of dynamic vibration absorbers attached to damped linear systems. J. Vib. Acoust. **124**(2), 284–295 (2002)
211. Kaveh, A., Fazam, M.F., and Maroofiazar, R.: Comparing H_2 and H_∞ algorithms for optimum design of tuned mass dampers under near-fault and far-fault earthquake motions. Period. Polytech. Civ. Eng. (2020)
212. Kaveh, A., Mahdavi, V.R.: Colliding bodies optimization: a novel meta-heuristic method. Comput. Struct. **139**, 18–27 (2014)
213. Ikago, K., Saito, K., Inoue, N.: Seismic control of single-degree-of-freedom structure using tuned viscous mass damper. Earthquak. Eng. Struct. Dynam. **41**(3), 453–474 (2012)
214. Smith, M.C.: Synthesis of mechanical networks: the inerter. IEEE Trans. Autom. Control **47**(10), 1648–1662 (2002)

215. Matta, E.: Performance of tuned mass dampers against near-field earthquakes. Struct. Eng. Mech. **39**(5), 621–642 (2011)
216. Giaralis, A., Taflanidis, A.: Optimal tuned mass-damper-inerter (TMDI) design for seismically excited MDOF structures with model uncertainties based on reliability criteria. Struct. Control Health Monit. **25**(2), e2082 (2018)
217. Zhao, Z., Zhang, R., Jiang, Y., Pan, C.: Seismic response mitigation of structures with a friction pendulum inerter system. Eng. Struct. **193**, 110–120 (2019)
218. Arakaki, T., Kuroda, H., Arima, F., Inoue, Y., Baba, K.: Development of seismic devices applied to ball screw: Part 1 Basic performance test of RD-series. AIJ J. Technol. Des. **5**(8), 239–244 (1999)
219. Kaveh, A., Farzam, M.F., Jalali, H.H., Maroofiazar, R.: Robust optimum design of a tuned mass damper inerter. Acta Mech. **231**(9), 3871–3896 (2020)
220. Ruiz, R., Taflanidis, A., Giaralis, A., Lopez-Garcia, D.: Risk-informed optimization of the tuned mass-damper-inerter (TMDI) for the seismic protection of multi-storey building structures. Eng. Struct. **177**, 836–850 (2018)
221. Marian, L., Giaralis, A.: Optimal design of a novel tuned mass-damper–inerter (TMDI) passive vibration control configuration for stochastically support-excited structural systems. Probab. Eng. Mech. **38**, 156–164 (2014)
222. Marian, L., Giaralis, A.: The tuned mass-damper-inerter for harmonic vibrations suppression, attached mass reduction, and energy harvesting. Smart Struct. Syst. **19**(6), 665–678 (2017)
223. Kaveh, A., Fahimi Farzam, M., Hojat Jalali, H.: Statistical seismic performance assessment of tuned mass damper inerter. Stru. Control Health Monit. e2602 (2020)

The Effect of SSI and Impulsive Motions on Optimum Active Controlled MDOF Structure

Serdar Ulusoy, Sinan Melih Nigdeli, and Gebrail Bekdaş

Abstract In this study, active controlled ten-story structures including soil-structure interaction (SSI) are investigated under different type of ground motions (near-fault and far ground motions) defined in FEMA P-695 to show the effect of earthquake records on SSI. In structures, where active tendons are located on each floor, the proportional-integral-derivative (PID) controllers are used. Optimum tuning of PID control parameters are calculated with the teaching-learning-based algorithm (TLBO) and harmony search algorithms (HSA) which are commonly used meta-heuristic algorithms. The properties such as swaying and rocking stiffness and damping of dense and soft soils were added to the foundation of the structures to consider the soil-structure interaction. The changes of structural reactions in two types of earthquake records are examined by decreasing or increasing these properties of soils by 10 and 20%. As a result, reduction and enhancement of properties of soft soil in structural responses are changed significantly, while these values are not changed much for dense soil.

Keywords Active structural control · Optimum tuning PID controllers · Impulsive motions · Soil-Structure interaction

S. Ulusoy (✉)
Department of Civil Engineering, Turkish-German University, 34820 Beykoz, Istanbul, Turkey
e-mail: serdar.ulusoy@tau.edu.tr

S. M. Nigdeli · G. Bekdaş
Department of Civil Engineering, Istanbul University-Cerrahpaşa, 34320 Avcılar, Istanbul, Turkey
e-mail: melihnig@istanbul.edu.tr

G. Bekdaş
e-mail: bekdas@istanbul.edu.tr

S. M. Nigdeli et al. (eds.), *Advances in Structural Engineering—Optimization*,
Studies in Systems, Decision and Control 326,
https://doi.org/10.1007/978-3-030-61848-3_6

1 Introduction

In the design of structures, the factors such as the topographic features of the region, fault distance and the magnitude of the earthquake are taken into consideration [1]. Therefore, the different control mechanisms are developed in order to protect the structures under the destructive effects of earthquakes. Among these control types, some passive control systems such as braced structural frames and reinforced concrete walls can be sufficient depending on the characteristics of the structure under far-fault of ground motions. Some passive control systems such seismic base isolation (especially in buildings such as hospitals, bridges etc.) and tuned mass damper (especially in high buildings) are used to reduce structural reactions under different type of ground motions. However, it is known from Northridge, California Earthquake of 17 January 1994 and Kobe, Japan Earthquake of 17 January 1995 that the structures with passive control systems are not enough to meet the ductility requirement of the structure [2] under the near-fault ground motions which are closer than 15 km to the fault distance. Therefore, the additional dampers or different control systems such as semi-active and active control are required to mitigate the large structural responses.

Under the influence of dynamic loads, some deformations occur on the foundation of structures. The design of the structures by neglecting these deformations may not reflect the real structural reaction. Especially in the structures on soft soil type, different results appear in comparison with fix-based structures. Therefore, soil-structure interaction should be taken into account. There are several studies about the structures considering soil-structure effect. 10-story simulated and 30-story real-life structures are investigated to detect the presence of SSI in structures from vibration records [3]. The impact of SSI was evaluated on traffic induced vibrations in structures [4] and under seismic loading both in the elastic and inelastic range of vibration for low-rise structures [5]. An efficient method, based on the Ritz concept is proposed for dynamic analysis of the response of multiple degrees of freedom (MDOF) structures considering SSI under seismic excitations [6]. It is determined that the fundamental frequency of structures depends on the soil–structure related rigidity; K_{ss} [7]. The soil-structures effect in four base-isolated bridges are examined by comparing identified and physical stiffness of sub structure components [8]. The change of structural reactions of elastomeric bearing in the base isolated bridge was observed for different type of soils [9]. The importance of SSI on the seismic response of base-isolated buildings and a high-rise building with a tuned mass damper (TMD) [10, 11] is investigated. The vibration control effectiveness of multiple tuned mass dampers (MTMD) and TMD in irregular structures are studied considering SSI [12]. An optimization approach (metaheuristic algorithms such as harmony search algorithm and bat algorithm) for optimum design of TMD implemented to seismic structures considering soil-structure interaction (SSI) is proposed [13]. The seismic performance of friction dampers for single degree of freedom system (SDOF) is affected by SSI effects [14]. The simple adaptive control (SAC) algorithm is used in conjunction with the magnetorheological (MR) damper to investigate the soil-structure interaction effects on the semi-active control of nonlinear structures [15].

SSI effects should be taken into account in the design of active control devices such as active tendon systems and active tuned mass damper (ATMD), especially for tall buildings located on soft site [16–19].

Metaheuristic algorithms which are divided into three different groups such as the swarm intelligence (flower pollination algorithm (FPA) [20], ant colony optimization (ACO) [21], artificial bee colony (ABC) [22], particle swarm optimization (PSO) [23] and bat algorithm (BA) [24]), evolutionary algorithms (genetic algorithm (GA) [25]) and the other metaheuristic algorithms (harmony search algorithm (HSA) [26], teaching learning based optimization (TLBO) [27] and Jaya algorithm (JA) [28]) are widely used in different fields of structural engineering to optimize the objective function. The metaheuristic-based methods such as BAT, FPA, TLBO, HS, JA, differential evolution, firefly, imperialist competitive algorithm etc. have been employed to determine the optimum parameters of seismic isolation, TMD, ATMD and active tendon systems under seismic excitation [29–39].

In this study, the effect of SSI is investigated under near and far-fault ground motions for active controlled MDOF structures located different types of dense and soft soils which have properties such as swaying and rocking stiffness and damping decreased or increased by 10 and 20%. The structures are actively controlled and the active tendon system with PID controller is used to mitigate the structural responses under seismic excitation. HSA and TLBO are utilized to determine optimum PID parameters which are effective in the performance of the active structural control. Also, time delay effect and control force limit are taken into account in the optimization process. Thus, both the stability problem in the structures was prevented and the capacity of the actuator in the appropriate range was determined.

2 Methodology

The active tendon-controlled structure and the equation of motion of this structure are presented in Fig. 1 and Eq. 1. a(t), v(t), x(t), [1] and u(t) represent the acceleration, velocity, displacement, unit and control signal vectors, respectively. [M], [C], [K] and [M^*] are the mass, damping, stiffness and acceleration mass. These matrices include a part of the sub-matrices such as mass matrix [M_f], damping matrix [C_f] and stiffness matrix of fixed based structure and [M_v], [M_z] matrices. [b] is the influence vector of the control signal, $a_g(t)$ is the ground acceleration, α is the angle of tendon with respect to the ground, k_c is the stiffness of tendon, m_i, c_i, k_i and I_i are the mass, damping, stiffness and mass moment of inertia of ith story for a n-story structure, m_0, I_0, x_0 and θ_0 are the mass, mass moment of inertia, displacement and rotation of foundation, z_i is the distance between base and ith story, c_r and c_s are rocking and swaying damping, k_r and k_s are rocking and swaying stiffness.

The control signal of the first floor $u_1(t)$ is obtained from Eq. 2 of PID controller, which compares the performance index (the displacement of the first floor of structures) with the reference signal assumed as zero to minimize the structural responses. After the determination of control signal vektor u(t) using Eq. 3, the forces $F_1(F -$

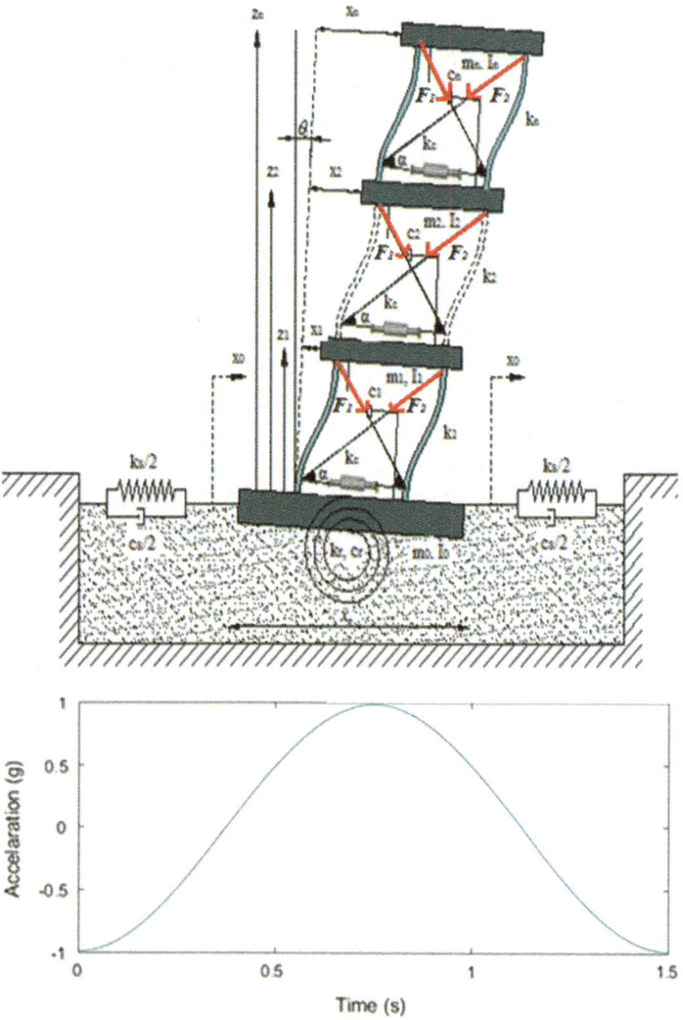

Fig. 1 The active-tendon controlled structures under directivity effect with 1.5 s periods and 230 cm/s PGV

$k_c u(t)$) and $F_2(F + k_c u\ (t))$ in the tendons occur in the dynamic state of the structure if it is assumed that the forces in the tendons are in the static state of structure F and there is a dynamic effect in the +x direction. Also, The PID parameters such as proportional gain (K_p), Integral Time (T_i) and derivative time (T_d) of the first floor are considered to provide the same control force for all stories to avoid confusion in its control algorithm.

$$[M]\,a(t) + [C]\,v(t) + [K]\,x(t) = -[M^*]\,a_g(t)\,[1] - (4k_c \cos\alpha)\,[b]\,u[t] \quad (1)$$

$$u_1(t) = K_P\left[e_1(t) + \frac{1}{T_i}\int_0^t e_1(t)dt + T_d\frac{de_1(t)}{dt}\right] \tag{2}$$

$$u(t) = [u_1 u_1 \ldots . u_1]^T nx1 \tag{3}$$

$$[M] = \begin{bmatrix} [M_f] & [M_v] & [M_z] \\ [M_v]^T & m_0 + \left(\sum_{i=1}^{n} m_i\right) & \left(\sum_{i=1}^{n} m_i z_i\right) \\ [M_z]^T & \left(\sum_{i=1}^{n} m_i z_i\right) & \left(\sum_{i=1}^{n} m_i z_i^2 + I_0 + \sum_{i}^{n} I_i\right) \end{bmatrix} \tag{4}$$

$$[M_f] = \begin{bmatrix} m_1 & & & & & & \\ & m_2 & & & & & \\ & & . & & & & \\ & & & . & & & \\ & & & & . & & \\ & & & & & . & \\ & & & & & & m_n \end{bmatrix}, \quad [M_v] = \begin{bmatrix} m_1 \\ m_2 \\ m_3 \\ . \\ . \\ m_n \end{bmatrix}, \quad [M_z] = \begin{bmatrix} m_1 z_1 \\ m_2 z_2 \\ m_3 z_3 \\ . \\ . \\ m_n z_n \end{bmatrix} \tag{5}$$

$$[M^*] = \begin{bmatrix} m_1 \\ m_2 \\ m_3 \\ . \\ . \\ m_n \\ m_0 + \left(\sum_{i=1}^{n} m_i\right) \\ \left(\sum_{i=1}^{n} m_i z_i\right) \end{bmatrix} \tag{6}$$

$$[C] = \begin{bmatrix} [C_f] & 0 & 0 \\ 0 & c_s & 0 \\ 0 & 0 & c_r \end{bmatrix} \tag{7}$$

$$[K] = \begin{bmatrix} [K_f] & 0 & 0 \\ 0 & k_s & 0 \\ 0 & 0 & k_r \end{bmatrix} \tag{8}$$

$$[C_f] = \begin{bmatrix} c_1+c_2 & -c_2 & & & & & \\ -c_2 & c_2+c_3 & -c_3 & & & & \\ & -c_3 & c_3+c_4 & -c_4 & & & \\ & & -c_4 & . & . & & \\ & & & . & . & . & \\ & & & & . & . & -c_n \\ & & & & & -c_n & c_n \end{bmatrix} \quad (9)$$

$$[K_f] = \begin{bmatrix} k_1+k_2 & -k_2 & & & & & \\ -k_2 & k_2+k_3 & -k_3 & & & & \\ & -k_3 & k_3+k_4 & -k_4 & & & \\ & & -k_4 & . & . & & \\ & & & . & . & . & \\ & & & & . & . & -k_n \\ & & & & & -k_n & k_n \end{bmatrix} \quad (10)$$

$$x(t) = [x_1\, x_2 \ldots x_{n-1}\, x_n] \quad (11)$$

$$v(t) = [v_1\, v_2 \ldots v_{n-1}\, v_n] \quad (12)$$

$$a(t) = [a_1\, a_2 \ldots a_{n-1}\, a_n] \quad (13)$$

$$[1] = [1\,1\,1\,1\,1]^T \quad nx1 \quad (14)$$

$$[b] = [1\,1\,1\,1\,1]^T \quad nx1 \quad (15)$$

PID controller parameters play an important role in calculating the control signal and reducing the structural reaction of active controlled structures according to this control signal. Therefore, numerical algorithms are needed to optimally tune these parameters. One of the numerical algorithms is metaheuristic algorithms. In this study, Harmony Search algorithm and Teaching-Learning Based Optimization are used to determine the Parameter of PID (design variables). The flowchart diagram of both algorithms is shown in Fig. 2.

Teaching-Learning Based Optimization which consists of two phases known as teacher and learner phase is developed by Rao et al. [27]. The mathematical expressions of teacher and learner phases are given in Eqs. 16 and 17, respectively.

$$x_i^{t+1} = x_i^t + rnd(1)(x_{\text{teacher}} - TFx_{\text{mean}}) \quad (16)$$

$$x_i^{t+1} = \left\{ \begin{array}{l} x_i^t + rnd(1)\left(x_j^t - x_k^t\right) iff\left(x_j^t\right) < f\left(x_k^t\right) \\ x_i^t + rnd(1)\left(x_k^t - x_j^t\right) iff\left(x_k^t\right) < f\left(x_j^t\right) \end{array} \right\} \quad (17)$$

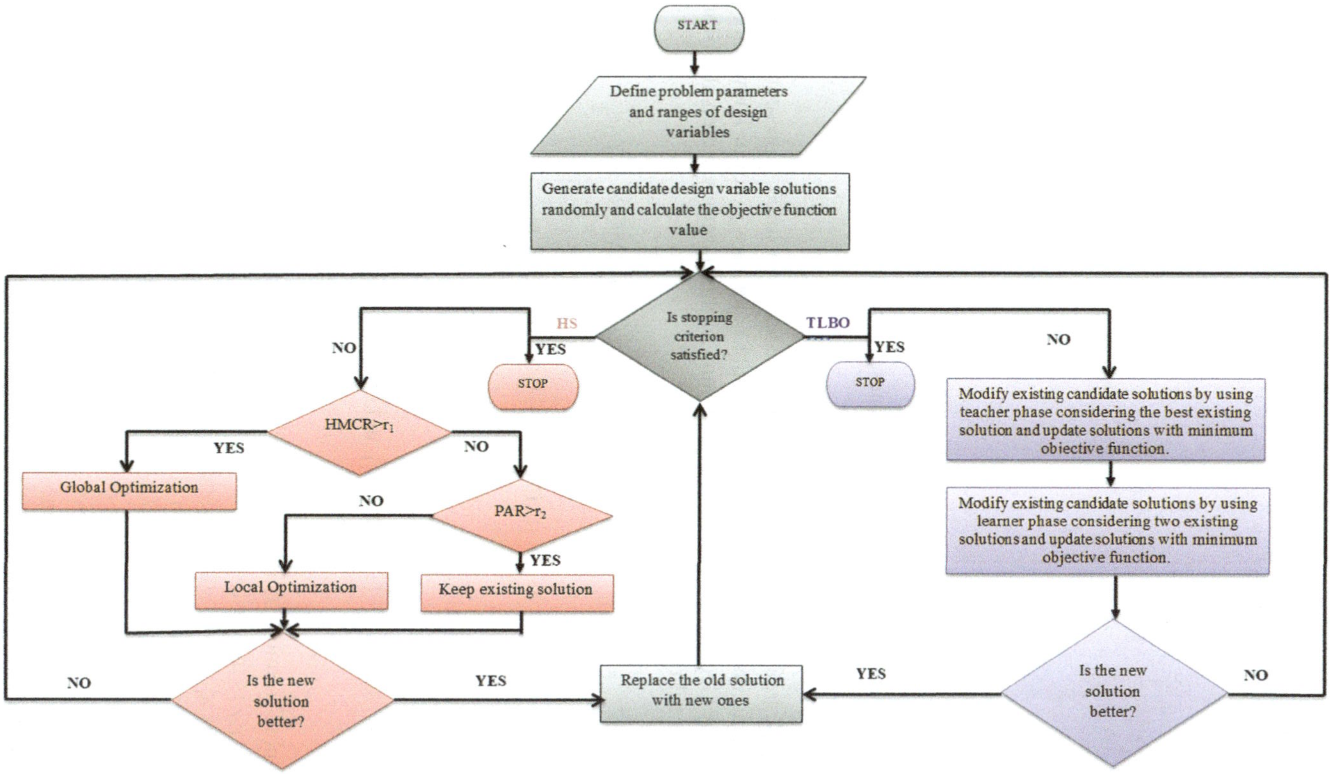

Fig. 2 The flowchart diagram of HSA and TLBO

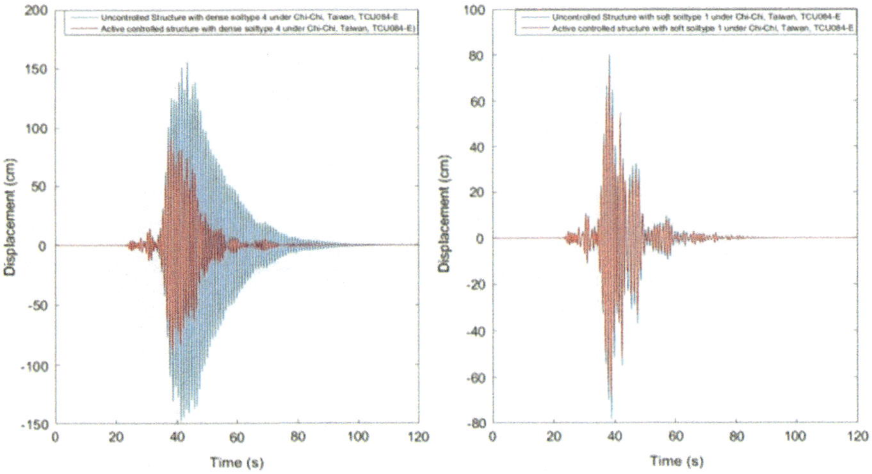

Fig. 3 The time history of maximum top displacement of structure with both soil types under critical excitation

Harmony Search Algorithm inspired by the musician who tried to present the best music to her audience with efforts such as popular notes, new notes or known notes is developed by Geem et al. [26]. The formulation of the Harmony search algorithm is described as follows:

$$x_i^{t+1} = x_{min} + rnd(1)\left(x_{max} - x_{min}\right) \quad \text{if HMCR} > r_1 \tag{18}$$

$$x_i^{t+1} = x_i^t \quad \text{if HMCR} \le r_1 \text{and PAR} > r_2 \tag{19}$$

$$x_i^{t+1} = x_{i+ni}^t \quad \text{if HMCR} \le r_1 \text{and PAR} \le r_2 \text{ (for discrete variables)} \tag{20}$$

$$x_i^{t+1} = x_i^t + b_w(r_3) \quad \text{if HMCR} \le r_1 \text{ and PAR} \le r_2 \text{ (for continuous variables)} \tag{21}$$

x_i^{t+1} is the new solution, x_i^t is existing solution, rnd(1) is random number between 0 and 1, $x_{teacher}$ is the best solution, TF is the teaching factor, x_{mean} is the average of all solutions x_j^t and x_k^t are two randomly chosen solutions, x_{max} and x_{min} are the maximum and minimum values of design variables, r_1, r_2 and r_3 are the random number, ni is the neighbourhood index and b_w is an arbitrary distance bandwidth. The numerical data in optimization process such as range of design variables (K_p, T_d and T_i), simulation time (ST), population number (PN) or Harmony Memory Size (HMS), iteration number (IN), control limit (CL) as a ratio of the total weight of the structure, Harmony Memory Considering Rate (HMCR), Pitch Adjusting Rate (PAR), Teaching Factor (TF) and the random number r_1, r_2, r_3 are given in Table 1.

Table 1 The numerical data in optimization process

Definition	Value	Unit
K_p	(−5)–(5)	–
T_1	(−5)–(5)	s
T	(−5)–(5)	s
HMCR	0.5	–
PAR	0.05	–
TF	1 or 2	–
IN	10,000	–
ST	20	s
PN or HMS	10	s
CL	10	%
r_1 and r_2	0–1	–
r_3	−1–1	–

Also, the directivity effect of near field ground motion is used in the optimization process.

The following equations of Makris [40], Cox and Ashford [1] and Sommerville [41] are used for the expression of directivity effect with 1.5 s periods and 230 cm/s peak ground velocity (PGV). T_p is the pulse period, V_p is the peak ground velocity, ω_p is the pulse frequency, ϕ is the directivity angle and M_w is the moment magnitude.

$$a_g(t) = \omega_p V_p \cos(\omega_p t) \quad 0 \leq t \leq T_p \tag{22}$$

$$\log T_p = -3.17 + 0.5 M_w \quad \text{for rock site} \tag{23}$$

$$\log T_p = -2.02 + 0.346 M_w \quad \text{for soft site} \tag{24}$$

$$\log V_p = 6.444 - 0.0187\phi - 5.022\log(M_w) \tag{25}$$

After determination of PID parameters of structures under directivity effect, the structural responses of active controlled structures with different soil types are investigated under far and near field ground motions defined in FEMA P-695 [42] and given in Tables 2 and 3.

3 Numerical Example

A ten-story shear building model with active tendon systems located on all floors is considered for numerical studies. The story properties of structure are presented in

Table 2 Far-field ground motion records

No	Date	Name	Component 1	Component 2
1	1994	Northridge	Northr/Mul009	Northr/Mul279
2	1994	Northridge	Northr/Los000	Northr/Los270
3	1999	Duzce, Turkey	Duzce/Bol000	Duzce/Bol090
4	1999	Hector Mine	Hector/Hec000	Hector/Hec090
5	1979	Imperial Valley	Impvall/H-Dlt262	Impvall/H-Dlt352
6	1979	Imperial Valley	Impvall/H-E11140	Impvall/H-E11230
7	1995	Kobe, Japan	Kobe/Nis000	Kobe/Nis090
8	1995	Kobe, Japan	Kobe/Shi000	Kobe/Shi090
9	1999	Kocaeli, Turkey	Kocaeli/Dzc180	Kocaeli/Dzc270
10	1999	Kocaeli, Turkey	Kocaeli/Arc000	Kocaeli/Arc090
11	1992	Landers	Landers/Yer270	Landers/Yer360
12	1992	Landers	Landers/Clw-Ln	Landers/Clw-Tr
13	1989	Loma Prieta	Lomap/Cap000	Lomap/Cap090
14	1989	Loma Prieta	Lomap/G03000	Lomap/G03090
15	1990	Manjil, Iran	Manjil/Abbar—L	Manjil/Abbar—T
16	1987	Superstition Hills	Superst/B-Icc000	Superst/B-Icc090
17	1987	Superstition Hills	Superst/B-Poe270	Superst/B-Poe360
18	1992	Cape Mendocino	Capemend/Rio270	Capemend/Rio360
19	1999	Chi-Chi, Taiwan	Chichi/Chy101-E	Chichi/Chy101-N
20	1999	Chi-Chi, Taiwan	Chichi/Tcu045-E	Chichi/Tcu045-N
21	1971	San Fernando	Sfern/Pel090	Sfern/Pel180
22	1976	Friuli, Italy	Friuli/A-Tmz000	Friuli/A-Tmz270

Table 4. The stiffness and angle to the ground of tendons are 36° and 372,100 N/m, respectively [43]. Also, the properties of different soil types are given in Table 5. These values are obtained by decreasing and increasing the values of the study of Liu et al. [44] by 10 or 20%. Thus, the change of the structural reactions of the active controlled structure with eight different soil types is examined. The optimum results (the optimum PID parameters) of active controlled structure with eight different soil types according to HSA and TLBO are given Table 6. The objective functions (the first floor displacement) obtained with two different algorithms are very close to each other. More successful results are emerged with the TLBO algorithm. The Code of the optimization process is written in Matlab and Simulink [45]. Fourth order Runge Kutta method with the step size $h = 0.001$ is used for dynamic analysis of active controlled structures. The time delay effect is taken into account as 20 ms similar to the study of Nigdeli and Boduroglu [46].

After the determination of PID parameters according to TLBO in Table 6, the percentage reductions of the maximum top displacements and total accelerations under the different ground motions defined in Tables 2 and 3 are presented in Figs. 4,

Table 3 Near-field ground motion records without pulse

No	Date	Name	Component 1	Component 2
1	1980	Northridge-01	LA—Sepulveda VA/0637-270	0637-360
2	1987	Loma Prieta	Bran/BRN000	BRN090
3	1999	Loma Prieta	Corralitos/CLS000	CLS090
4	1992	Cape Mendocino	Cape Mendocino/CPM000	CPM090
5	1979	Gazli, USSR	Karakyr/GAZ000	GAZ090
6	1979	Imperial Valley-06	Bonds Corner/H-BCR140	H-BCR230
7	1999	Imperial Valley-06	Chihuahua/H-CHI012	H-CHI282
8	1992	Denali, Alaska	TAPS Pump Sta. #10/PS10047	PS10317
9	1992	Nahanni, Canada	Site 1/S1010	S1280
10	1994	Nahanni, Canada	Site 2/S2240	S2330
11	1989	Northridge-01	Northridge—Saticoy/STC090	STC180
12	1994	Chi-Chi, Taiwan	TCU067/TCU067-E	TCU067-N
13	1999	Chi-Chi, Taiwan	TCU084/TCU084-E	TCU084-N
14	1999	Kocaeli, Turkey	Yarimca/YPT060	YPT330

Table 4 The story properties of the structure

Story	m_i (t)	k_i (N/m)	c_i (Ns/m)	I_i (kgm^2)	z_i (m)
1	529	1,206,400,000	4,632,600	22,248,500	3.50
2	529	1,206,400,000	4,632,600	22,248,500	7.00
3	529	1,206,400,000	4,632,600	22,248,500	10.5
4	529	1,206,400,000	4,632,600	22,248,500	14.0
5	529	1,206,400,000	4,632,600	22,248,500	17.5
6	529	1,206,400,000	4,632,600	22,248,500	21.0
7	529	1,206,400,000	4,632,600	22,248,500	24.5
8	529	1,206,400,000	4,632,600	22,248,500	28.0
9	529	1,206,400,000	4,632,600	22,248,500	31.5
10	529	1,206,400,000	4,632,600	22,248,500	35.0

5, 6, 7. In this case, the simulation time for these ground motions is 120 s. The maximum displacement values of the uncontrolled structure model under the far and near field ground motions are 53.38 (for dense soil type 1 under Northridge, Mul009) and 155.83 cm (for dense soil type 4 under Chi-Chi, Taiwan, TCU084-E), respectively. The total acceleration values are 26.96 (for dense soil type 4 under Northridge, Mul009) and 82.90 m/s^2 (for dense soil type 4 under Chi-Chi, Taiwan, TCU084-E). In the active controlled structure, the maximum displacements were reduced to 35.75 and 134.50 cm for the far and near field ground motions. Likewise, the total acceleration values were decreased to 19.04 and 50.02 m/s^2. While the maximum top displacement and total acceleration values of the uncontrolled or

Table 5 The properties of different soil types

Soil type	c_s (Ns/m)	c_r (Ns/m)	k_s (N/m)	k_r (N/m)
Dense soil type 1	1.584×10^9	1.380×10^{11}	6.900×10^{10}	2.292×10^{13}
Dense soil type 2	1.452×10^9	1.265×10^{11}	6.325×10^{10}	2.101×10^{13}
Dense soil type 3	1.188×10^9	1.035×10^{11}	5.175×10^{10}	1.719×10^{13}
Dense soil type 4	1.056×10^9	0.920×10^{11}	4.600×10^{10}	1.528×10^{13}
Soft soil type 1	2.628×10^8	2.712×10^{10}	2.292×10^9	9.036×10^{11}
Soft soil type 2	2.409×10^8	2.486×10^{10}	2.101×10^9	8.283×10^{11}
Soft soil type 3	1.971×10^8	2.034×10^{10}	1.719×10^9	6.777×10^{11}
Soft soil type 4	1.752×10^8	1.808×10^{10}	1.528×10^9	6.024×10^{11}

Table 6 The optimum PID parameters of active controlled structures with different soil type according to HSA and TLBO

Type of soils		Dense soil				Soft soil			
		K_p	T_d	T_i	X (cm)	K_p	T_d	T_i	X (cm)
Type 1	HS	−2.0548	0.7744	0.0637	9.59	−3.0856	0.2615	0.5159	9.89
	TLBO	−2.0510	0.7868	0.0616	9.59	−3.0870	0.2617	0.5168	9.89
Type 2	HS	−2.0912	0.7495	0.0675	9.59	−3.1417	0.2382	0.6317	10.03
	TLBO	−2.0485	0.7876	0.0615	9.59	−3.2453	0.2308	0.6000	10.02
Type 3	HS	−2.2266	0.3352	1.3862	9.83	−3.4226	0.1920	0.7209	10.34
	TLBO	−2.0270	0.8016	0.0596	9.60	−3.4978	0.1919	0.6522	10.33
Type 4	HS	−2.0553	0.7719	0.0637	9.61	−3.3091	0.1982	0.7488	10.52
	TLBO	−2.0228	0.8027	0.0595	9.60	−3.5462	0.1908	0.5713	10.51

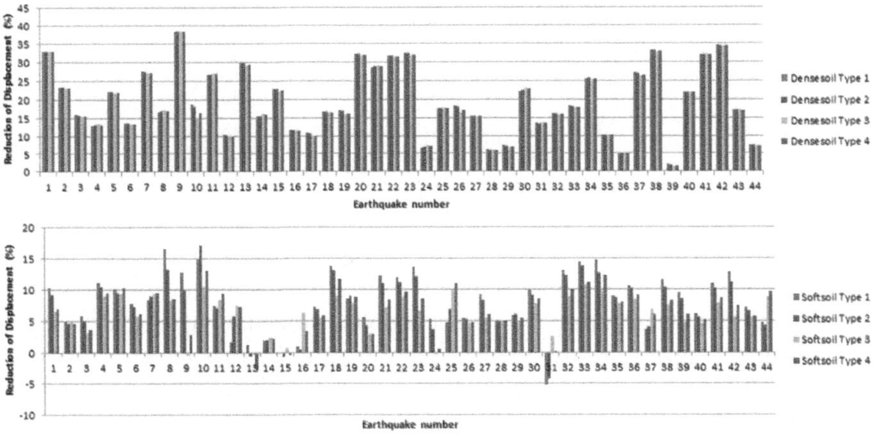

Fig. 4 The percentage reduction of maximum top displacement of structures with different soil types under far-field ground motions according to TLBO

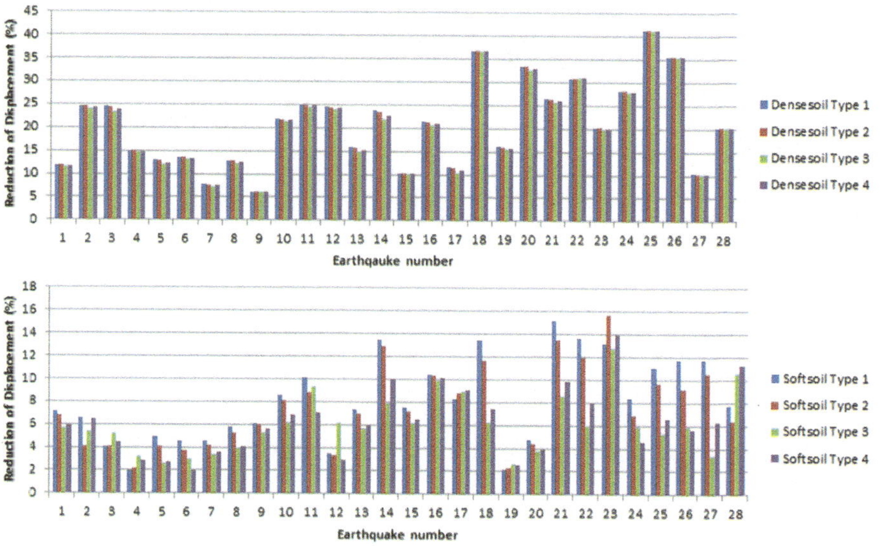

Fig. 5 The percentage reduction of maximum top displacement of structures with different soil types under near-field ground motions according to TLBO

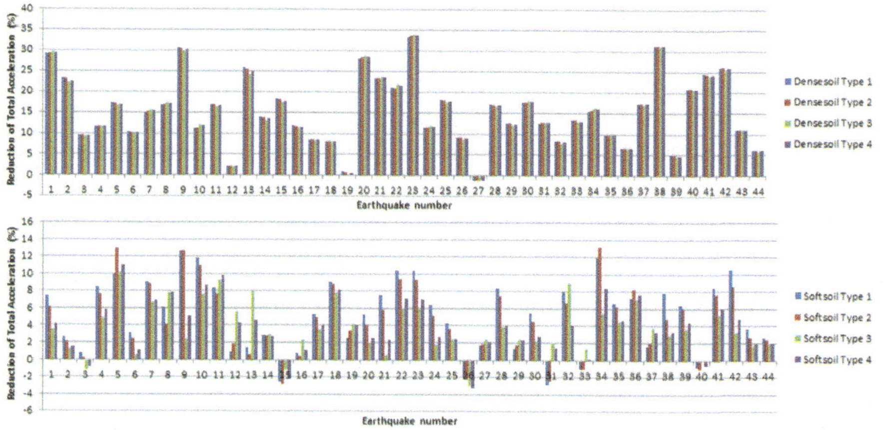

Fig. 6 The percentage reduction of maximum total acceleration of structures with different soil types under far-field ground motions according to TLBO

controlled structure do not differ in four different dense soil types, they differ in four different soft soil types. This means that the structural responses of the active controlled structure located on soft soil types are more affected than the structure with dense soil significantly. There is a decrease in the maximum top displacement of the structure in almost all records (near and far field ground motions). The percentage reduction of the top displacement of structure ranges from 1.52 to 38.52% for the

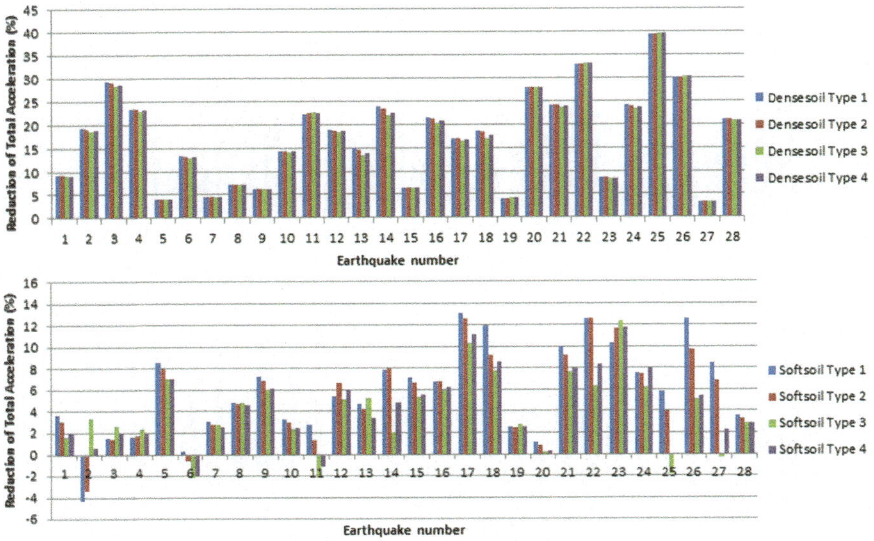

Fig. 7 The percentage reduction of maximum total acceleration of structures with different soil types under near-field ground motions according to TLBO

far field ground motions and from 2.22 to 41.48% for the near field ground motions. It is observed that the active controlled structure located on soft soil types has not performed the function of the active control system because of the increase of the top displacement of structure under three far field ground motions such as such as Kobe (Nis000 and Shi00) and Superstition Hills (B-Icc000). The maximum structural responses of the uncontrolled and active controlled structures under the near and far field ground motions are shown in Tables 7 and 8, respectively. Also, the

Table 7 The maximum structural responses of active controlled structure under near field ground motions

Structure under near field-ground motions	Maximum displacement (cm)		Maximum total acceleration (m/s^2)	
	Uncontrolled	Controlled	Uncontrolled	Controlled
Structure with dense soil type 1	155.32	90.89	81.99	49.58
Structure with dense soil type 2	155.43	91.00	82.17	49.66
Structure with dense soil type 3	155.70	91.40	82.62	49.89
Structure with dense soil type 4	155.83	91.26	82.90	50.02
Structure with soft soil type 1	80.12	71.23	43.07	40.54
Structure with soft soil type 2	76.37	69.00	40.72	39.08
Structure with soft soil type 3	66.57	62.17	34.62	34.62
Structure with soft soil type 4	60.31	57.10	30.83	31.38

Table 8 The maximum structural responses of active controlled structure under far field ground motions

Structure under far field-ground motions	Maximum displacement (cm)		Maximum total acceleration (m/s^2)	
	Uncontrolled	Controlled	Uncontrolled	Controlled
Structure with dense soil type 1	53.38	35.75	26.93	19.04
Structure with dense soil type 2	53.35	35.73	26.94	19.03
Structure with dense soil type 3	53.26	35.71	26.95	19.00
Structure with dense soil type 4	53.20	35.67	26.96	18.97
Structure with soft soil type 1	29.82	26.73	16.32	15.11
Structure with soft soil type 2	28.72	26.06	15.75	14.78
Structure with soft soil type 3	26.28	24.45	14.44	13.84
Structure with soft soil type 4	25.82	23.67	13.65	13.16

displacement time histories of structure with both soil types under critical excitation are given in Fig. 3.

4 Conclusion

In this study, the active controlled structure located on different soil type such as soft and dense soil is investigated under far and near field ground motions to evaluate the effect of SSI and impulsive motions on the active control system. Teaching-learning based optimization and harmony search algorithm are used to determine the parameters of PID controller. The conclusions about the effect of SSI and impulsive motions on active controlled structure are as follows:

(1) The values of objective function for two metaheuristic algorithms (TLBO and HAS) are very close to each other. Thus, these algorithms are very suitable for the use of the optimization problem of active control devices.
(2) There are no significant changes in the structural responses of uncontrolled and active controlled structure located on four different dense soil types under the various ground motions. For this reason, active control systems are applicable in structures with dense soil types.
(3) The soft soil types and the ground motions affect the performance of the active tendon systems in reducing of the structural responses. In this case, passive systems can be chosen instead of active tendon systems, especially in the far field ground motions which do not cause high structural responses.

References

1. Cox, K.E., Ashford, S.A.: Characterization of large velocity pulses for laboratory testing. Report 2002/22. Pacific Earthquake Engineering Research Center, University of California, Berkeley, California, USA (2002)
2. Chopra, A.K., Chintanapakdee, C.: Comparing response of SDF systems to near-fault and far-fault earthquake motions in the context of spectral regions. Earthquake Eng. Struct. Dynam. **30** (2001)
3. Şafak, E.: Detection and identification of soil-structure interaction in buildings from vibration recordings. J. Struct. Eng. **121**(5), 899–906 (1995)
4. François, S., Pyl, L., Masoumi, H.R., Degrande, G.: The influence of dynamic soil–structure interaction on traffic induced vibrations in buildings. Soil Dynam. Earthq. Eng. **27**(7), 655–674 (2007)
5. Dutta, S.C., Bhattacharya, K., Roy, R.: Response of low-rise buildings under seismic ground excitation incorporating soil–structure interaction. Soil Dynam. Earthq. Eng. **24**(12), 893–914 (2004)
6. Chopra, A.K., Gutierrez, J.A.: Earthquake response analysis of multistorey buildings including foundation interaction. Earthquake Eng. Struct. Dynam. **3**(1), 65–77 (1974)
7. Khalil, L., Sadek, M., Shahrour, I.: Influence of the soil–structure interaction on the fundamental period of buildings. Earthquake Eng. Struct. Dynam. **36**(15), 2445–2453 (2007)
8. Chaudhary, M.T.A., Abe, M., Fujino, Y.: Identification of soil–structure interaction effect in base-isolated bridges from earthquake records. Soil Dynam. Earthq. Eng. **21**(8), 713–725 (2001)
9. Thakkar, S.K., Maheshwari, R.: Study of seismic base isolation of bridge considering soil structure interaction (1995)
10. Spyrakos, C.C., Koutromanos, I.A., Maniatakis, C.A.: Seismic response of base-isolated buildings including soil–structure interaction. Soil Dynam. Earthq. Eng. **29**(4), 658–668 (2009)
11. Wang, J.F., Lin, C.C.: Seismic performance of multiple tuned mass dampers for soil–irregular building interaction systems. Int. J. Solids Struct. **42**(20), 5536–5554 (2005)
12. Liu, M.Y., Chiang, W.L., Hwang, J.H., Chu, C.R.: Wind-induced vibration of high-rise building with tuned mass damper including soil–structure interaction. J. Wind Eng. Ind. Aerodyn. **96**(6–7), 1092–1102 (2008)
13. Bekdaş, G., Nigdeli, S.M.: Metaheuristic based optimization of tuned mass dampers under earthquake excitation by considering soil-structure interaction. Soil Dynam. Earthq. Eng. **92**, 443–461 (2017)
14. Sanghai, S.S., Pawade, P.Y.: Effectiveness of friction dampers on seismic response of structure considering soil-structure interaction. Građevinar. **72**(01), 33–44 (2020)
15. Amini, F., Bitaraf, M., Nasab, M.S.E., Javidan, M.M.: Impacts of soil-structure interaction on the structural control of nonlinear systems using adaptive control approach. Eng. Struct. **157**, 1–13 (2018)
16. Nazarimofrad, E., Zahrai, S.M.: Fuzzy control of asymmetric plan buildings with active tuned mass damper considering soil-structure interaction. Soil Dynam. Earthq. Eng. **115**, 838–852 (2018)
17. Lin, C.C., Chang, C.C., Wang, J.F.: Active control of irregular buildings considering soil–structure interaction effects. Soil Dynam. Earthq. Eng. **30**(3), 98–109 (2010)
18. Nazarimofrad, E., Zahrai, S.M.: Seismic control of irregular multistory buildings using active tendons considering soil–structure interaction effect. Soil Dynam. Earthq. Eng. **89**, 100–115 (2016)
19. Enrique Luco, J.: A simple model for structural control including soil–structure interaction effects. Earthquake Eng. Struct. Dynam. **27**(3), 225–242 (1998)
20. Yang, X.-S.: Flower pollination algorithm for global optimization. In: Unconventional Computation and Natural Computation, Lecture Notes in Computer Science, vol. 7445, pp. 240–249 (2012)

21. Dorigo, M., Maniezzo, V., Colorni, A.: The ant system: Optimization by a colony of cooperating agents. IEEE Trans. Syst. Man Cybern. B. **26**, 29–41 (1996)
22. Karaboga, D., Basturk, B.: On the performance of artificial bee colony (ABC) algorithm. Appl. Soft Comput. **8**(1), 687–697 (2008)
23. Kennedy, J., Eberhart, R.C.: Particle swarm optimization. In: Proceedings of IEEE International Conference on Neural Networks, no. IV, Nov 27–Dec 1, pp. 1942–1948. Perth Australia (1995)
24. Yang, X.S.: A new metaheuristic bat-inspired algorithm. Nature inspired cooperative strategies for optimization (NICSO 2010), 65–74 (2010)
25. Holland, J.H.: Adaptation in Natural and Artificial Systems. University of Michigan Press, Ann Arbor, Michigan (1975)
26. Geem, Z.W., Kim, J.H., Loganathan, G.V.: A new heuristic optimization algorithm: harmony search. Simulation **76**, 60–68 (2001)
27. Rao, R.V., Savsani, V.J., Vakharia, D.P.: Teaching–learning-based optimization: a novel method for constrained mechanical design optimization problems. Comput. Aided Des. **43**(3), 303–315 (2011)
28. Rao, R.: Jaya: A simple and new optimization algorithm for solving constrained and unconstrained optimization problems. Int. J. Ind. Eng. Comput. **7**(1), 19–34 (2016)
29. Skandalos, K., Afshari, H., Hare, W., Tesfamariam, S.: Multi-objective optimization of inter-story isolated buildings using metaheuristic and derivative-free algorithms. Soil Dynam. Earthq. Eng. **132**, 106058 (2020)
30. Çerçevik, A.E., Avşar, Ö., Hasançebi, O.: Optimum design of seismic isolation systems using metaheuristic search methods. Soil Dynam. Earthq. Eng. **131**, 106012 (2020)
31. Quaranta, G., Marano, G.C., Greco, R., Monti, G.: Parametric identification of seismic isolators using differential evolution and particle swarm optimization. Appl. Soft Comput. **22**, 458–464 (2014)
32. Bekdaş, G., Nigdeli, S.M., Yang, X.S.: A novel bat algorithm based optimum tuning of mass dampers for improving the seismic safety of structures. Eng. Struct. **159**, 89–98 (2018)
33. Yucel, M., Bekdaş, G., Nigdeli, S.M., Sevgen, S.: Estimation of optimum tuned mass damper parameters via machine learning. J. Build. Eng. **26**, 100847 (2019)
34. Bekdaş, G., Kayabekir, A.E., Nigdeli, S.M., Toklu, Y.C.: Tranfer function amplitude minimization for structures with tuned mass dampers considering soil-structure interaction. Soil Dynam. Earthq. Eng. **116**, 552–562 (2019)
35. Kayabekir, A.E., Bekdaş, G., Nigdeli, S.M., Geem, Z.W.: Optimum design of PID controlled active tuned mass damper via modified harmony search. Appl. Sci. **10**(8), 2976 (2020)
36. Katebi, J., Shoaei-Parchin, M., Shariati, M., Trung, N.T., Khorami, M.: Developed comparative analysis of metaheuristic optimization algorithms for optimal active control of structures. Eng. Comput. 1–20 (2019)
37. Ulusoy, S., Niğdeli, S.M., Bekdaş, G.: Optimization of PID controller parameters for active control of single degree of freedom structures. Challenge **5**(4), 130–140 (2019)
38. Ulusoy, S., Bekdaş, G., Nigdeli, S.M.: Active structural control via metaheuristic algorithms considering soil-structure interaction. Struct. Eng. Mech. **75**(2), 175 (2020)
39. Ulusoy, S., Nigdeli, S.M., Bekdaş, G.: Novel metaheuristic-based tuning of PID controllers for seismic structures and verification of robustness. J. Build. Eng. 101647 (2020)
40. Makris, N.: Rigidity-plasticity-viscosity: can electrorheological dampers protect base-isolated structures from near-source ground motions? Earthquake Eng. Struct. Dynam. **26**, 571–591 (1997)
41. Sommerville, P.G.: Magnitude scaling of the near fault rupture directivity pulse. Phys. Earth Planet. Inter. **137**, 201–212 (2003)
42. FEMA P-695: Quantification of Building Seismic Performance Factors. Federal Emergency Management Agency, Washington DC (2009)
43. Chung, L.L., Reinhorn, A.M., Soong, T.T.: Experiments on active control of seismic structures. J. Eng. Mech. **114**(2), 241–256 (1988). https://doi.org/10.1061/(asce)0733-9399
44. Liu, M.Y., Chiang, W.L., Hwang, J.H., Chu, C.R.: Wind-induced vibration of high-rise building with tuned mass damper including soil- structure interaction. J. Wind Eng. Ind. Aerodyn. (2008). https://doi.org/10.1016/j.jweia.2007.06.034

45. MathWorks Inc.: MATLAB R2015b. Natick, MA, USA (2015)
46. Nigdeli, S.M., Boduroglu, B.H.: Active tendon control of torsionally irregular structures under near-fault ground motion excitation. Comput. Aided Civil Infrastruct. Eng. **28**, 718–736 (2013). https://doi.org/10.1111/mice.12046

Metaheuristic Algorithms for Optimal Design of Truss Structures

Ali Mortazavi and Vedat Togan

Abstract In today's extremely developing and competitive world exploiting from a limited amount of available resources to maximize the profits and advantages becomes one of the most challenging efforts of human beings. For instance, in the engineering design of many systems, by implementing a number of restriction criteria, achieving a feasible state under proper economic conditions is targeted. Optimization methods put forward a proper approach to solve these types of problems. Generally, optimization deals with the selection of the best candidate (mostly with regard to some constraints) from a set(s) of available alternatives. Different types of quantitative optimization problems arise in different disciplines from most of the engineering areas to economics. Therefore, the developing solution techniques and algorithms have been the focus of engineering science and applied mathematics for a while.

Keywords Metaheuristic methods · Truss structures · Size and layout optimization

1 Introduction

In today's extremely developing and competitive world exploiting from a limited amount of available resources to maximize the profits and advantages becomes one of the most challenging efforts of human beings. For instance, in the engineering design of many systems, by implementing a number of restriction criteria, achieving a feasible state under proper economic conditions is targeted [1–4]. Optimization methods put forward a proper approach to solve these types of problems. Generally, optimization deals with the selection of the best candidate (mostly with regard to some constraints) from a set(s) of available alternatives. Different types of quantitative

A. Mortazavi (✉)
Graduate School of Natural and Applied Sciences, Ege University, Izmir, Turkey
e-mail: ali.mortazavi.phd@gmail.com

V. Togan
Civil Engineering Department, Karadeniz Technical University, Trabzon, Turkey
e-mail: togan@ktu.edu.tr

S. M. Nigdeli et al. (eds.), *Advances in Structural Engineering—Optimization*,
Studies in Systems, Decision and Control 326,
https://doi.org/10.1007/978-3-030-61848-3_7

optimization problems arise in different disciplines from most of the engineering areas to economics. Therefore, the developing solution techniques and algorithms have been the focus of engineering science and applied mathematics for a while [5–7].

The optimization techniques can be divided into two main groups as deterministic and non- deterministic methods. In the past, and with the absence of today's modern computers, deterministic methods form the fundamental platform for solving most of the optimization problems. The deterministic approaches require the continuous objective function and its first and perhaps higher-order gradients information to determine their search direction and step sizes. These methods are rapid and accurate, but they have two main drawbacks. Firstly, they require a continuous or at least partial-continuous objective function while finding such a function and acquisition of its gradient(s) information for most of the complex engineering problems is very difficult or even impossible. Secondly, the performance of these methods highly depends on the starting point of the optimization process [8]. Such that, if the algorithm is launched from an improper location in the search domain the process can be easily trapped into nears local minima. Such a condition may often be encountered in many engineering problems with non-convex and non-smooth search spaces. Under these circumstances employing an alternative method(s) would seem like a proper solution.

In this regard, the metaheuristic techniques are the numeric approaches which remove the aforementioned limitations. These techniques do not require any continuous objective function and/or its different order(s) gradient information. These methods are generally inspired by natural rules, physical principles, or even the social behaviors of humans. These methods have simple mathematical formulations that are based on a directed-stochastic search pattern. Such a formulation makes these methods be able to provide both exploration and exploitation search behaviors to roll over the promising region(s) of the search domain through the affordable computational time [9–12]. Due to their numeric essence, metaheuristic methods do not require to impose any extra simplification about the original optimization problem, and this leads these methods to be used for solving various types of problems. Such that, the successful applications of these methods are reported in different fields, such as physics, engineering, chemistry, economics, social sciences, marketing, robotics, and arts [13, 14].

It should be noted that although the metaheuristic techniques as the global optimizer tool can be employed on solving different types of problems, they do not guarantee to find a global solution ever. Indeed, they have the ability to discover promising feasible solutions for a certain problem. Yet, a well-designed metaheuristic technique usually can provide a solution so closed to the exact optimum. These algorithms work based on the iterative process that in each iteration current solutions candidates (i.e. agents) conducting toward a new location to improve the best previously found solution. During this process, the initial population is iteratively improved and when the termination condition(s) is met the best agent of the population is selected as the final optimal solution. [15, 16].

In the current chapter, the main concern is devoted to the structural optimization of the truss systems. Truss structures are constructed via interlocking axial elements in a geometric pattern. In these systems load transmission is occurred via converting external loads to axial tensions among the struts. Trusses provide rigid and lightweight load-carrying systems which are widely used for different purposes in the field of sutural engineering. Their weight advantage causes them to show a reliable dynamic behavior under transient loads (e.g. earthquake) thus they are suitable to span large areas with few (or even without) interior supports. Also, the construction time for truss systems in comparison with other privilege systems (i.e. frame systems) is less and this affirmatively affects the construction cost. On the other hand, from the architectural point of view, the truss structures provide esthetic appearance and by a proper design they can be installed as expose facade. In the current chapter there are five different metaheuristic techniques are described and applied to weight optimization of these structures. The methods are listed as Firefly Algorithm (FA), Teaching and Learning Based Optimization (TLBO), Ion Motion Optimization (IMO), Drosophila food-Search Optimization (DSO) and Interactive Search Algorithm (ISA).

2 Optimization Algorithms

In the current section, applied optimization methods are briefly described. For each algorithm an illustrative pseudo code is, also, addressed.

2.1 Firefly Algorithm (FA)

The firefly algorithm (FA) is originally introduced by Yang [17], this search algorithm of this method imitates the fireflies' behavior in the real world. This algorithm models the flashing phenomenon of fireflies and its impact on their own species. To get a practical mathematical model, some simplifications and idealization have been made on their complex interaction behavior. these idealizations are as follows:

(a) All fireflies are assumed to be unisex species; on the other words all of them irrespective to their gender are able to be attracted to each other.
(b) The amount of attraction is directly related to the amount of brightness. Such that, for a pair of fireflies, the brighter one is more attractive, thus the dimmer firefly will move toward the shinier one. It should be noted that the distance between them also affects the attractiveness level. Indeed, if they are far apart, they will not affect each other. In this condition, the distant firefly would perfume a random search.
(c) The optimization objective function's landscape highly affects the effect of the fireflies' brightness (e.g. in foggy weather fireflies may not recognize each other even from close distances).

Based on these assumptions the attraction level of pair of firefly algorithm is mathematically formulated as bellow:

$$\beta(r_{ij}) = \beta_0 e^{-\gamma r_{ij}^2} \tag{1}$$

in which, r_{ij} designates the distance between a twosome of fireflies. This distance is defined as second norm (i.e. Euclidean distance) as $r_{ij} = \left\| \boldsymbol{X}_i - \boldsymbol{X}_j \right\|$. Also, γ and β_0 indicate the light absorption coefficient and attractiveness level for the $r_{ij} = 0$ condition, respectively. Consequently, based on the given definition the updating process FA method is mathematically expressed and below:

$$\begin{aligned} \Delta\mathbf{X}_i &= \beta_0 e^{-\gamma_{ij}^2}\left(\mathbf{X}_i - \mathbf{X}_j\right) + \alpha\left(rand - \frac{1}{2}\right) \\ \mathbf{X}_i^{t+1} &= \mathbf{X}_i^t + \Delta\mathbf{X}_i \end{aligned} \tag{2}$$

in which, superscript t and $t + 1$ demonstrate the current and updated locations of the ith firefly, respectively. Also, α is random number selected from [0,1] interval with uniform probability, and 'rand' provides a vector of random number uniformly distributed in [0,1] interval, and $\beta_0 = 1$ [17]. To provide more illustration the pseudo code for FA is addressed as follows (Table 1).

Table 1 The pseudo code for FA

Initialize algorithm internal parameters;
Generate random population of fireflies
Light intensity at $\mathbf{X}_i$ is determined by $f(\mathbf{X}_i)$
while (termination conditions are not met)
for (each particle i)
for (each particle j)
Adjust attractiveness using Eq. 1
For each firefly,
if *$(f(\mathbf{X}_j) > f(\mathbf{X}_i))$,*
firefly i should move towards firefly j
end
Evaluate updated solutions
end
end

2.2 Teaching and Learning Based Optimization (TLBO) Algorithm

Teaching and Learning Based Optimization (TLBO) has been obtained based on the knowledge transfer process in the conventional educational system (e.g. classroom) by Rao et al. [14]. This algorithm applies two different search patterns to navigate the agents, they are named as the teaching phase and learning phase. The teaching phase works based on the knowledge transfer between teacher and students. Since the teacher is accepted as the most knowledgeable person in the class, in the teaching phase the best agent of the population is selected as the teacher, and all other agents of the population (i.e. students) are conducted toward the best agent (i.e. teacher). The learning phase of TLBO is based on pairwise knowledge sharing of agents. After these two phases if the location of any agent is improved it is accepted otherwise it is rejected. So, this method does not apply any memory stick. TLBO performs these two phases in each iteration thus in each iteration twice of population size the objective function evaluations (OFEs) are required. Based on this information TLBO is mathematically formulated as follows:

Teaching phase:

$$\mathbf{X}^{(new,i)} = \mathbf{X}^i + r(\mathbf{X}_{teacher} - T_F\mathbf{X}_{mean}) \tag{3.1}$$

$$\begin{cases} \mathbf{X}^i = \mathbf{X}^{(new,i)} & \text{if } f\left(\mathbf{X}^{(new,i)}\right) < f\left(\mathbf{X}^i\right) \\ \mathbf{X}^i = \mathbf{X}^i & \text{otherwise} \end{cases}$$

Learning phase:

$$\begin{aligned} \mathbf{X}_i^{new} &= \mathbf{X}_i + r.\left(\mathbf{X}_i - \mathbf{X}_j\right) \text{ if } f(\mathbf{X}_i) \le f\left(\mathbf{X}_j\right) \\ \mathbf{X}_i^{new} &= \mathbf{X}_i + r.\left(\mathbf{X}_j - \mathbf{X}_i\right) \text{ otherwise} \end{aligned} \tag{3.2}$$

$$\begin{cases} \mathbf{X}^i = \mathbf{X}^{(new,i)} & \text{if } f\left(\mathbf{X}^{(new,i)}\right) < f\left(\mathbf{X}^i\right) \\ \mathbf{X}^i = \mathbf{X}^i & \text{otherwise} \end{cases}$$

The subscripts of *i* and *j* indicate the ith and jth agents of the population, and $\mathbf{X}_{mean}$ is the mean agent which is obtained via averaging of all agents of the population. It is mathematically defined as below:

$$\mathbf{X}_{mean} = \left[m\left(\sum_{j=1}^{np} x_j^1\right), m\left(\sum_{j=1}^{np} x_j^2\right), \ldots, m\left(\sum_{j=1}^{np} x_j^i\right), \ldots, m\left(\sum_{j=1}^{np} x_j^{nd}\right) \right] \tag{4}$$

in which, *np* designates the number of the population; *nd* shows the problem dimension and *m*(.) returns the mean value of inputted variables. To describe the mechanism of TLBO its pseudo code is provided as follows (Table 2).

Table 2 The pseudo code for TLBO

```
Generate random population of agents (i.e. students)
while (not termination condition)
Select the best agent of the population as the teachers
  for (each agent i)
      perform teaching phase based on Eq. (3.1)
      evaluate updated agent
      if new location of the agent is improved
         hold it
      else
         reject it
      end
  end
  for (each agent i)
      select random j^th agent which (i≠j)
      perform learning phase based on Eq. (3.2)
      evaluate updated agent
      if new location of the agent is improved
         hold it
      else
         reject it
      end
  end
end
```

Teaching Phase

Learning Phase

2.3 Drosophila Food-Search Optimization (DSO)

Drosophila Food-search Optimization (DSO) is a population-based algorithm which imitates the food search behavior of the Drosophila Melanogaster insect. This method has been introduced by Das and Singh [18]. The formulation of this method so called Redundant Search (RD) is mathematically are formulated as follow:

$$\begin{aligned}
&U_{i,k} = V_{i,k} + \left|V_{r3,k} - V_{r4,k}\right| \\
&W_{i,k} = V_{i,k} + \left|V_{r3,k} - V_{r4,k}\right| fork = r1\, and\, r2; \\
&\text{for } j \neq r1 \text{ and } j \neq r2,\ U_{i,k} = V_{i,j} \text{ and } W_{i,j} = V_{i,j} \\
&V_{i,j}^{'} = \text{Min}\{f(V_{i,j}), f(U_{i,j}), f(W_{i,j})\},\quad i = 1, 2, \ldots, P \text{ and } j = 1, 2, \ldots, D
\end{aligned} \tag{5}$$

where, $i \in \{1, 2, \ldots, p\}$ and $j \in \{1, 2, \ldots, D\}$ which D and p are the problem dimension and population size, respectively. $r1, r2 \in [1, D]$ are two random integers. Also, the current and updated agent's location are respectively shown with $V_{i,k}$ and $V_{i,j}^{'}$. If the relative improvement for each agent in any iteration is lower than one percent, a neighborhood search is performed as formulated below:

Table 3 The pseudo code for DSO

generate random population of agents (i.e. insects)
evaluate fitness of each agent in the populations
while *(not termination condition)*
use tournament selection
for *(each agent i)*
make the Redundant Search using Eq. (5)
calculate objective function of each agent
save best string the best agent
if *the improvement level of any agent is less than 1%,*
apply neighborhood search using Eq. (6)
end
if *the neighborhood agent is improved the solution*
replace it with old one
end
end
end

$$X_{neighbor} = 0.5\frac{(R_2^2 - R_3^2)f(R_1) + (R_3^2 - R_1^2)f(R_2) + (R_1^2 - R_2^2)f(R_3)}{(R_2 - R_3)f(R_1) + (R_3 - R_1)f(R_2) + (R_1 - R_2)f(R_3)} \tag{6}$$

in which $f(\cdot)$ designates objective function value for desired agent, and R_1, R_2 and R_3 are randomly selected agents among the population so that $R_1 \neq R_2 \neq R_3$.

Based on given information the pseudo code for DSO is defined as below (Table 3):

2.4 *Ions Motion Optimization (IMO)*

Ion motion algorithm simulated repulsion and attraction forces of free ions (i.e. cations and anions) in nature. This algorithm applies two different navigation patterns based on ion motions in the liquid and solid environments [19]. Notably, IMO does not employ any random coefficient in the liquid phase. So, in this phase, all agents (ions) movements are not stochastic. Also, since the initial population is divided into two main groups (i.e. anions and cations) the initial population size should be selected as an even number. Subsequently, the IMO both liquid and solid phases are addressed as follow:

$$\begin{aligned}
&\underline{\textit{Liquid phase}}\\
&A_{ij} = A_{ij} + AF_{ij} \times (Cbest_j - A_{ij})\\
&C_{ij} = C_{ij} + CF_{ij} \times (Abest_j - C_{ij})\\
&\text{where,}\\
&AF_{ij} = \frac{1}{1 + e^{-0.1/AD_{ij}}}\\
&CF_{ij} = \frac{1}{1 + e^{-0.1/CD_{ij}}}\\
&\underline{\textit{Solid phase}}\\
&A_{ij} = A_{ij} + AF_{ij} \times (Cbest_j - A_{ij})\\
&C_{ij} = C_{ij} + CF_{ij} \times (Abest_j - C_{ij})\\
&\text{where,}\\
&AF_{ij} = \frac{1}{1 + e^{-0.1/AD_{ij}}}\\
&CF_{ij} = \frac{1}{1 + e^{-0.1/CD_{ij}}}\\
&\underline{\textit{Solid phase}}\\
&\text{if } \left(CbestFit \geq \frac{CworstFit}{2} \quad and \quad AbestFit \geq \frac{AworstFit}{2}\right)\\
&\quad \text{if } rand() > 0.5\\
&\qquad A_i = A_i + \Phi_1 \times (Cbest - 1)\\
&\qquad C_i = C_i + \Phi_2 \times (Abest - 1)\\
&\quad \text{else}\\
&\qquad A_i = A_i + \Phi_1 \times (Cbest)\\
&\qquad C_i = C_i + \Phi_2 \times (Abest)\\
&\quad \text{end}\\
&\quad \text{if } rand() < 0.05\\
&\qquad Reinitialized\ A_i\ and\ C_i\\
&\quad \text{end}\\
&\text{end}
\end{aligned} \tag{7}$$

in which, $AD_{ij} = |A_{ij} - Cbest_j|$, $CD_{ij} = |D_{ij} - Abest_j|$ are defined as Euclidean distance between i^{th} agent and best anion and cation, respectively. The pseudo code for IMO is shown as follows (Table 4).

2.5 Interactive Search Algorithm (ISA)

Interactive Search Algorithm (ISA) is the hybrid optimization technique which combines the integrated Particle Swarm Optimization (iPSO) and Teaching and Learning Based Optimization (TLBO) methods [20]. This method applies two different navigation schemes so-called tracking and interacting search patterns. In the tracking pattern, all particles (agents) are conducted toward the three main locations in the search domain. These locations are spotted by weighted particle and the best particle of the colony and location stored in ISA memory. In the interacting pattern, particles try to improve their location via the pairwise knowledge sharing process. The ISA is mathematically formulated as follows:

Table 4 The pseudo code for IMO

generate random population of agents (i.e. ions)
while (not termination condition)
evaluate the objective function of all agents (i.e. ions)
select the best and worst Anion and Cation
select force factor using Eq. (8)
perform the liquid phase based on Eq. (8)
If $\left(CbestFit \geq \frac{CworstFit}{2} \quad and \quad AbestFit \geq \frac{AworstFit}{2}\right)$
Perform solid phase motion based on the Eq. (8)
end
end

$$\left.\begin{aligned} &\text{if } \tau \geq 0.3 \\ &{}^{t+1}V_i = \omega_i \times {}^{t}V_i + \left(\varphi_{1i} \odot \left({}^{t}X_j^{P} - {}^{t}X_i\right)\right) \\ &\qquad + \left(\varphi_{2i} \odot \left({}^{t}X^{G} - {}^{t}X_j^{P}\right)\right) \\ &\qquad + \left(\varphi_{3i} \odot \left({}^{t}X^{W} - {}^{t}X_j^{P}\right)\right) \end{aligned}\right\} Tracking \tag{8}$$

$$\left.\begin{aligned} &\text{if } \tau < 0.3 \\ &{}^{t+1}V_i = r_i^{T} \odot \left(X_i - X_j\right) \quad \text{if} \quad f(X_i) < f(X_j) \\ &{}^{t+1}V_i = r_i^{T} \odot \left(X_j - X_i\right) \quad \text{if} \quad f(X_i) \geq f(X_j) \end{aligned}\right\} Interacting$$

$${}^{t+1}X_i = {}^{t}X_i + {}^{t+1}V_i$$

In which, superscripts of $t + 1$ and t show the updated and current condition of the desired corresponding variable. Also, ω_i is the inertia weight and φ_{1i}, φ_{2i}, φ_{3i} are random vectors which their components are uniformly selected from [0,1]. Also, $\odot$ indicates the Hadamard multiplication and i and j are shown the ith and jth agent of the population, respectively. for more illustration the pseudo code for ISA method is give in Table 5.

3 Structural Optimization

Structural optimization generally deals with finding an optimal design of load-carrying systems. Since the weight of the structure plays an important role from both

Table 5 The pseudo code for ISA

```
generate random population of agents (i.e. particles)
while (not termination condition)
    for (each particle i)
      calculate tendency factor
        if  ≥ 0.3
          perform tracking search using Eq. (8)
        else
          perform interacting search using Eq. (8)
        end
      select the best particle
    end
end
```

economical and mechanical aspects, weight minimization of the structural system is mostly defined as the main objective function of a structural optimization problem. For instance, the lighter structure not only has a lower construction cost but also shows better performance under the earthquake loads. On the other hand, to find a feasible solution there are several constraints that should be considered during the optimization process. These constraints can include dynamic (e.g. natural frequencies), static (e.g. stress and displacement), and/or stability (e.g. buckling) criteria.

Different features of the structure can be taken into account as the main decision variables of the problem; two most important of them which mostly used in truss structure optimization problems are size and shape parameters. Size optimization, deals with selecting optimal cross sections for the structural members while the shape optimization tries to optimize the nodal coordinates of the joints of the structural system. Size variables can be selected either from continuous or discrete variable sets. Based on given information a structural optimization is mathematically formulated as follows:

$$
\begin{aligned}
&\text{find} \quad \mathbf{X} = \{x_1, \ldots, x_n\} \\
&\text{to minimize: } \mathrm{f}(\mathbf{X}) = \sum_{e=1}^{ne} \mathrm{L_e} \rho_e \mathrm{A_e} \\
&\text{subjected to: } \mathrm{g_k}(\mathbf{X}) \leq 0 \qquad \mathrm{k} = 1 \ldots \mathrm{nc} \\
&\qquad\qquad\qquad \mathrm{x_{min,i}} \leq \mathrm{x_i} \leq \mathrm{x_{max,i}}
\end{aligned}
\tag{9}
$$

in this formulation, **X** denotes the design vector, and f(**X**) returns total weight of the structure. L_e, ρ_e, and A_e designate the length, material density and cross-sectional area of the e*th* member, respectively. $g_k(X)$ shows the k*th* constraint of the problem while

nc indicate the total number of imposed constraints. Also, $x_{min,i}$ and $x_{max,i}$ demonstrate the lower and upper bounds for ith design variable. In the current chapter, the numeric examples are selected such that involve different type of variables (i.e. discrete and continuous variables) subjected to dynamic, static, and stability constraints, for both size and shape optimization of truss structures.

4 Numerical Examples

In this section, four truss structures' weights are minimized using explained methods. The obtained results are compared with each other through the illustrative tables. The standard deviation (Std.) of the process, achieved mean, and the best weights and Number of Structure Analyses (NSAs) are provided for each example. All algorithms are coded in Matlab® platform. In these examples, different properties of structural systems (e.g. discrete and continuous variables and dynamic and static constraints) are considered. To provide more clarity on the properties of tested problems, their features are abbreviated in Table 6. The parameters setting for applied algorithms are given in Table 7. The stop criteria are 1000 * D (D = problem dimension) or 50 null iterations (i.e. iterations without improvements) which ever achieved faster. The

Table 6 Properties of the test problems

Property \ Example	A 160-bar pyramid truss	A 120-bar dome structure	A 582-bar spatial truss tower	A 18-bar planar truss
	1	2	3	4
Continuous variables		✓		✓
Discrete variables	✓		✓	
Static constraint(s)	✓		✓	✓
Dynamic constraint(s)		✓		
Stability constraint(s)	✓		✓	✓
Size optimization	✓	✓	✓	✓
Layout optimization				✓

Table 7 Parameters setting for the utilized algorithms

Algorithm	Parameter setting
FA [17]	$\alpha = 0.25, \beta = 0.2, \gamma = 1$
TLBO [14]	$T_F = \text{round}[1 + \text{rand}(0, 1)\{2 - 1\}]$
DSO [18]	–
IMO [19]	–
ISA [21]	$\tau = 0.3$

population size is considered as 10, 10, 35 and 15 for first, second, third and fourth examples, respectively.

4.1 A 160-Bar Pyramid Truss

As the first example weight minimization of a 160-bar pyramid spatial truss structure is considered. The schematic view for this structure is shown in Fig. 1. The pyramid's base part is in the square form which its each sides length is 52ft (16 m), it's the highest node is located at 5.25ft from the basement. The structure consists 160 struts. The variables for size optimization are selected from a discrete set of pipes (hollow circular) profiles addressed in AISC database. These profiles are given in Table 8. The stress in the members are limited considering the stability criteria given in AISC-ASD provisions for hollow sections. In all principal directions the nodal displacement is limited up to 1.75 in for all nodes of the truss.

All nodes of the structure are subjected to the −1.92 kips (−8.53 kN) in vertical direction. The structural members maintaining symmetry of the system are put into 7 independent groups. The corresponding grouping is shown on the Fig. 1. Found optimal solutions by all methods are given in Table 9, based on this table except IMO all other methods can nearly find the same optimal solution while ISA with the lowest required number of structural analyses shows the fastest convergence rate among all other methods. Based on standard deviation (Std.) values both TLBO and ISA demonstrate the most stable behavior among all other cited techniques.

4.2 A 120-Bar Dome Structure

As shown in Fig. 2, the 120-bar dome structure is selected as the next example. For this structure, the elasticity modulus and density of applied material are 210 GPa and 7971.81 kg/m^3, respectively. As shown in Fig. 2, maintaining symmetry, the structure's members are categorized into 7 independent groups. 3000 kg, 500 kg and 100 kg non-structural concentrated masses are attached to the nodes 1, 2–13 and 14–37, respectively. First and second natural frequencies of the system are considered as the dynamic constraints of the problem such that $\omega_1 \geq 9$ Hz and $\omega_2 \geq 11$ Hz conditions should be satisfied. Sizing variables are selected from a continuous search domain with upper and lower bounds of 129.3 and 1 cm^2, respectively. Obtained optimal results for all applied methods are reported in Table 10. Corresponding natural frequencies are also given in Table 11. According to given data ISA can find the lightest structure, while considering the standard deviation values, TLBO shows the most stable optimization process among other tested methods. Also, none of methods violate the natural frequency constraints of the problem.

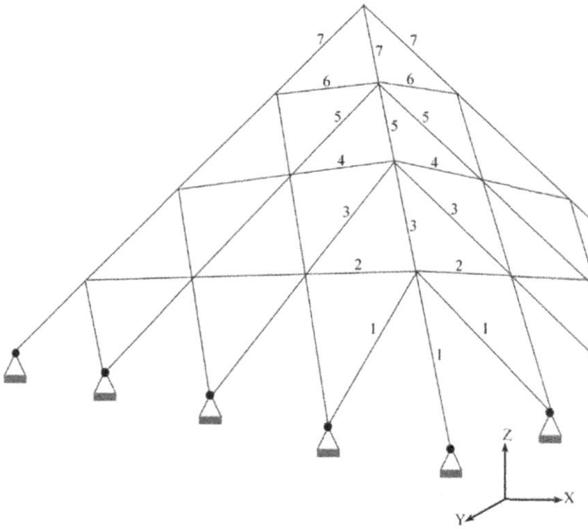

a) 3-D view

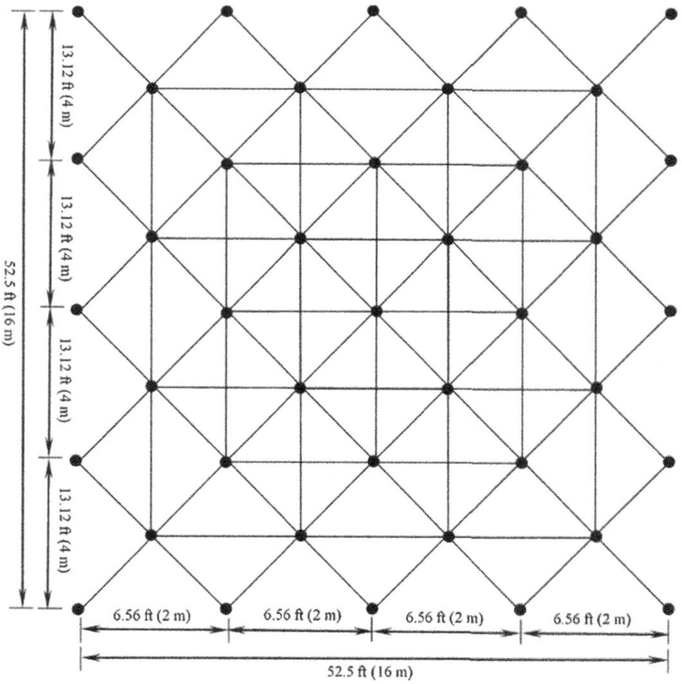

b) Front view

Fig. 1 The 160-bar pyramid structure

Table 8 Cross sectional specifications

Number	Section	Area (in^2)	Number	Section	Area (in^2)
1	PIPE1/2Std	0.2500	20	PIPE3-1/2XS	3.6800
2	PIPE1/2XS	0.3200	21	PIPE2-1/2XXS	4.0300
3	PIPE3/4Std	0.3330	22	PIPE5Std	4.3000
4	PIPE3/4XS	0.4330	23	PIPE4XS	4.4100
5	PIPE1Std	0.4940	24	PIPE3XXS	5.4700
6	PIPE1XS	0.6390	25	PIPE6Std	5.5800
7	PIPE1-1/4Std	0.6690	26	PIPE5XS	6.1100
8	PIPE1-1/2Std	0.7990	27	PIPE4XXS	8.1000
9	PIPE1-1/4XS	0.8810	28	PIPE8Std	8.4000
10	PIPE2Std	1.0700	29	PIPE6XS	8.4000
11	PIPE1-1/2XS	1.0700	30	PIPE5XXS	11.3000
12	PIPE2XS	1.4800	31	PIPE10Std	11.9000
13	PIPE2-1/2Std	1.7000	32	PIPE8XS	12.8000
14	PIPE3Std	2.2300	33	PIPE12Std	14.6000
15	PIPE2-1/2XS	2.2500	34	PIPE6XXS	15.6000
16	PIPE2XXS	2.6600	35	PIPE10XS	16.1000
17	PIPE3-1/2Std	2.6800	36	PIPE12XS	19.2000
18	PIPE3XS	3.0200	37	PIPE8XXS	21.3000
19	PIPE4Std	3.1700			

Table 9 Comparison of optimum designs for 160-bar pyramid

Sizing variables	Optimal cross sections				
	FA	TBLO	DSO	IMO	ISA
A1	PIPE2Std	PIPE2Std	PIPE2Std	PIPE2Std	PIPE2Std
A2	PIPE1-1/4Std	PIPE1-1/4Std	PIPE1-1/4Std	PIPE1-1/4Std	PIPE1-1/4Std
A3	PIPE2Std	PIPE2Std	PIPE2Std	PIPE2Std	PIPE2Std
A4	PIPE1-1/4Std	PIPE1-1/4Std	PIPE1-1/4Std	PIPE1-1/4Std	PIPE1-1/4Std
A5	PIPE2Std	PIPE2Std	PIPE2Std	PIPE2Std	PIPE2Std
A6	PIPE1-1/4Std	PIPE1-1/4Std	PIPE1-1/4Std	PIPE1-1/4Std	PIPE1-1/4Std
A7	PIPE2Std	PIPE2Std	PIPE2Std	PIPE2Std	PIPE2Std
Best weight (lb)	6148.35	6148.35	6148.35	6338.31	6148.35
Mean weight (lb)	6154.91	6148.35	6152.35	6378.35	6148.35
OFEs	2200	1350	1510	880	810
Std. Dev.	1.54	0.00	0.77	15.98	0.00

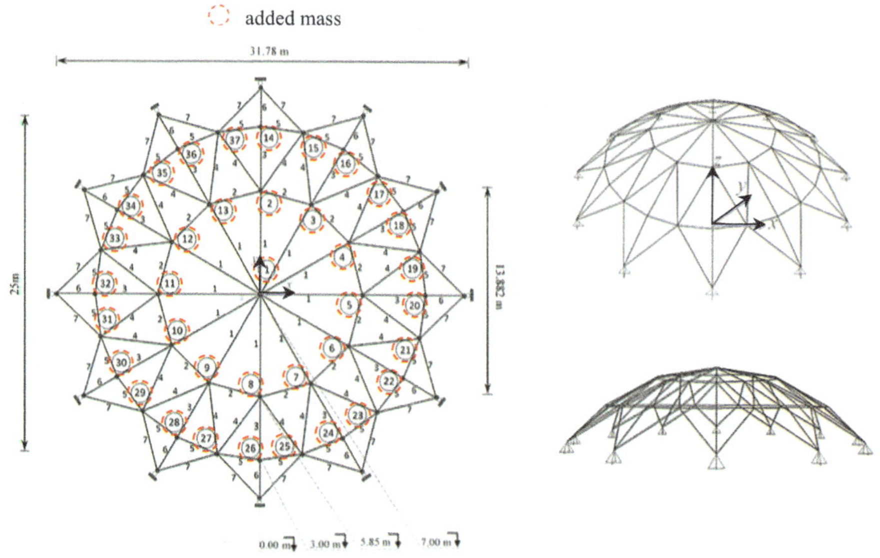

Fig. 2 The 120-bar dome structure

Table 10 The optimal result for 120-bar dome

Design variables (area cm^2)	FA	TLBO	DSO	IMO	ISA
A1	20.0325	19.513	19.571	19.698	19.509
A2	38.2935	40.391	41.148	41.684	40.391
A3	11.7403	10.607	11.439	11.206	10.606
A4	21.9118	21.142	21.315	21.300	21.137
A5	10.200	9.805	10.094	9.588	9.813
A6	10.9328	11.778	12.514	12.739	11.779
A7	14.6337	14.816	15.080	15.179	14.819
Best weight (kg)	8789.50	8707.322	8886.91	8890.63	8707.40
Mean weight (kg)	8799.21	8707.523	8893.33	8923.28	8709.28
Std.	4.73	1.45	13.44	48.88	1.92
NSA	10,500	5050	5850	3880	4880

Table 11 Attained first four natural frequencies (Hz) of the 120-bar dome

Frequency number	FA	TLBO	DSO	IMO	ISA
1	9.0001	9.0000	9.0000	9.0020	9.0000
2	11.0007	11.0000	11.0000	11.0030	11.0000
3	11.0053	11.0000	11.0000	11.0030	11.0000
4	11.0129	11.0100	11.0100	11.0070	11.0100

4.3 A 582-Bar Spatial Truss Tower

As shown in Fig. 3 a 582-bar spatial truss structure weight minimization is considered in this section. Maintaining the symmetry, structural members are divided into 32 different groups. The tower is subjected to the following load cases:

I. All nodes are horizontally subjected to −6.75 kips concentrated force (in z-direction)
II. All nodes are horizontally subjected to 1.12 kips concentrated force (in x-direction)
III. All nodes are horizontally subjected to 1.12 kips concentrated force (in y-direction)

As addressed in Table 12, sizing variables are selected from discrete set of predefined w-shape profiles and based on this table the upper and lower bounds of cross-sectional areas are 215.0 in^2 (1387.09 cm^2) and 6.16 in^2 (39.74 cm^2), respectively. For all nodes the displacement is restricted up to 3.15 in (8 cm) in all principal directions. According to AISC-ASD89 for all structure's members the stresses should be restricted as follows:

$$\begin{cases} \sigma_i^+ = 0.6F_y & \sigma_i \geq 0 \\ \sigma_i^- & \sigma_i < 0 \end{cases} \tag{10}$$

where σ_i^+ and σ_i^- designate tensile and compressive stresses, respectively and for compressive stress:

$$\sigma_i^- \begin{cases} \left[\left(1 - \frac{\lambda_i^2}{2C_c^2}\right)F_y \Big/ \left(\frac{5}{3} + \frac{3\lambda_i}{8C_c} - \frac{\lambda_i^3}{8C_c^3}\right)\right] & for\ \lambda_i < C_c \\ \frac{12\pi^2 E}{23\lambda_i^2} & for\ \lambda_i \geq C_c \end{cases} \tag{11}$$

Based on above formulation compressive members' allowable stress depends on their slenderness ratio (λ). The critical slenderness ratio is shown by C_c and it is defined as below:

$$C_c = \sqrt{\frac{2\pi^2 E}{F_y}} \tag{12}$$

Based on the AISC-ASD code, maximum allowable slenderness ratios should be limited up to 300 and 200 for tensile and compressive members, respectively. Slenderness calculation and its restriction are addressed in follows:

$$\lambda_i = \frac{k_i l_i}{r_i} \leq \begin{cases} 300\ for\ tension\ members \\ 200\ for\ compression\ members \end{cases} \tag{13}$$

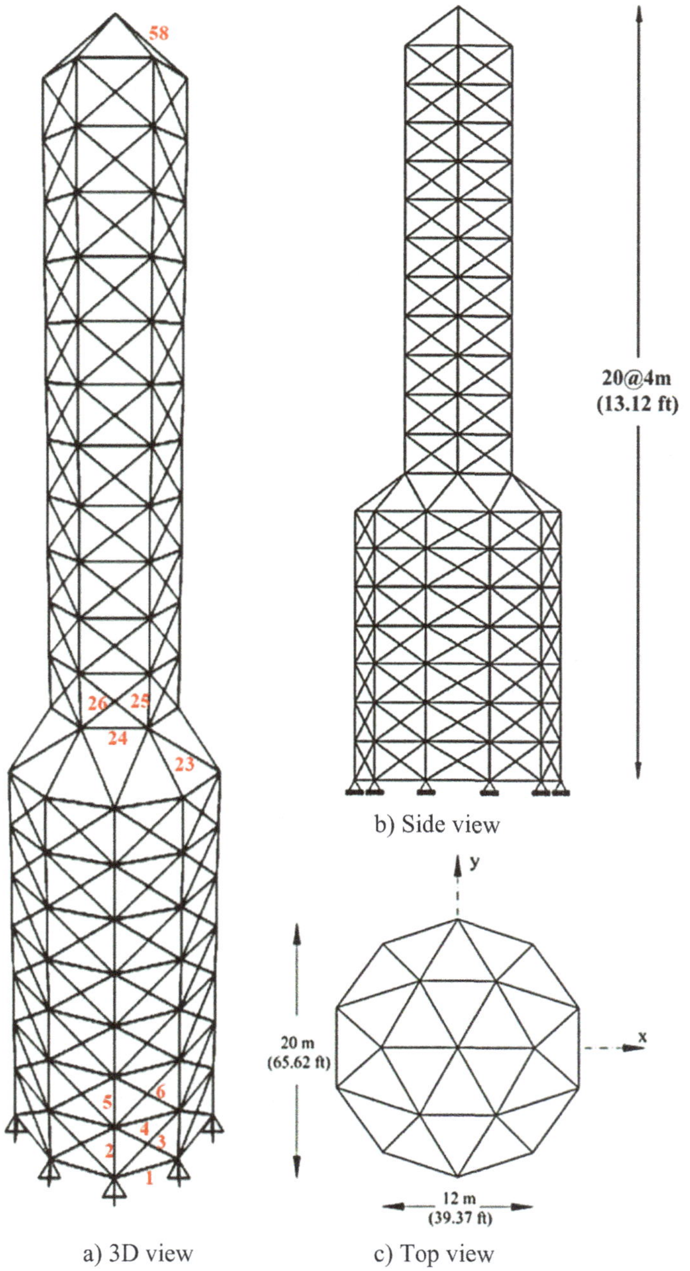

a) 3D view b) Side view c) Top view

Fig. 3 The 582-bar spatial truss system

Table 12 Discrete cross-sections available for the sizing variables of the 582-bar tower problem

W27 × 178	W21 × 122	W18 × 50	W14 × 455	W14 × 74	W12 × 136	W10 × 77
W27 × 161	W21 × 111	W18 × 46	W14 × 426	W14 × 68	W12 × 120	W10 × 68
W27 × 146	W21 × 101	W18 × 40	W14 × 398	W14 × 61	W12 × 106	W10 × 60
W27 × 114	W21 × 93	W18 × 35	W14 × 370	W14 × 53	W12 × 96	W10 × 54
W27 × 102	W21 × 83	W16 × 100	W14 × 342	W14 × 48	W12 × 87	W10 × 49
W27 × 94	W21 × 73	W16 × 89	W14 × 311	W14 × 43	W12 × 79	W10 × 45
W27 × 84	W21 × 68	W16 × 77	W14 × 283	W14 × 38	W12 × 72	W10 × 39
W24 × 162	W21 × 62	W16 × 67	W14 × 257	W14 × 34	W12 × 65	W10 × 33
W24 × 146	W21 × 57	W16 × 57	W14 × 233	W14 × 30	W12 × 58	W10 × 30
W24 × 131	W21 × 50	W16 × 50	W14 × 211	W14 × 26	W12 × 53	W10 × 26
W24 × 117	W21 × 44	W16 × 45	W14 × 193	W14 × 22	W12 × 50	W10 × 22
W24 × 104	W18 × 119	W16 × 40	W14 × 176	W12 × 336	W12 × 45	W8 × 67
W24 × 94	W18 × 106	W16 × 36	W14 × 159	W12 × 305	W12 × 40	W8 × 58
W24 × 84	W18 × 97	W16 × 31	W14 × 145	W12 × 279	W12 × 35	W8 × 48
W24 × 76	W18 × 86	W16 × 26	W14 × 132	W12 × 252	W12 × 30	W8 × 40
W24 × 68	W18 × 76	W14 × 730	W14 × 120	W12 × 230	W12 × 26	W8 × 35
W24 × 62	W18 × 71	W14 × 665	W14 × 109	W12 × 210	W12 × 22	W8 × 31
W24 × 55	W18 × 65	W14 × 605	W14 × 99	W12 × 190	W10 × 112	W8 × 28
W21 × 147	W18 × 60	W14 × 550	W14 × 90	W12 × 170	W10 × 100	W8 × 24
W21 × 132	W18 × 55	W14 × 500	W14 × 82	W12 × 152	W10 × 88	W8 × 21

in which l_i, r_i and λ_i designate length, radius of gyration, and the slenderness ratio for the ith member, respectively. According to AISC-ASD 1989 code, for compressive members if the given restriction for slenderness ratio is violated maximum compressive stress should be limited up to $\left(\frac{12\pi^2 E}{23\lambda_i^2}\right)$.

Obtained optimal results for current example is provided in Table 13. Based on the given results ISA can find lightest structure. Based on the reported Number of Structural Analyses (NSAs) DSO shows the fastest convergence rate. Eventually, the announced standard deviation (Std.) indicate the TLBO demonstrates the most stable behavior during the whole optimization process. The results indicate that each method can show different affirmative and negative behaviors on certain problems.

4.4 A 18-Bar Planar Truss

In the current section as shown in Fig. 4 an 18-bar planar cantilever truss system is optimized. Both size and layout parameters of the system are considered as the decision variables. The size variables are selected from discrete set defined as {2.0, 2.25, ..., 21.5, 21.75} in^2. The elasticity modulus and density of the material

Table 13 Optimal results for the 582-bar spatial tower

Design variables	Optimal cross sections				
	FA	TLBO	DSO	IMO	ISA
1	W8 × 21	W8 × 21	W8 × 24	W8 × 21	W8 × 21
2	W12 × 79	W24 × 84	W12 × 72	W12 × 79	W18 × 76
3	W8 × 28	W8 × 21	W8 × 28	W8 × 24	W8 × 21
4	W10 × 60	W24 × 62	W12 × 58	W10 × 60	W12 × 65
5	W8 × 24	W8 × 21	W8 × 24	W8 × 24	W8 × 21
6	W8 × 21	W8 × 21	W8 × 24	W8 × 21	W8 × 21
7	W10 × 68	W16 × 57	W10 × 49	W8 × 48	W14 × 48
8	W8 × 24	W8 × 21	W8 × 24	W8 × 24	W8 × 21
9	W8 × 21	W8 × 21	W8 × 24	W8 × 21	W8 × 21
10	W14 × 48	W12 × 53	W12 × 40	W10 × 45	W12 × 50
11	W12 × 26	W8 × 21	W12 × 30	W8 × 24	W8 × 21
12	W21 × 62	W10 × 77	W12 × 72	W10 × 68	W16 × 77
13	W18 × 76	W21 × 83	W18 × 76	W14 × 74	W12 × 79
14	W12 × 53	W21 × 57	W10 × 49	W8 × 48	W8 × 48
15	W14 × 61	W18 × 76	W14 × 82	W18 × 76	W12 × 79
16	W8 × 40	W8 × 21	W8 × 31	W8 × 31	W8 × 21
17	W10 × 54	W10 × 22	W14 × 61	W8 × 21	W16 × 67
18	W12 × 26	W18 × 55	W8 × 24	W16 × 67	W8 × 21
19	W8 × 21	W8 × 21	W8 × 21	W8 × 24	W8 × 21
20	W14 × 43	W8 × 21	W12 × 40	W8 × 21	W8 × 48
21	W8 × 24	W14 × 30	W8 × 24	W8 × 40	W8 × 21
22	W8 × 21	W8 × 21	W14 × 22	W8 × 24	W8 × 21
23	W10 × 22	W8 × 21	W8 × 31	W8 × 21	W12 × 26
24	W8 × 24	W8 × 21	W8 × 28	W10 × 22	W8 × 21
25	W8 × 21	W8 × 21	W8 × 21	W8 × 24	W8 × 21
26	W8 × 21	W8 × 21	W8 × 21	W8 × 21	W8 × 21
27	W8 × 24	W10 × 22	W8 × 24	W8 × 21	W8 × 21
28	W8 × 21	W8 × 21	W8 × 28	W8 × 24	W8 × 21
29	W8 × 21	W8 × 21	W16 × 36	W8 × 21	W8 × 21
30	W6 × 25	W8 × 31	W8 × 24	W8 × 21	W8 × 21
31	W10 × 33	W8 × 21	W8 × 21	W8 × 24	W8 × 21
32	W8 × 28	W12 × 22	W8 × 24	W8 × 24	W8 × 21
Best volume (m^3)	21.83	20.30	22.06	22.39	20.07
Mean volume (m^3)	22.32	20.65	23.41	24.98	20.31
Std. (m^3)	1.67	0.34	1.03	2.66	0.35
NSAs	6405	8540	2415	17,500	5810

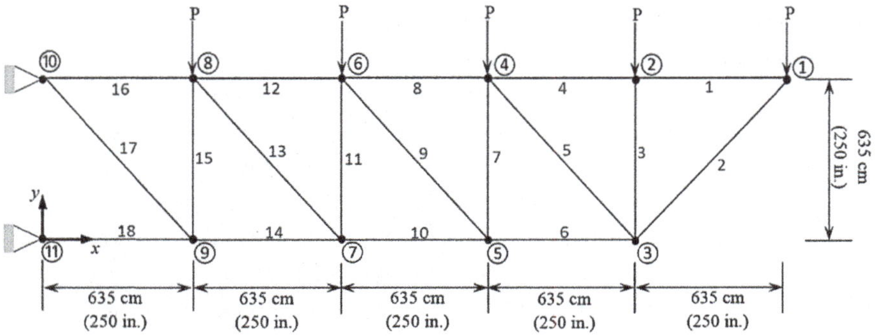

Fig. 4 A 18-bar cantilever truss

are E = 10 Msi (68,947.5728 MPa) and ρ = 0.1 lb/in^3 (2768 kg/m^3), respectively. Allowable stress for all members is limited up to ±20 ksi (±137.8951 MPa) for tensile and compressive members. Also, the Euler buckling criterion should be satisfied as $\sigma_{eb} \leq \frac{\alpha E A_e}{L_e^2}$, where α = 1, E, A_e and L_e are eth member's material elasticity modulus, cross-sectional area and length, respectively. The layout variables are limited as, 775 in ≤ X_3 ≤ 1225 in, 525 in ≤ X_5 ≤ 925 in, 275 in ≤ X_7 ≤ 725 in, 25 in ≤ X_9 ≤ 475 in, 225 in ≤ Y_3, Y_5, Y_7, Y_9 ≤ 245 in. The members are put into four independent groups, these are $G1 = \{1, 4, 8, 12, 16\}$, $G2 = \{2, 6, 10, 14, 18\}$, $G3 = \{3, 7, 11, 15\}$, $G4 = \{5, 9, 13, 17\}$. The nodes of 1, 2, 4, 6 and 8 are subjected to the P = −10 kips vertical loads.

For more illustration, the final layout for the optimal structure of the lightest found solution is schematically given in Fig. 5. Obtained optimal results for layout and size variables, statistical data, and convergence information for this example is given in Table 14. Based on these data, ISA finds the lightest structure while DSO providing the least standard deviation provides the most stabile behavior through the optimization process. Number of required structural analyses indicate that ISA shows the fastest convergence rate among all other algorithms.

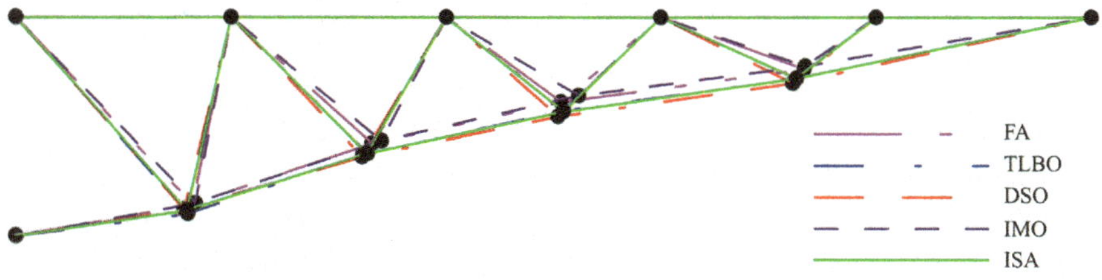

Fig. 5 Schematic of optimal layout for 18-bar planar truss structure

Table 14 Optimal results for the 18-bar spatial tower

Design variables	FA	TLBO	DSO	IMO	ISA
X_3	916.4975	906.9373	903.100	917.4475	907.2491
Y_3	190.5241	179.8866	174.301	193.7899	179.8671
X_5	916.4975	637.0087	630.300	654.3243	636.7873
Y_5	152.9217	142.6170	136.300	159.9436	141.8271
X_7	649.4695	408.6414	402.101	424.4821	407.9442
Y_7	105.425	94.1563	90.4900	108.5779	94.0559
X_9	205.4255	199.6503	195.3001	208.4691	198.7897
Y_9	36.4252	25.3657	30.6001	37.6349	29.5157
G_1	14.25	12.25	12.65	12.75	12.50
G_2	11.75	17.50	17.22	18.50	17.50
G_3	6.00	5.75	6.17	4.75	5.75
G_4	8.00	4.25	3.55	3.25	3.75
Best volume (m^3)	4520.99	4528.7969	4515.7	4530.7	4512.365
Mean volume (m^3)	4582.66	4531.99	4530.12	4588.98	4529.98
Std. (m^3)	25.87	14.1	12.87	29.76	12.91
NSAs	6810	9510	10,500	5100	4500

References

1. Mortazavi, A.: Large-scale structural optimization using a fuzzy reinforced swarm intelligence algorithm. Adv. Eng. Softw. **142**, 102790 (2020)
2. Moloodpoor, M., Mortazavi, A., Ozbalta, N.: Thermal analysis of parabolic trough collectors via a swarm intelligence optimizer. Sol. Energy **181**, 264–275 (2019)
3. Mortazavi, A., Toğan, V.: Sizing and layout design of truss structures under dynamic and static constraints with an integrated particle swarm optimization algorithm. Appl. Soft Comput. **51**, 239–252 (2017)
4. Toğan, V., Mortazavi, A.: Sizing optimization of skeletal structures using teaching-learning based optimization. Optim. Control: Theor. Appl. **7**, 12 (2017)
5. Salomon, R.: Re-evaluating genetic algorithm performance under coordinate rotation of benchmark functions. A survey of some theoretical and practical aspects of genetic algorithms. Biosystems **39**, 263–278 (1996)
6. Grandhi, R.: Structural optimization with frequency constraints—a review. AIAA J **31**, 2296–2303 (1993)
7. Ghosh, A., Mallik, A.K.: Theory of Mechanisms and Machines. Affiliated East-West Press, New Delhi/Madras (1988)
8. Mortazavi, A.: Size and layout optimization of truss structures with dynamic constraints using the interactive fuzzy search algorithm. Eng. Optim. 1–23 (2020)
9. Mortazavi, A.: Comparative assessment of five metaheuristic methods on distinct problems. Dicle Univ. J. Eng. **10**, 879 (2019)
10. Tejani, G.G., Savsani, V.J., Patel, V.K., Mirjalili, S.: Truss optimization with natural frequency bounds using improved symbiotic organisms search. Knowl.-Based Syst. **143**, 162–178 (2018)
11. Pavithr, R.S., Gursaran, : Quantum Inspired Social Evolution (QSE) algorithm for 0-1 knapsack problem. Swarm Evol. Comput. **29**, 33–46 (2016)

12. Bekdaş, G., Nigdeli, S.M., Yang, X.-S.: Sizing optimization of truss structures using flower pollination algorithm. Appl. Soft Comput. **37**, 322–331 (2015)
13. Venkata Rao R., Saroj, A.: A self-adaptive multi-population based Jaya algorithm for engineering optimization, Swarm Evol. Comput. **37**, 1–26 (2017)
14. Rao, R.V., Savsani, V.J., Vakharia, D.P.: Teaching-learning-based optimization: a novel method for constrained mechanical design optimization problems. Comput.-Aided Design **43**, 303–315 (2011)
15. Patel, V.K., Savsani, V.J.: Heat transfer search (HTS): a novel optimization algorithm. Inf. Sci. **324**, 217–246 (2015)
16. Hasançebi, O., Teke, T., Pekcan, O.: A bat-inspired algorithm for structural optimization. Comput. Struct. **128**, 77–90 (2013)
17. Yang, X.-S.: Firefly algorithms for multimodal optimization. In: Watanabe, O., Zeugmann, T. (eds.) Stochastic Algorithms: Foundations and Applications, pp. 169–178. Springer, Heidelberg (2009)
18. Das, K.N., Singh, T.K.: Drosophila food-search optimization. Appl. Math. Comput. **231**, 566–580 (2014)
19. Javidy, B., Hatamlou, A., Mirjalili, S.: Ions motion algorithm for solving optimization problems. Appl. Soft Comput. **32**, 72–79 (2015)
20. Mortazavi, A., Toğan, V., Nuhoğlu, A.: Interactive search algorithm: a new hybrid metaheuristic optimization algorithm. Eng. Appl. Artif. Intell. **71**, 275–292 (2018)
21. Arora, S., Singh, S.: Butterfly optimization algorithm: a novel approach for global optimization. Soft. Comput. **23**, 715–734 (2019)

Total Potential Optimization Using Hybrid Metaheuristics: A Tunnel Problem Solved via Plane Stress Members

Yusuf Cengiz Toklu, Gebrail Bekdaş, Aylin Ece Kayabekir, Sinan Melih Nigdeli, and Melda Yücel

Abstract Total Potential Optimization using Metaheuristic Algorithms (TPO/MA) is an alternative structural analysis method starting with the same principles of the Finite Element Method (FEM). In TPO/MA as in FEM, the structure at hand is divided into finite parts. In these parts, if FEM is to be used, the equilibrium equations are written in a matrix form in local coordinates, then they are combined to give a matrix equation valid for the totality of the structure. The final step is solving this matrix equation to find the displacements in the structure. On the other hand, in TPO/MA, potential energies of the elements are written, then they are summed for the totality of the structure, yielding a functional to be minimized to find the equilibrium position according to the minimum potential energy principle. This minimization gives the displacements of the structure. That is why this method can also be called Finite Element Method with Energy Minimization (FEMEM). FEM is very efficient for linear systems, but for nonlinear systems the matrix obtained depends on material properties, displacements, and loads, i.e. it is not constant and may be ill-conditioned. This makes solutions difficult and sometimes impossible. It has been shown in the literature that TPO/MA can overcome these difficulties much more easily, and can solve problems that cannot be solved by FEM. In this study, TPO/MA is applied to tunnel problems with plane stress properties. Minimization process is applied to

Y. C. Toklu
Department of Civil Engineering, Beykent University, 34398 Sariyer, Istanbul, Turkey
e-mail: cengiztoklu@gmail.com

G. Bekdaş · A. E. Kayabekir · S. M. Nigdeli (✉) · M. Yücel
Department of Civil Engineering, Istanbul University - Cerrahpaşa, 34320 Avcılar, Istanbul, Turkey
e-mail: melihnig@istanbul.edu.tr

G. Bekdaş
e-mail: bekdas@istanbul.edu.tr

A. E. Kayabekir
e-mail: ecekayabekir@gmail.com

M. Yücel
e-mail: melda.yucel@yahoo.com.tr

S. M. Nigdeli et al. (eds.), *Advances in Structural Engineering—Optimization*, Studies in Systems, Decision and Control 326,
https://doi.org/10.1007/978-3-030-61848-3_8

several metaheuristic algorithms and hybrid ones, which are then compared with each other as to accuracy and precision.

Keywords Total potential optimization using metaheuristic algorithms · Finite element method with energy minimization · Harmony search · Flower pollination algorithm · Hybrid algorithms · Plane stress analysis

1 Introduction

For the analysis of structures, finite element method (FEM) is the most widely used one. By constructing a matrix equation including stiffness matrix, loads and deflections, the solution of unknown deflections can be found by solving the matrix system. This procedure works well for linear systems, but in nonlinear systems, the situation changes because additional iterative techniques become necessary [1–4]. Another technique can be forwarded by considering the principle that the total potential energy of a structural system becomes minimum in the equilibrium condition. In that case, if the total potential energy is minimized like an optimization problem, the deflections defined as design variables can be found. This theory is not limited to the type behavior of the structure, and the nonlinear problems can be solved directly.

According to this theory, Toklu proposed analysis of trusses trough energy minimization by using an adaptive local search for problems with material and geometrical nonlinearities [5]. Then, Toklu et al. proposed music inspired metaheuristic algorithm called harmony search (HS) for minimization of total potential energy [6]. The same methodology using total potential optimization using metaheuristic algorithms (TPO/MA) was also proposed for large truss structures with linear elastic members that have large deflections as to the existence of more than one solution which cannot be solved via FEM [7]. Temur et al. investigated the material nonlinearity of truss structures through TPO/MA by using teaching-learning-based optimization (TLBO) [8]. A truss-like structure called tensegric structures and used in architectural works, bridges, covering large areas including habitats in outer space was analyzed via TPO/MA using genetic algorithms [9]. Also, cable net systems were analyzed by using TPO/MA employing HS [10]. To solve problems with high number of freedoms, hybrid metaheuristic algorithms JAYA algorithm (JA), flower pollination algorithm (FPA) and TLBO [11]. Differently than truss like structures, structural members generated by plane-stress members were analyzed via TPO/MA for linear [12, 13] and nonlinear systems [14].

In the area of heuristic calculations, solving complex problems will be more important in the future than the metaphor of the employed algorithm [15]. For that reason, the best heuristic method combining several features of known algorithm is searched in this study instead of generating a new algorithm with a new name using the existing features of the other algorithms. The novel part of the work done in the study is to find best hybrid method for the plane stress analysis. In that case, it is possible to solve complex plane stress problems with high number of freedoms since plane stress analysis problem may have a lot of design variables compared to truss systems. It depends on the presentation of plane stress member with meshing elements. Thus, the stress problem can be analyzed in detail. In the study, a tunnel analysis example with 300 design variables which is double of the biggest truss system solved by TPO/MA [11].

2 The Total Potential Energy of Plane Stress Members

The structural systems are considered with plane-stress members by different meshing options. By meshing the systems are to triangular elements shown as Fig. 1. If the nodal displacements of ith node are respectively u_i and v_i at the nodes in x and y directions, the displacement fields such as u(x, y) and v(x, y) are respectively written as Eqs. (1) and (2) for consideration of a linear variation of the displacement over the triangle. In these equations, C_1–C_4 are constants, and these constants can be written according the strains if partial derivatives of displacement field are taken respect to x and y. These strains such as ε_x, ε_y and γ defining the normal strain in x and y direction and shear strain are respectively given as Eqs. (3)–(5).

$$u(x, y) = u_i + C_1 x + C_2 y \tag{1}$$

$$v(x, y) = v_i + C_3 x + C_4 y \tag{2}$$

$$\varepsilon_x = \frac{\partial u}{\partial x} = C_1 \tag{3}$$

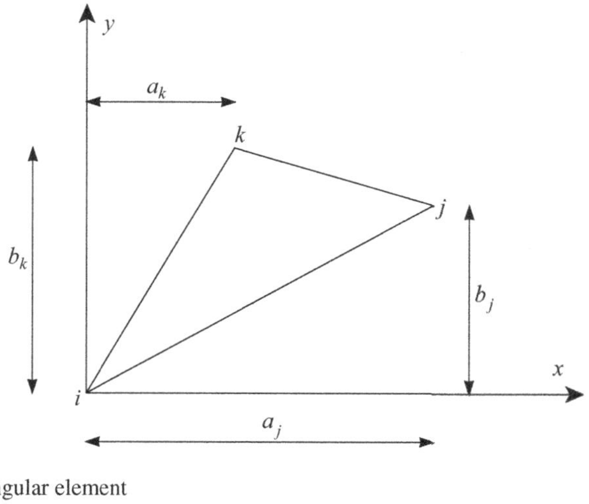

Fig. 1 The triangular element

$$\varepsilon_y = \frac{\partial v}{\partial y} = C_4 \tag{4}$$

$$\gamma = \frac{\partial u}{\partial y} + \frac{\partial v}{\partial x} = C_2 + C_3 \tag{5}$$

In the case of where the ith node is the origin point (0, 0), the nodal displacement of nodes i, j and k are respectively written as Eqs. (6)–(8) for defining distances; a_j, a_k, b_j and b_k shown in Fig. 1.

$$u(0,0) = u_i,\ \dot{v}(0,0) = v_i \tag{6}$$

$$u(a_j, b_j) = u_j,\ v(a_j, b_j) = v_j \tag{7}$$

$$u(a_k, b_k) = u_k,\ v(a_k, b_k) = v_k \tag{8}$$

As a matrix vector equation system, the relations between the nodal displacements are shown as Eq. (9).

$$\begin{bmatrix} u_j \\ v_j \\ u_k \\ v_k \end{bmatrix} = \begin{bmatrix} a_j & b_j & 0 & 0 \\ 0 & 0 & a_j & b_j \\ a_k & b_k & 0 & 0 \\ 0 & 0 & a_k & b_k \end{bmatrix} \begin{bmatrix} C_1 \\ C_2 \\ C_3 \\ C_4 \end{bmatrix} + \begin{bmatrix} u_i \\ v_i \\ u_i \\ v_i \end{bmatrix} \tag{9}$$

Solving matrix Eq. (9), the constants; C_1–C_4 can be found as:

$$C_1 = \frac{b_k(u_j - u_i)}{a_j b_k - a_k b_j} + \frac{b_j(u_k - u_i)}{a_k b_j - a_j b_k} \tag{10}$$

$$C_2 = \frac{a_k(u_j - u_i)}{a_k b_j - a_j b_k} + \frac{a_j(u_k - u_i)}{a_j b_k - a_k b_j} \tag{11}$$

$$C_3 = \frac{b_k(v_j - v_i)}{a_j b_k - a_k b_j} + \frac{b_j(v_k - v_i)}{a_k b_j - a_j b_k} \tag{12}$$

$$C_4 = \frac{a_k(v_j - v_i)}{a_k b_j - a_j b_k} + \frac{a_j(v_k - v_i)}{a_j b_k - a_k b_j} \tag{13}$$

After C_1, C_2, C_3, and C_4 are determined, strains can be calculated from Eqs. (3)–(5).

For a system, the strain density (e) is written as Eq. (14), where the normal stress in x and y direction and shear stress are respectively defined as σ_x, σ_y and τ, and these stresses are formulated as Eqs. (15)–(17) for a linear stress-strain relationship of two-dimensional planed stress. In these equations, E and ν are constant parameters respect to the material of the system and named as elasticity modulus and Poisson's ratio, respectively.

$$e = \int_{\varepsilon=0}^{\varepsilon} \sigma d\varepsilon = \frac{1}{2}\left(\sigma_x \varepsilon_x + \sigma_y \varepsilon_y + \tau\gamma\right) \tag{14}$$

The right-hand side of Eq. (14) is valid for linear systems where stresses and strains are uniform over an element. For nonlinear structures, the integral in the equation should be evaluated.

$$\sigma_x = \frac{E}{1 - \nu^2}(\varepsilon_x + \nu\varepsilon_y) \tag{15}$$

$$\sigma_y = \frac{E}{1 - \nu^2}(\varepsilon_y + \nu\varepsilon_x) \tag{16}$$

$$\tau = \frac{E}{1 - \nu^2}\left(\frac{1 - \nu}{2}\gamma\right) \tag{17}$$

Strain energy density of an element, e, can be calculated from Eq. (14) using strains from Eqs. (3)–(5) and stresses from Eqs. (15)–(17). The strain energy of a single element (U_i) is written as Eq. (18), if the strain energy density of the element and volume of element labelled as m are e_m and V_m, respectively.

$$U_i = e_E V_m \tag{18}$$

The volume of the element m is written as Eq (19) which is the multiplication of thickness (t) and area of the triangle element.

$$V_m = \frac{(a_j b_k - a_k b_j)t}{2} \tag{19}$$

To find the total strain energy of the system, the strain energies of all elements are summed for a system consisting of n elements (Eq. (20).

$$U = \sum_{m=1}^{n} U_m \tag{20}$$

The total potential energy of the system (Π_p) is found by subtracting the work done by external forces from the total strain energy. If the number of nodes is p and the point loads are defined as P_{xi} and P_{yi} for the forces applied ith node in x and y directions, Π_p can be written as Eq. (21). For the minimum values Π_p, the system is in the static equilibrium.

$$\Pi_p = U - \sum_{i=1}^{p} (P_{xi}u_i + P_{yi}v_i) \tag{21}$$

3 The Analysis Methodology

By using TPO/MA, the analysis problems are handled as an optimization problem in which total potential energy is minimized to find the unknown nodal displacements taken as design variables. The application area of TPO/MA is increased by adding plates and plate-like structures. Now, it is the time to demonstrate the ability of the method on complex systems that cannot be directly handled and solving big structural systems with a high number of freedoms.

The steps of the TPO/MA for plane-stress problems until on optimization problem that considers the minimization of total potential energy are as follows:

- Divide the system into triangle defined by nodal coordinates and have a nodal displacement of each node.
- Calculate the strain energy of each triangle according to Eq. (18) by considering fixed nodes.
- Calculate the total strain energy according to Eq. (20) by summing the energy value of all members.
- Calculate the work done by external forces and find the total potential energy according to Eq. (21).

Then, the problem is handled as an optimization problem and the nodal displacement (u_i and v_i for 1 to the number of free nodes; p) are assigned with candidate values according to the rules of the employed metaheuristic algorithm. The number of design variables is equal to s which is calculated according to Eq. (22) where p defines the number of fixed nodes and pf defines the number of fixed displacements.

$$s = 2p - p_f \tag{22}$$

All hybrid algorithms proposed in the study are two phase algorithms and a general flowchart of TPO/MA for a two-phase algorithm is shown as Fig. 2.

3.1 Hybrid Methods

Algorithms used in this study are harmony search (HS), flower pollination (FPA), teaching learning-based optimization (TLBO) and Jaya algorithm (JA). JA was developed by Rao [16] is a single-phase algorithm and it have a great convergence ability in optimization problems including analysis via TPO/MA. Despite of that, JA may

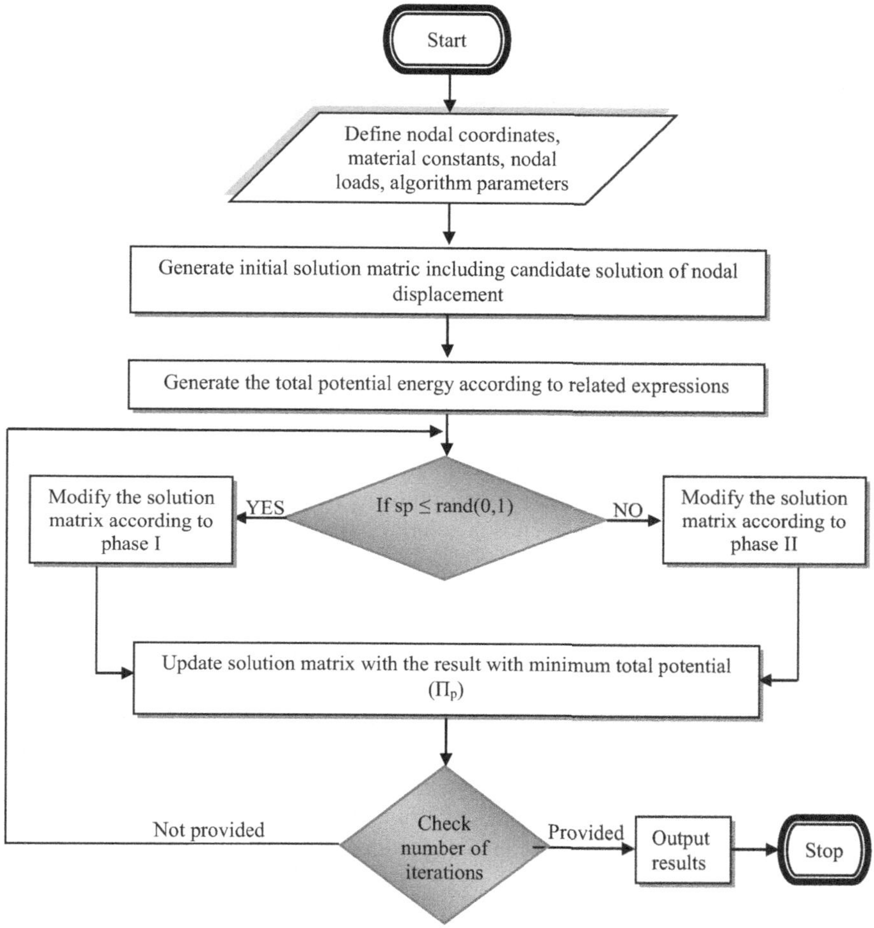

Fig. 2 The flowchart of the TPO/MA for two-phase metaheuristic algorithms

get trapped in a local optimum for a big optimization problem since it is a single-phase algorithm. For that reason, a second phase is added to the JA by hybridizing it with the other algorithms.

The formulation of JA on providing of a new candidate solution (X_{new}) is given as Eq. (23).

$$X_{new} = X_{old} + \text{rand}(0,\ 1)(g^* - X_{old}) - \text{rand}(0,\ 1)(g^w - X_{old}) \tag{23}$$

The existing solution (X_{old}) is modified by using the solutions with minimum (g^*) and maximum (g^w) total potential energy values. A random number between 0 and 1 is shown as rand(0, 1). The good convergence ability is provided by using the best and worst solutions (g^* and g^w).

The trials made during this study has shown that JA may get trapped into a local optimum for problems with a high number of variables. For that reason, four different hybrid algorithms are proposed in this study.

In the first hybrid algorithm, the music inspired metaheuristic algorithm called harmony search (HS) is combined with JA. HS is developed by Geem et al. [17] and it imitates the improvising performance of a composer to gain attention of the audience. This algorithm is a two phase one including global and local search phase. In global phase, random solutions are produced by using all solution ranges. In local phase, one of the existing solutions is used and neighborhood values are scanned. In the developed hybrid algorithm, the local search phase of HS was used as phase II. The formulation of the phase II is shown as in Eq. (24).

$$X_{new} = X_{ran} + PAR(2rand(0, 1) - 1)X_{ran} \tag{24}$$

In Eq. (24), a randomly chosen candidate solution within the existing ones is shown with X_{ran} and the parameter "pitch adjusting rate" (PAR) is a ratio controlling the acceptable range around the chosen solution. This first hybrid method is named as the Jaya Algorithm with HS (JA-HS).

As the second hybrid algorithm, JA is combined with the student phase of TLBO. TLBO imitates the education process including teacher and student phases [18]. In the teacher phase, the education process conducted by a teacher is formalized by using the best solution of the existing ones. This phase is similar to single phase of JA since the best solution is also used. For that reason, the student phase of TLBO using two randomly chosen solutions (X_{rand1} and X_{rand2}) from the existing ones are used as the second phase. The student phase or learner phase is formulized as seen in Eq. (25). The equation is updated according to objective function solutions ($f(X_{rand1})$, $f(X_{rand2})$). This hybrid algorithm is named as Jaya Algorithm with Probabilistic Student Phase (JA1SP).

$$X_{new} = \begin{cases} X_{old} + rand(0, 1)(X_{rand1} - X_{rand2});\ f(X_{rand1}) > f(X_{rand2}) \\ X_{old} + rand(0, 1)(X_{rand2} - X_{rand1});\ f(X_{rand1}) < f(X_{rand2}) \end{cases} \tag{25}$$

As the third hybrid algorithm, the rand(0, 1) value of JA phase of JA1SP is updated with a random number generated according to a Levy distribution (L) as seen in Eq. (26). Thus, three algorithms such as JA, TLBO and FPA are combined in an algorithm called Jaya Algorithm with Levy Flight and Probabilistic Student Phase (JALSP). Levy distribution is used in the global pollination phase of FPA developed by Yang [19]. In FPA, the Levy distribution is used because of the imitation of movement of pollinators. By using this hybrid algorithm, the performance of the method is tested on different random distribution than linear distribution.

$$X_{new} = X_{old} + L(g^{*} - X_{old}) - L(g^{w} - X_{old}) \tag{26}$$

Lastly, a hybrid algorithm combining student phase of TLBO and local search of HS (SPHS) was used. In the formulations of the phase using the local search of HS, the best existing solution is used instead of a randomly chosen solution. The student phase of TLBO already uses two randomly chosen solutions. For that reason, the best solution is used for the other phase as seen in Eq. (27).

$$X_{new} = g^{*} + PARg^{*}(2rand(0, 1) - 1) \tag{27}$$

For all algorithms, the feature of choosing a phase in an iteration used in HS and FPA by using parameters such as harmony memory considering rate (HMCR) or switch probability (sp) is used. In that case, all hybrid algorithms include features of more than two algorithms.

4 Numerical Examples

The method is presented for a tunnel structure, but a cantilever beam example is also given to show the importance on using hybrid metaheuristic algorithms. In that case, hybrid metaheuristic algorithms are needed for problems when the number of design variables creates an obstacle for single metaheuristic algorithms in search of the best solution and in reaching a high level of robustness. All solutions were compared with finite element method (FEM) results. Harmony memory considering rate (HMCR) or switch probability (sp) was taken as 0.5 in all cases. PAR was taken as 0.1 in all analyses.

4.1 Cantilever Beam Under a Concentered Load

The cantilever beam shown as Fig. 3 was tested for two different meshing options. The first option given as Fig. 4 has 15 free nodes, the number of variables is 30. As seen in Table 1, single metaheuristic algorithms are effective to find the sane solution with FEM.

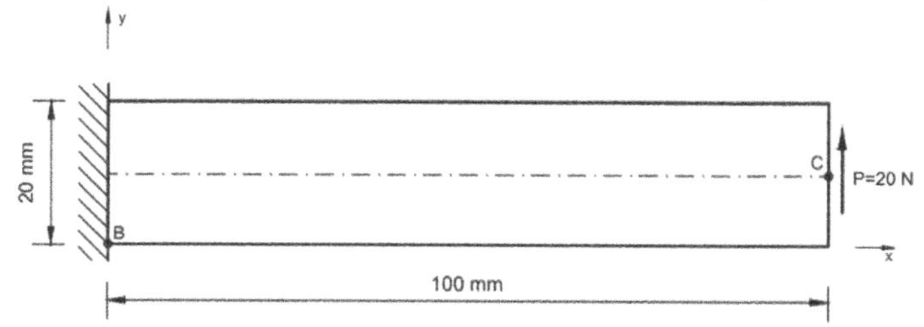

Fig. 3 Structure 1: cantilever beam [20, 21]

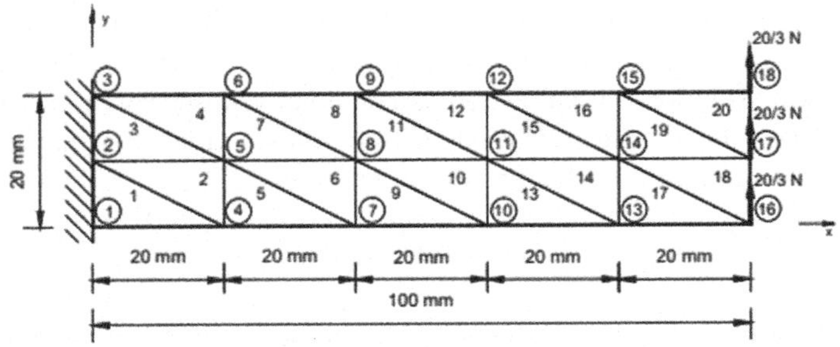

Fig. 4 Mathematical model prepared for the analyses of cantilever beam (option 1) [12]

When the system was investigated for the second option (Fig. 5), with 200 variables, the single algorithms are not effective to find the same results with FEM. As seen in Table 2, it is possible to find the exact solution by using JA1SP.

4.2 Tunnel System

The tested example is a tunnel system shown in Fig. 6 [22]. Considering its symmetry, only half of the system which has 150 nodes are analyzed (see Fig. 7). In that case, when the fixed displacements are known, 265 design variables are considered in the study. For nodes 145–150, −400,000 N external loads in y direction are taken.

The shaded area in Fig. 6 and the bold members in Fig. 7 represents concrete members while the other part of the system is rock. The elasticity modulus and Poisson ratio of the system are 33,000 N/mm^2 and 0.2, respectively for the concrete, while these values are respectively 100 N/mm^2 and 0.3 for rock. The thickness of the tunnel system is 1 m.

As seen in Table 3, JA1SP is the best method in minimization of the energy. The worst hybrid algorithm is SPHS, but it has similar displacements with JA1SP. The displacement plots of JA1SP and SPHS are given as Figs. 8 and 9, respectively. The match with the FEM results are clearly seen.

5 Conclusion and Future Studies

TPO/MA is an alternative method that can be applied to analysis of general structural systems. This has been proved until now with applications on trusses, cable net systems, tensegric structure and plates. The capacity of solving problem via TPO/MA can be increased by using the hybrid metaheuristic method. As presented in this study, while single metaheuristics are effective in solving a cantilever beam problem within 30 variables, hybrid method can solve the option with frequent meshing with 200

Table 1 Solutions for the cantilever beam (option 1) [12]

Node	TPOMA (FPA)		TPOMA (JA)		TPOMA (HS)		TPOMA (HHS)		TPOMA (TLBO)		FEM	
	Δx (mm)	Δy (mm)	Δx (mm)	Δy (mm)	Δx (mm)	Δy (mm)	Δx (mm)	Δy (mm)	Δx (mm)	Δy (mm)	Δx (mm)	Δy (mm)
4	0.19369	0.25151	0.19369	0.25151	0.20715	0.27089	0.01926	0.02499	0.19369	0.25151	0.1937	0.2515
5	0.00059	0.24176	0.00059	0.24176	−0.00003	0.26078	0.00006	0.02402	0.00059	0.24176	0.0006	0.2418
6	−0.19912	0.25858	−0.19912	0.25858	−0.21326	0.27713	−0.01980	0.02569	−0.19912	0.25858	−0.1991	0.2586
7	0.34701	0.84814	0.34701	0.84814	0.36992	0.90709	0.03452	0.08432	0.34701	0.84814	0.347	0.8481
8	−0.00314	0.8425	−0.00314	0.8425	−0.0046	0.90067	−0.00031	0.08375	−0.00314	0.8425	−0.0031	0.8425
9	−0.35408	0.85627	−0.35408	0.85627	−0.37917	0.91413	−0.03522	0.08512	−0.35408	0.85627	−0.3541	0.8563
10	0.4547	1.70967	0.4547	1.70967	0.48581	1.83084	0.04525	0.17003	0.4547	1.70967	0.4547	1.7097
11	−0.00576	1.70704	−0.00576	1.70704	−0.00686	1.82991	−0.00057	0.16976	−0.00576	1.70704	−0.0058	1.707
12	−0.46619	1.71762	−0.46619	1.71762	−0.50155	1.8422	−0.04639	0.17082	−0.46619	1.71762	−0.4662	1.7176
13	0.51858	2.74696	0.51858	2.74696	0.55003	2.93925	0.05162	0.27326	0.51858	2.74696	0.5186	2.747
14	−0.0084	2.74754	−0.0084	2.74754	−0.01011	2.94063	−0.00083	0.27331	−0.0084	2.74754	−0.0084	2.7475
15	−0.53422	2.75474	−0.53422	2.75473	−0.57002	2.94848	−0.05317	0.27403	−0.53422	2.75473	−0.5342	2.7547
16	0.53861	3.87428	0.53861	3.87428	0.57192	4.13881	0.05362	0.38549	0.53861	3.87428	0.5386	3.8743
17	−0.01054	3.87495	−0.01054	3.87495	−0.01274	4.14002	−0.00105	0.38556	−0.01054	3.87495	−0.0105	3.8749
18	−0.55939	3.88219	−0.55939	3.88219	−0.59412	4.14664	−0.05569	0.38628	−0.55939	3.88219	−0.5594	3.8822
Π_p (Nmm)	−388		−388		−386		−388		−388		−388	
Analysis number	999103		672375		999343		531810		987341		–	

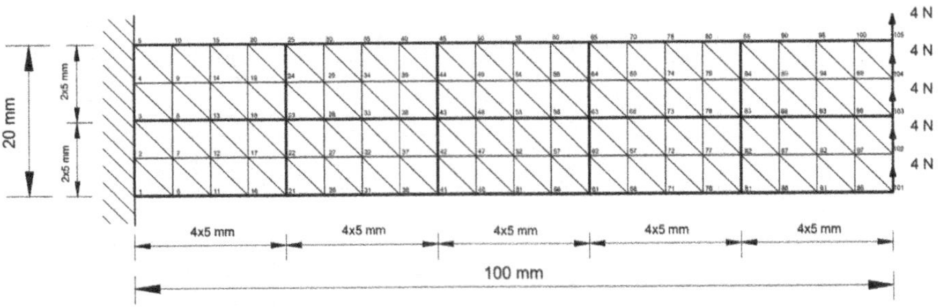

Fig. 5 Mathematical model prepared for the analyses of cantilever beam (option 2) [13]

Table 2 Solutions for the cantilever beam (option 2) [13]

	TPOMA (FPA)	TPOMA (JA)	TPOMA (TLBO)	TPOMA (JA1SP)	FEM
Π_p (Nmm)	−84.3	−5.7	−51.4	−84.5	−84.5

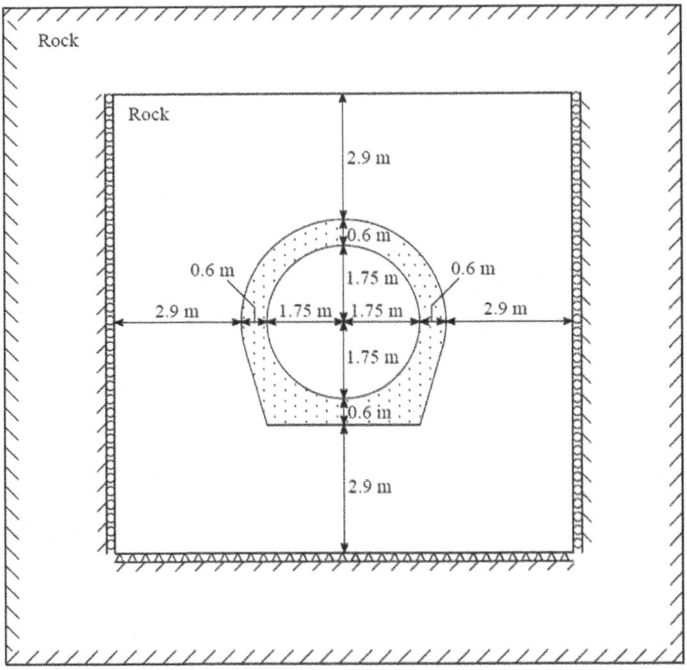

Fig. 6 The tunnel system

variables. Also, the hybrid ones are effective to solve the tunnel system with 265 design variables.

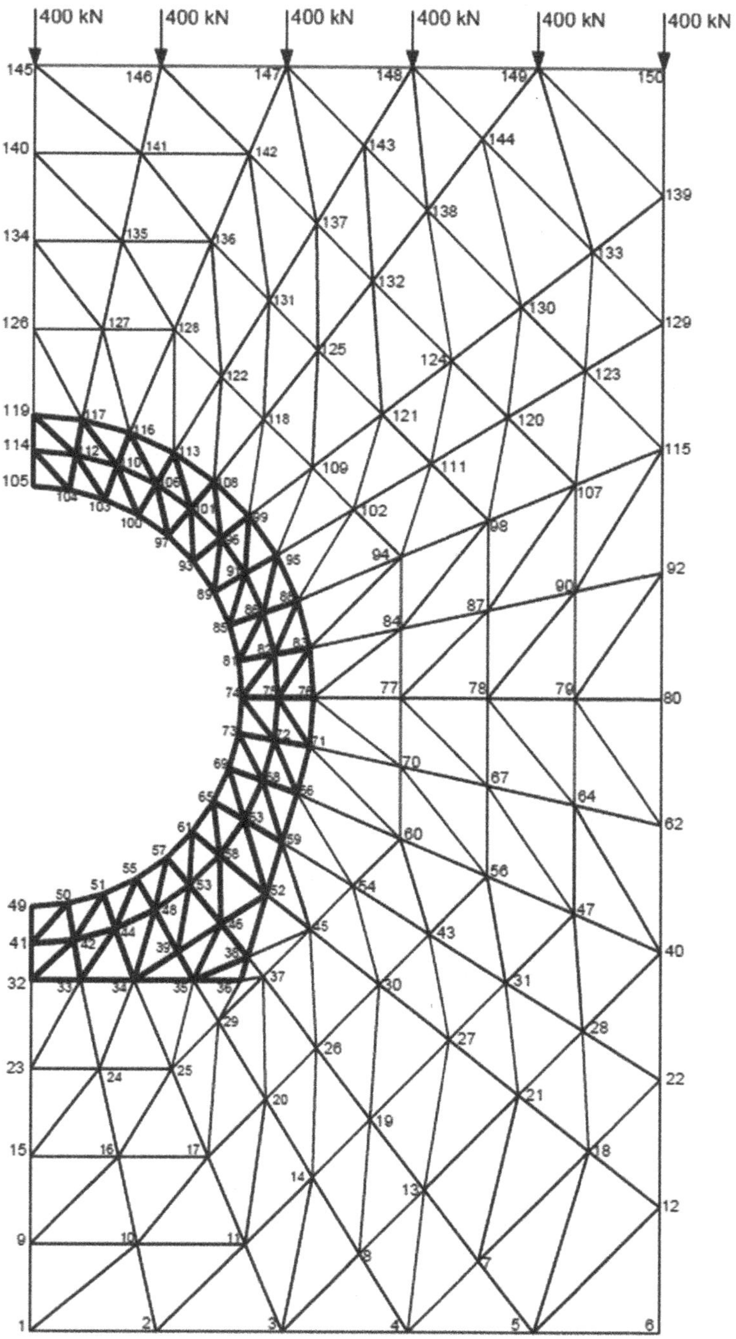

Fig. 7 Analysis model of the tunnel system

Table 3 Solutions for the tunnel system

	TPOMA (JA1SP)	TPOMA (JALSP)	TPOMA (SPHS)	TPOMA (JA-HS)	FEM
Π_p (Nm)	−40339.39	−40255.739	−40241.99	−40308.64	−39791.31

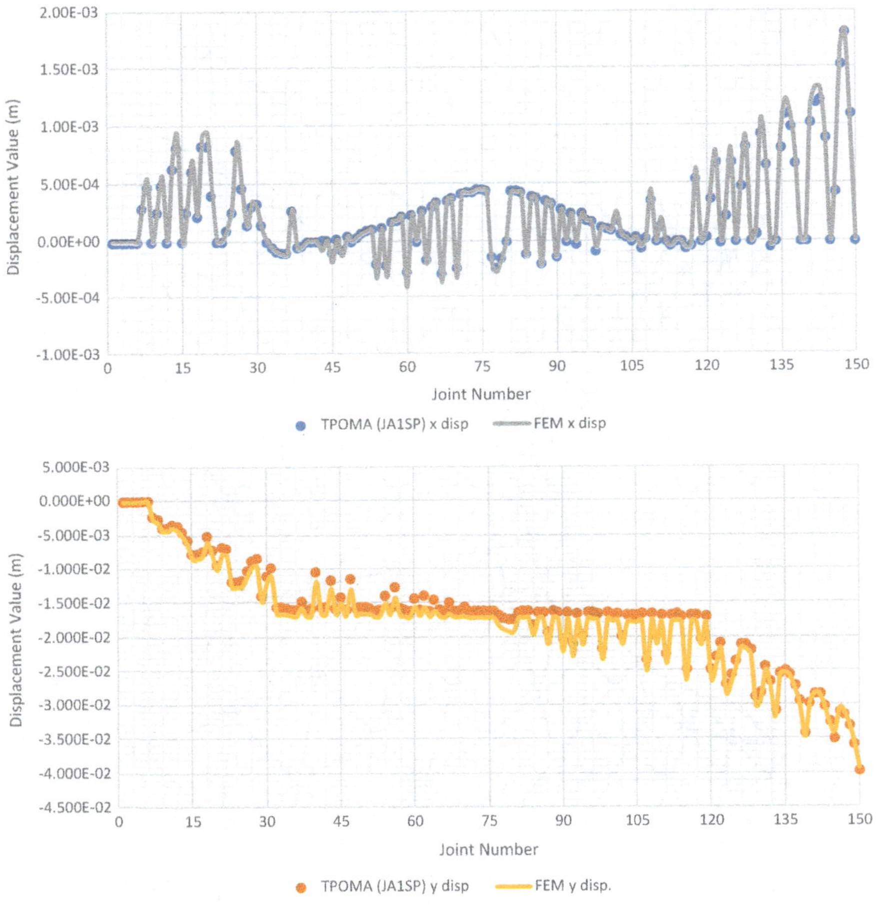

Fig. 8 Nodal displacement of structure 5 for both directions (JA1SP)

It is still needed to investigate various structures with various combinations of metaheuristics, since a proposed algorithm may outperform another one in an example, while it is vice versa for the second example according to no-free lunch theorem [23].

The proposed four hybrid algorithms were tested on tunnel problem presented as Structure 5 and all hybrid algorithms are seen to be effective in finding lower energy value than FEM. In that case, TPO/MA has proved itself more effective in

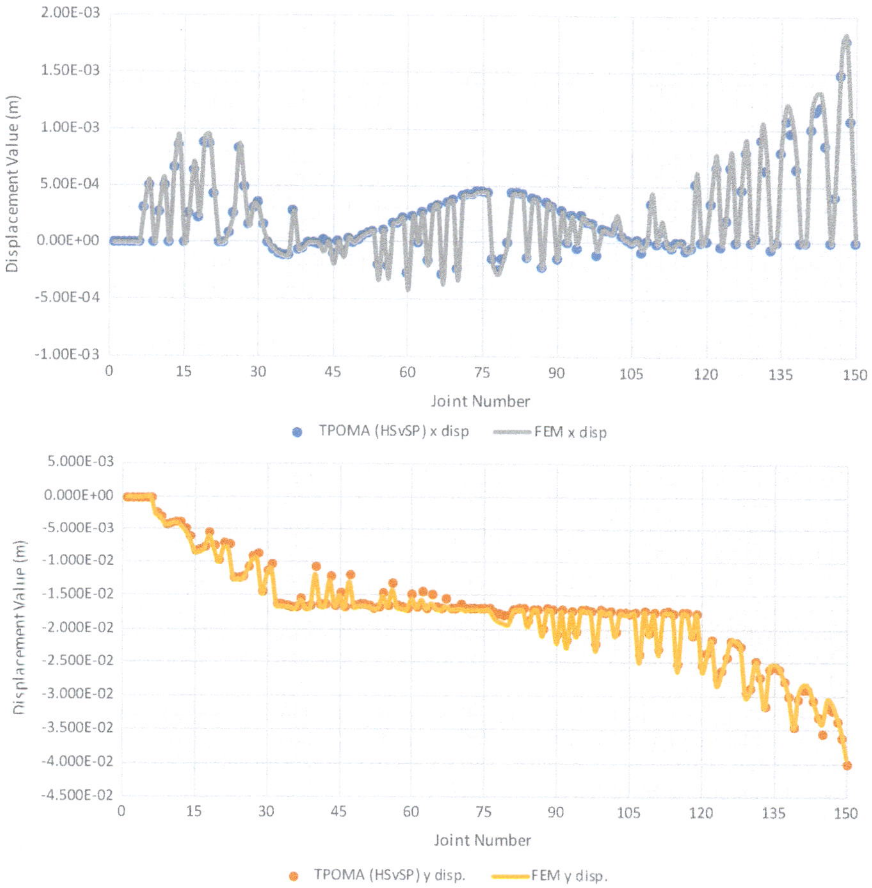

Fig. 9 Nodal displacement of structure 4 for both directions (SPHS)

second-order analysis. The algorithm JA1SP is the best one in minimization of the total potential energy.

References

1. Levy, R., Spillers, W.R.: Analysis of Geometrically Nonlinear Structures, 2nd edn. Kluwer Academic Publishers, Dordtrecht (2003)
2. Saffari, H., Fadaee, M.J., Tabatabaei, R.: Nonlinear analysis of space trusses using modified normal flow algorithm. J. Struct. Eng. **134**(6), 998–1005 (2008)
3. Krenk, S.: Non-linear Modeling and Analysis of Solids and Structures. Cambridge University Press, Cambridge (2009)
4. Greco, M., Menin, R.C.G., Ferreira, I.P., Barros, F.B.: Comparison between two geometrical nonlinear methods for truss analyses. Struct. Eng. Mech. **41**(6), 735–750 (2012)

5. Toklu, Y.C.: Nonlinear analysis of trusses through energy minimization. Comput. Struct. **82**(20–21), 1581–1589 (2004)
6. Toklu, Y.C., Bekdas, G., Temur, R.: Analysis of trusses by total potential optimization method coupled with harmony search. Struct. Eng. Mech. **45**(2), 183–199 (2013)
7. Toklu, Y.C., Temur, R., Bekdaş, G.: Computation of Non-unique Solutions for Trusses Undergoing Large Deflections. Int. J. Comput. Methods **12**(3), 1550022 (2015)
8. Temür, R., Bekdaş, G., Toklu, Y.C.: Total potential energy minimization method in structural analysis considering material nonlinearity. Challenge J. Struct. Mech. **3**(3), 129–133 (2017)
9. Toklu, Y.C., Uzun, F.: Analysis of tensegric structures by total potential optimization using metaheuristic algorithms. J. Aerosp. Eng. **29**(5), Sept 04016023 (2016)
10. Toklu, Y.C., Bekdaş, G., Temur, R.: Analysis of cable structures through energy minimization. Struct. Eng. Mech. **62**(6), 749–758 (2017)
11. Bekdas, G., Kayabekir, A.E., Nigdeli, S.M., Toklu, Y.C.: Advanced energy based analyses of trusses employing hybrid metaheuristics. Struc. Des. Tall Spec. Build. **28**(9), e1609 (2019)
12. Kayabekir, A.E., Toklu, Y.C., Bekdaş, G., Nigdeli, S.M., Yücel, M., Geem, Z.W.: A novel hybrid harmony search approach for the analysis of plane stress systems via total potential optimization. Appl. Sci. **10**(7), 2301 (2020)
13. Toklu, Y.C., Bekdaş, G., Kayabekir, A.E., Nigdeli, S.M., Yucel, M.: Total potential optimization using metaheuristics: analysis of cantilever beam via plane-stress members. In: 6th International Conference on Harmony Search, Soft Computing and Applications (ICHSA 2020), 16–17 July. Istanbul, Turkey (2020)
14. Toklu, Y.C., Kayabekir, A.E., Bekdaş, G., Nigdeli, S.M., Yücel, M.: Analysis of plane stress systems via total potential optimization method considering non-linear behavior. J. Struc. Eng. **146**(11). https://doi.org/10.1061/(asce)st.1943-541x.0002808
15. Sörensen, K., Sevaux, M., Glover, F.: A history of metaheuristics. Handbook of Heuristics, 1–18 (2018)
16. Rao, R.: Jaya: A simple and new optimization algorithm for solving constrained and unconstrained optimization problems. Int. J. Ind. Eng. Comput. **7**(1), 19–34 (2016)
17. Geem, Z.W., Kim, J.H., Loganathan, G.V.: A new heuristic optimization algorithm: harmony search. Simulation **76**, 60–68 (2001)
18. Rao, R.V., Savsani, V.J., Vakharia, D.P.: Teaching–learning-based optimization: a novel method for constrained mechanical design optimization problems. Comput. Aided Des. **43**(3), 303–315 (2011)
19. Yang, X.S.: Flower pollination algorithm for global optimization. In: Unconventional Computation and Natural Computation, pp. 240–249 (2012)
20. Cook, R.D.: Finite Element Modelling for Stress Analysis. John Wiley & Sons, Inc., USA, 0-471-10774-31995
21. Gazi H.: Sonlu Elemanlar Yönemi ile Çalışan Bir Bilgisayar Programının Geliştirilmesi ve Düzlem Gerilme—Düzlem Deformasyon Problemlerinin Analizi, MSc Thesis, Istanbul University, Turkey
22. Weawer, W., Johnston, P.R.: Finite Element for Structural Analysis. Prentice—Hall, Inc., New Jersey (1984)
23. Wolpert, D., Macready, W.G.: No free lunch theorems for optimization. IEEE Trans. Evol. Comput. **1**(1), 67–82 (1997)

Buckling Analysis and Stacking Sequence Optimization of Symmetric Laminated Composite Plates

Celal Cakiroglu and Gebrail Bekdaş

Abstract Symmetric laminated composite plates are frequently used in structural design due to their high strength to weight ratio. The stacking sequence of these laminates is known to have a major impact on the performance of these structural members. As a measure of performance often the buckling load of a laminated composite plate is used. In this study the buckling load of a laminated composite plate is obtained using analytical techniques. In order to optimize the structural performance a newly developed and very efficient meta-heuristic technique called the Jaya algorithm has been utilized. The fiber orientation angles of the plies as well as the ply thicknesses are regarded as continuous valued design variables and allowed to vary randomly while keeping the symmetry of the configuration which resulted in dispersed symmetric plates. CFRP is chosen as the plate material due to being the most frequently used composite material in the industry. The material properties and plate aspect ratio are treated as constant values. It was shown that allowing the fiber orientation angles to take values other than the commonly used 0°, ± 45°, 90° combined with proper thickness sequence can greatly enhance the structural performance.

Keywords Buckling · Optimization · Laminated composite plates

1 Introduction

Fiber reinforced composite materials found broad application in structural engineering due to their superior stiffness and weight properties compared to more traditional materials such as aluminum. Particularly these materials found application

C. Cakiroglu (✉)
Department of Civil Engineering, Turkish-German University, Sahinkaya Cad 86, Istanbul 34820, Turkey
e-mail: cakiroglu@tau.edu.tr

G. Bekdaş
Department of Civil Engineering, Istanbul University, Cerrahpaşa, Istanbul 34310, Turkey
e-mail: bekdas@istanbul.edu.tr

S. M. Nigdeli et al. (eds.), *Advances in Structural Engineering—Optimization*, Studies in Systems, Decision and Control 326,
https://doi.org/10.1007/978-3-030-61848-3_9

in the laminated composite plates due to the advantages they provide in terms of weight reduction and stiffness increase of aircraft panels. Due to the high slenderness of these structural members, their buckling behaviour has been extensively researched. Barakat et al. [1] optimized CFRP(carbon fiber reinforced polymer) and boron/epoxy laminates to achieve the maximum buckling load using a sequential linear programming technique. Similarly, De Almeida et al. [2, 3] worked on the buckling load maximization of laminates. In [2, 3] a well established meta-heuristic algorithm called harmony-search algorithm and genetic algorithms were used for the optimization. The buckling load was computed using finite element analysis. Another meta-heuristic algorithm used in the optimization of laminates is the ant colony algorithm which was applied by Abachizadeh et al. [4, 5]. In [4, 5] a multi-objective optimization scheme was applied to symmetric hybrid laminates where the fundamental frequency and weight of the laminate were optimized simultaneously. Similarly, Wang et al. [6] applied a modified ant colony algorithm to the buckling load maximization problem. The ant colony algorithm was also applied by Aymerich and Serra [7, 8] to the buckling load optimization problem of a laminate with a constant ply thickness sequence.

The current study deals with the buckling load maximization of a rectangular symmetric laminate laminate made of nine layers. The number of layers, the total plate thickness, the side lengths as well as the mechanical properties of the plate are based on the study carried out by Führer [9]. In this current study the buckling loads were calculated using analytical methods and later used in the Jaya optimization scheme. The following sections give an overview of these analytical techniques together with their theoretical background.

Fiber reinforced laminae have different elasticity properties in the fiber direction and in the transverse directions. Therefore they are classified as orthotropic materials [10]. In order to define the mechanical properties of fiber reinforced laminae, two different moduli of elasticity are defined in the fiber direction and the transverse direction. These moduli are denoted with E_1 and E_2 respectively. The shear modulus is denoted with G_{12}. In the following text the fiber direction and the transverse direction are shown with 1 and 2 respectively. In order to describe the stress–strain relationship of a lamina we need the Poisson's ratio which relates strains in the transverse direction to the strains in the longitudinal direction and vice versa. The in-plane strains and stresses shown in Fig. 1 are related to each other as shown in Eq. (1) [11].

$$\nu_{12} = \frac{\varepsilon_2}{\varepsilon_1},\ \nu_{21} = \frac{\varepsilon_1}{\varepsilon_2},\ \varepsilon_1 = \frac{\sigma_1}{E_1} - \nu_{21}\frac{\sigma_2}{E_2},\ \varepsilon_2 = \frac{\sigma_2}{E_2} - \nu_{12}\frac{\sigma_1}{E_1} \tag{1}$$

$$\tau_{12} = \gamma_{12} G_{12}$$

The relationships in Eq. (1) can be combined in matrix form to give Eq. (2),

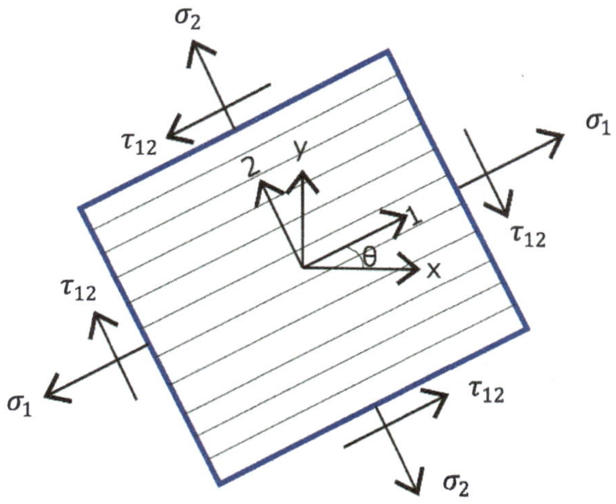

Fig. 1 Fibre direction and transverse direction of a lamina are denoted with the indices 1 and 2 respectively

$$\begin{Bmatrix} \varepsilon_1 \\ \varepsilon_2 \\ \gamma_{12} \end{Bmatrix} = \begin{bmatrix} \frac{1}{E_1} & -\frac{\nu_{21}}{E_2} & 0 \\ -\frac{\nu_{12}}{E_1} & \frac{1}{E_2} & 0 \\ 0 & 0 & \frac{1}{G_{12}} \end{bmatrix} \begin{Bmatrix} \sigma_1 \\ \sigma_2 \\ \tau_{12} \end{Bmatrix} \tag{2}$$

The matrix in Eq. (2) is called the compliance matrix and usually denoted with [S]. By taking the inverse of [S] we can obtain the stiffness matrix [Q] and the stress strain relationship in Eq. (3),

$$\begin{Bmatrix} \sigma_1 \\ \sigma_2 \\ \tau_{12} \end{Bmatrix} = \begin{bmatrix} \frac{E_1}{1-\nu_{12}\nu_{21}} & \frac{\nu_{21}E_1}{1-\nu_{12}\nu_{21}} & 0 \\ \frac{\nu_{12}E_2}{1-\nu_{12}\nu_{21}} & \frac{E_2}{1-\nu_{12}\nu_{21}} & 0 \\ 0 & 0 & G_{12} \end{bmatrix} \begin{Bmatrix} \varepsilon_1 \\ \varepsilon_2 \\ \gamma_{12} \end{Bmatrix} \tag{3}$$

The stiffness matrix [Q] in Eq. (3) denotes the stress–strain relationship with respect to the 1–2 axes. If there is an angle θ between the fiber direction and the x-axis as shown in Fig. 1, then the stiffness matrix needs to be tranformed to the x–y coordinate system. This can be done with the help of the transformation matrices. The transformation of stresses between coordinate systems is shown in Eq. (4) [11],

$$\begin{Bmatrix} \sigma_x \\ \sigma_y \\ \tau_{xy} \end{Bmatrix} = \begin{bmatrix} \cos^2\theta & \sin^2\theta & -2\sin\theta\cos\theta \\ \sin^2\theta & \cos^2\theta & 2\sin\theta\cos\theta \\ \sin\theta\cos\theta & -\sin\theta\cos\theta & \cos^2\theta - \sin^2\theta \end{bmatrix} \begin{Bmatrix} \sigma_1 \\ \sigma_2 \\ \tau_{12} \end{Bmatrix} \tag{4}$$

The transformation matrix in Eq. (4) is denoted with [T]. The transformation of strains can be done in the same way. If we denote all vectors and matrices with respect to the x–y coordinate system with a bar, we obtain Eq. (5)

$$\begin{gathered}[\overline{\sigma}] = [T][\sigma] = [T][Q][\varepsilon] = [T][Q]\left[T^{-1}\right][\overline{\varepsilon}] \\ [T][Q]\left[T^{-1}\right] = \left[\overline{Q}\right] \Rightarrow [\overline{\sigma}] = \left[\overline{Q}\right][\overline{\varepsilon}]\end{gathered} \tag{5}$$

Figure 2 shows the normal and shear stress resultants acting on the cross section of an infinitesimal rectangle taken from a laminated composite plate. The stress resultants in Fig. 2 are computed through the integration of normal and shear stresses over the thickness of the laminate as in Eq. (6) [11]. The notation used in the numbering of plies in a general laminated composite plate can be seen in Fig. 3.

$$\begin{gathered}\begin{Bmatrix} N_x \\ N_y \\ N_{xy} \end{Bmatrix} = \sum_{k=1}^{N} \int_{z_{k-1}}^{z_k} \begin{Bmatrix} \sigma_x \\ \sigma_y \\ \tau_{xy} \end{Bmatrix}_k dz = \sum_{k=1}^{N} \left[\overline{Q}\right]_k \left(\begin{Bmatrix} \varepsilon_x^0 \\ \varepsilon_y^0 \\ \gamma_{xy}^0 \end{Bmatrix} \int_{z_{k-1}}^{z_k} dz + \begin{Bmatrix} \kappa_x \\ \kappa_y \\ \kappa_{xy} \end{Bmatrix} \int_{z_{k-1}}^{z_k} z dz \right) \\ \begin{Bmatrix} M_x \\ M_y \\ M_{xy} \end{Bmatrix} = \sum_{k=1}^{N} \int_{z_{k-1}}^{z_k} \begin{Bmatrix} \sigma_x \\ \sigma_y \\ \tau_{xy} \end{Bmatrix}_k z dz = \sum_{k=1}^{N} \left[\overline{Q}\right]_k \left(\begin{Bmatrix} \varepsilon_x^0 \\ \varepsilon_y^0 \\ \gamma_{xy}^0 \end{Bmatrix} \int_{z_{k-1}}^{z_k} z dz + \begin{Bmatrix} \kappa_x \\ \kappa_y \\ \kappa_{xy} \end{Bmatrix} \int_{z_{k-1}}^{z_k} z^2 dz \right)\end{gathered} \tag{6}$$

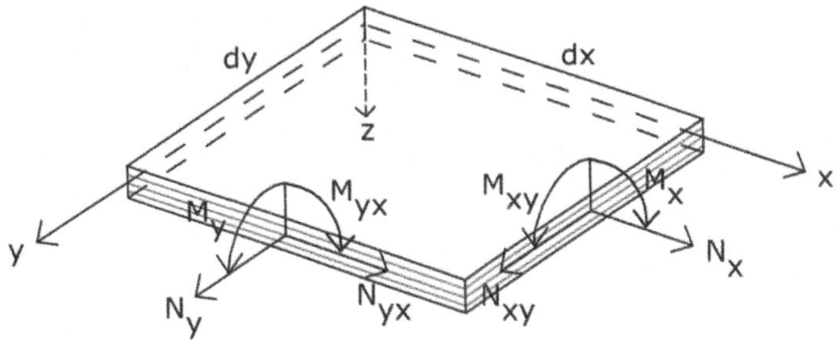

Fig. 2 Forces and moments on a laminate cross-section [11]

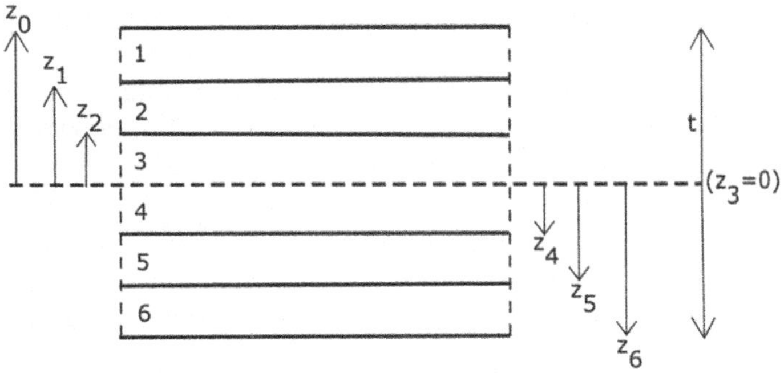

Fig. 3 Numbering convention for layers and their z-coordinates in a laminate with 6 layers [11]

Equation (6) shows the integration of the resultant forces for a general laminate with N layers where $\left[\varepsilon^0\right]$ is the vector of strain values at z = 0 and $[\kappa]$ is the vector of curvatures that satisfies the relationship in Eq. (7).

$$\left\{\begin{array}{c}\varepsilon_x\\ \varepsilon_y\\ \gamma_{xy}\end{array}\right\}=\left\{\begin{array}{c}\varepsilon_x^0\\ \varepsilon_y^0\\ \gamma_{xy}^0\end{array}\right\}+z\left\{\begin{array}{c}\kappa_x\\ \kappa_y\\ \kappa_{xy}\end{array}\right\}=\left\{\begin{array}{c}\frac{\partial u_0}{\partial x}\\ \frac{\partial v_0}{\partial y}\\ \frac{\partial u_0}{\partial y}+\frac{\partial v_0}{\partial x}\end{array}\right\}-z\left\{\begin{array}{c}\frac{\partial^2 w}{\partial x^2}\\ \frac{\partial^2 w}{\partial y^2}\\ 2\frac{\partial^2 w}{\partial x\partial y}\end{array}\right\} \tag{7}$$

In Eq. (7) u_0 and v_0 denote the displacements of the middle surface of the laminate (z = 0) in x and y directions respectively whereas w is the plate displacement in z-direction. Combining Eq. (6) and Eq. (7) gives the matrix equation in Eq. (8) [12]

$$\left\{\begin{array}{c}N_x\\ N_y\\ N_{xy}\\ M_x\\ M_y\\ M_{xy}\end{array}\right\}=\left[\begin{array}{cccccc}A_{11}&A_{12}&A_{13}&B_{11}&B_{12}&B_{13}\\ A_{21}&A_{22}&A_{23}&B_{21}&B_{22}&B_{23}\\ A_{31}&A_{32}&A_{33}&B_{31}&B_{32}&B_{33}\\ B_{11}&B_{12}&B_{13}&D_{11}&D_{12}&D_{13}\\ B_{21}&B_{22}&B_{23}&D_{21}&D_{22}&D_{23}\\ B_{31}&B_{32}&B_{33}&D_{31}&D_{32}&D_{33}\end{array}\right]\left\{\begin{array}{c}\varepsilon_x^0\\ \varepsilon_y^0\\ \gamma_{xy}^0\\ \kappa_x\\ \kappa_y\\ \kappa_{xy}\end{array}\right\} \tag{8}$$

In Eq. (8) the A_{ij}, B_{ij} and D_{ij} parts of the matrix are called the extension stiffness matrix, the extension-bending coupling matrix and the bending stiffness matrix respectively. The equations for the computation of these matrices are given in Eq. (9) where N denotes the total number of layers [12].

$$\begin{aligned}A_{ij}&=\sum_{k=1}^{N}\left(\overline{Q_{ij}}\right)_k(z_k-z_{k-1})\\ B_{ij}&=\frac{1}{2}\sum_{k=1}^{N}\left(\overline{Q_{ij}}\right)_k\left(z_k^2-z_{k-1}^2\right)\\ D_{ij}&=\frac{1}{3}\sum_{k=1}^{N}\left(\overline{Q_{ij}}\right)_k\left(z_k^3-z_{k-1}^3\right)\end{aligned} \tag{9}$$

2 Analytical Techniques for the Computation of the Buckling Load of Orthotropic Symmetric Plates

Due to their high slenderness laminated composite plates are prone to buckling under compression. Figure 4 shows a general load case for a laminated composite plate under compressive and shear forces. T_x, T_y and T_{xy} denote the compressive line loads

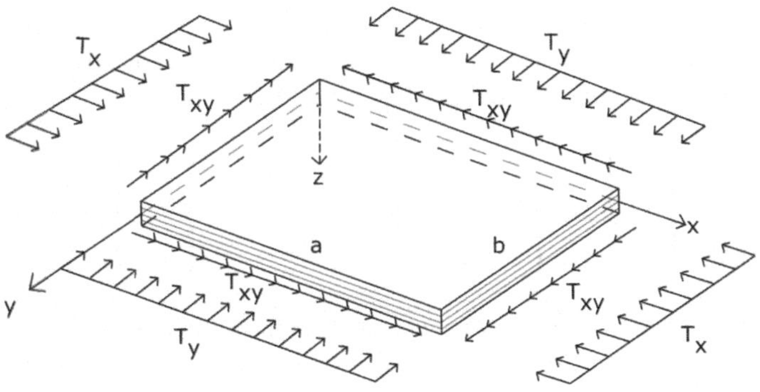

Fig. 4 Line loads acting on a laminate

in x and y directions and the shear line load respectively. The long and short side lengths of the plate are denoted with a and b.

The buckling equation for orthotropic plates is given in Eq. (10) [13],

$$D_1\frac{\partial^4 w}{\partial x^4}+2D_3\frac{\partial^4 w}{\partial x^2\partial y^2}+D_2\frac{\partial^4 w}{\partial y^4}+T_x\frac{\partial^2 w}{\partial x^2}+T_y\frac{\partial^2 w}{\partial y^2}+2T_{xy}\frac{\partial^2 w}{\partial x\partial y}=0 \quad (10)$$

where $D_1 = D_{11}$, $D_2 = D_{22}$, $D_3 = D_{12} + 2D_{33}$. In case of uniaxial compression in x direction the T_{xy} and T_y terms are equal to zero which gives us the uniaxial compression buckling equation in Eq. (11),

$$D_1\frac{\partial^4 w}{\partial x^4}+2D_3\frac{\partial^4 w}{\partial x^2\partial y^2}+D_2\frac{\partial^4 w}{\partial y^4}+T_x\frac{\partial^2 w}{\partial x^2}=0 \quad (11)$$

For the boundary conditions with all edges simply supported, the plate deflection w can be expressed as in Eq. (12) where m, n are positive integers and $\lambda_m = \frac{\pi m}{a}$, $\lambda_n = \frac{\pi n}{b}$ [13].

$$w(x,y)=\sum_m\sum_n w_{mn}\sin\lambda_m x\cdot\sin\lambda_n y \quad (12)$$

By inserting Eq. (12) into Eq. (11) we obtain Eq. (13),

$$T_x=\frac{\pi^2}{b^2}\left(D_1\left(\frac{mb}{a}\right)^2+D_2\frac{n^4}{\left(\frac{mb}{a}\right)^2}+2D_3n^2\right) \quad (13)$$

It follows that for any m, n values corresponding to a deflected shape of the plate there exists a line load T_x that satisfies the equilibrium of the deflected plate the minimum value of which is the critical buckling load T_x^c. Clearly, T_x increases with n therefore for the calculation of T_x^c, $n = 1$ should be inserted into Eq. (13) which gives us Eq. (14).

$$T_x = \frac{\pi^2}{b^2}\left(D_1\left(\frac{mb}{a}\right)^2 + D_2\frac{n^4}{\left(\frac{mb}{a}\right)^2} + 2D_3n^2\right) \quad (14)$$

Equation (14) shows that besides the bending stiffness matrix, the critical buckling load also depends on the aspect ratio (a/b) of the plate, the half wave number m and the side length perpendicular to the direction of T_x. Also, for any given value of m, T_x is a nonlinear function of the aspect ratio.

3 Methods and Results

To visualize the variation of T_x with respect to m and a/b we calculate the critical buckling load of a nine layered symmetric laminate with the material properties given in [9]. The ply angle sequence used in [9] is [45°, –45°, 0°, 90°, 0°, 90°, 0°, –45°, 45°]. The mechanical properties of a carbon fibre reinforced composite (CFRP) lamina as used in [9] are listed in Table 1.

The short side length of the plate in [9] is 150 mm and the total thickness is 2.25 mm with all layers having equal thickness.

Figure 5 shows the variation of T_x with the aspect ratio for m values between 1 and 4. For any given aspect ratio between 1 and 3, the critical line load T_x corresponds to the first intersection point of a straight line starting from and perpendicular to the a/b axis with one of the curves in Fig. 5. Using this technique it can be found that the critical load occurs at m = 1, m = 2 and m = 3 for the aspect ratios 1, 2 and 3 respetively. Each one of the curves in Fig. 5 corresponds to the minimum line load for a certain range of a/b. For example, the curve of m = 1 gives us T_x^c for $a/b \leq 1.5$ and the curve of m = 2 gives T_x^c for $1.5 \leq a/b \leq 2.6$.

Figure 6 shows the T_x function for a wider range of a/b values. It can be observed that for large aspect ratios the graphs of all m values are close to each other. The T_x^c value can be approximated by equating the first derivative of T_x to zero [13]. Let $\lambda = (mb/a)^2$, then Eq. (14) can be written as in Eq. (15)

Table 1 Material properties of CFRP plates [9]

Mechanical property	Carbon fibre (CFRP)
E_1 (N/mm^2)	157,000
E_2([N/mm^2)	8500
G_{12} (N/mm^2)	4200
ν_{12}	0.35

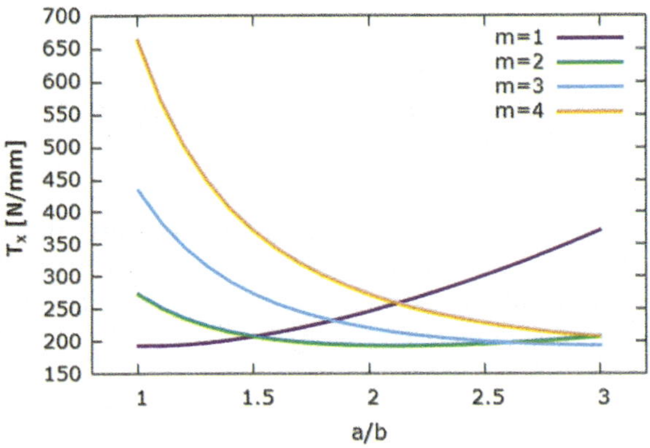

Fig. 5 The variation of the critical line load with respect to the aspect ratio

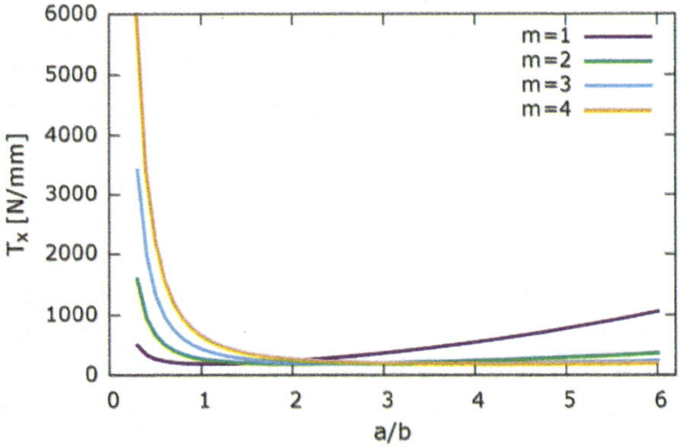

Fig. 6 The variation of the critical line load with respect to the aspect ratio

$$T_x = \frac{\pi^2}{b^2}\left(D_1\lambda + \frac{D_2}{\lambda} + 2D_3\right) \tag{15}$$

Setting the first derivative of T_x with respect to λ equal to zero we obtain Eq. (16) for the critical buckling load T_x^c.

$$T_x^c = \frac{2\pi^2}{b^2}\sqrt{D_1 D_2}\left(1 + \frac{D_3}{\sqrt{D_1 D_2}}\right) \tag{16}$$

3.1 Stacking Sequence Optimization

The Jaya optimization algorithm was developed by Rao [14] as a general-purpose optimization technique that can function without any algorithm specific parameters. Since algorithms that depend on algorithm-specific parameters such as the ant colony optimization or evolution programming require proper tuning of these parameters, they are only accessible to experienced users. Therefore, the parameter free nature of the Jaya algorithm is a great advantage for practicing design engineers who do not have the necessary expertise in optimization techniques.

In the process of optimizing the stacking sequence in this study the fiber orientation angles and the layer thicknesses of the laminate are chosen to be the design variables. These variables are allowed to vary as continuous valued variables within predefined constraints. One of these constraints is the minimum allowable thickness of a ply which is set to 0.1 mm due to the fabrication limitations. Other constraints were placed on the total plate thickness and the maximum thickness of an individual ply as 2.25 mm and 0.5 mm respectively. The fiber orientation angles were allowed to take any value between –90° and 90°. The Jaya algorithm is initiated with the random generation of a population of candidate solution vectors which constitute a population. The population size can be determined by the user and in this study, the population contains ten different design vectors with eighteen design variables each. To each member of the population a buckling load is assigned using the analytical methods described in the previous section. Once to each solution vector the corresponding buckling load is assigned, the solution vectors are ranked according to their performances. The best and worst performing design vectors are identified and the entire population goes through a Jaya iteration process which can be described with Eq. (17)

$$x_i^{t+1} = x_i^t + r_1\left(g^* - \left|x_i^t\right|\right) - r_2\left(g^w - \left|x_i^t\right|\right) \tag{17}$$

In Eq. (17) i is the index showing the position of a variable in a design vector whereas t shows the Jaya iteration number. g^* and g^w are the variables with the index i in the best and worst performing design vectors respectively. r_1 and r_2 are random numbers between zero and one. After each Java iteration a design vector is replaced with a newly generated design vector only if the newly generated design vector performs better than the old design vector. The Jaya iteration in Eq. (17) is repeated for a predetermined number of times or until a condition defined by the user has been satisfied. A flow chart of the Jaya iteration process can be seen in Fig. 7.

The buckling load of a rectangular laminated CFRP composite plate with the aspect ratio of three was maximized using the Jaya algorithm. After every Jaya iteration the buckling loads corresponding to the best and worst performing design vectors are recorded. The development of these buckling loads is shown in Fig. 8 for the first 50 iterations. In Fig. 8 the purple line shows the buckling loads of the best performing design vectors after each iteration and the green line shows the buckling loads of the worst performing vectors. Clearly, after 25 iterations all design vectors

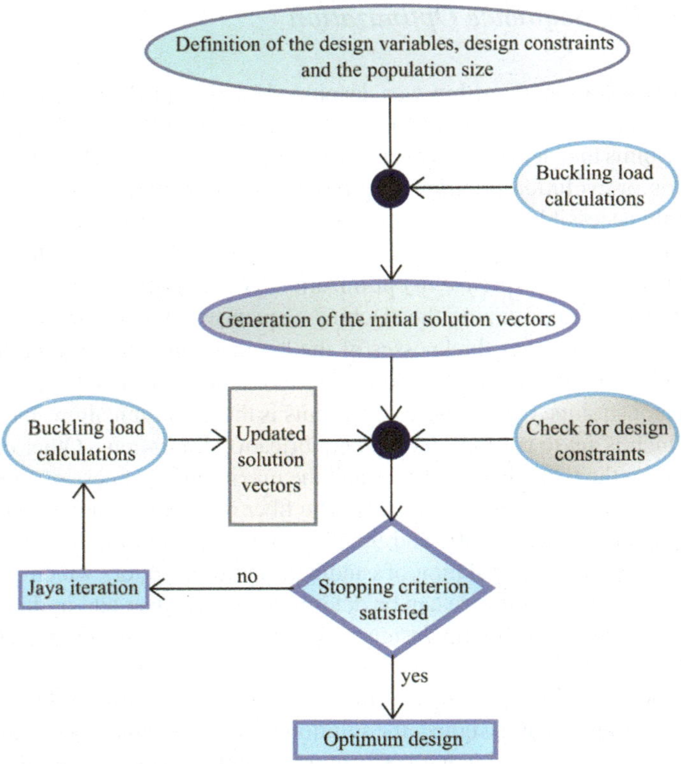

Fig. 7 Flow chart of the Jaya algorithm

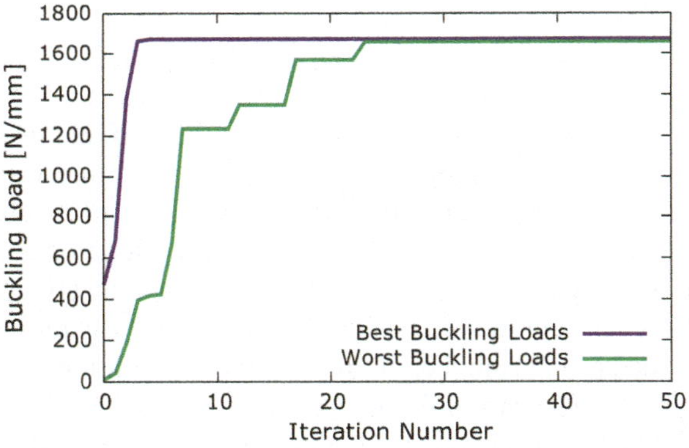

Fig. 8 Development of the buckling loads in the first 50 Jaya iterations

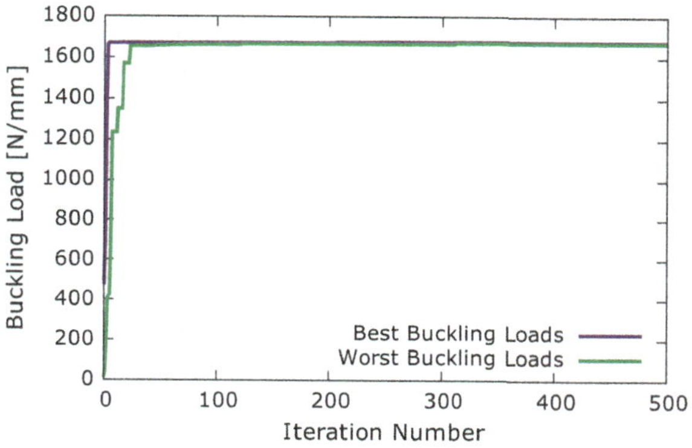

Fig. 9 Development of the buckling loads in the first 50 Jaya iterations

were updated in such a way that the performance differences between them became negligible. Figure 9 shows the development of the buckling loads throughout 500 iterations. It can be observed that no significant improvements in the buckling loads due to Jaya iterations could be achieved after the initial 50 iteration steps. The best performing fiber angle sequence after 500 iterations was [42.75°, –30.08°, –74.10°, 35.98°, 69.60°, 35.98°, –74.10°, –30.08°, 42.75°] whereas the best performing ply thickness sequence was [0.50, 0.10, 0.10, 0.34, 0.14, 0.34, 0.10, 0.10, 0.50] where the unit of thickness is mm. The maximum buckling load that could be achieved after 500 Jaya iterations was 1680 N/mm which is nearly ten times higher than the buckling load corresponding to the stacking sequence [45°, –45°, 0°, 90°, 0°, 90°, 0°, –45°, 45°] with equal ply thicknesses used in [9].

4 Conclusion

Laminated composite plates are highly efficient structural members due to their superior strength to weight ratio. The performance of these structural members is largely determined by several parameters such as the sequence of the ply thicknesses and the fiber orientation angles. One of the most frequently used structural performance indicators for laminated composite plates is the buckling load. This indicator is commonly used in the literature as a maximization objective in optimization studies related to laminates. This study presented an analytical method that predicts the buckling load of a symmetric laminated composite plate. Also, a novel meta-heuristic optimization technique called the Jaya algorithm was presented. The fiber orientation angles and ply thicknesses of a laminate are treated as continuous valued design variables of the stacking sequence optimization problem. It was demonstrated that by allowing the fiber orientation angles to take values other than 0°, ± 45°, 90° the structural performance can be greatly enhanced. To obtain the optimum stacking sequence configuration for any given plate geometry and material properties the availability of efficient optimization algorithms such as the Jaya algorithm plays a crucial role. Future studies in this field can include material properties and aspect ratio as design variables in the optimization process. Another possible direction of future research in this field can include the application of different optimization techniques to the stacking sequence optimization and a comparison of their performances. In addition to buckling under uniaxial compressive line load also, buckling due to shear forces and temperature gradients can be counted among possible research directions related to the optimization of laminated composite plates.

References

1. Barakat, S., Abu-Farsakh, G.: The use of an energy-based criterion to determine optimum configurations of fibrous composites. Compos. Sci. Technol. **59**, 1891–1899 (1999)
2. De Almeida, F.S.: Stacking sequence optimization for maximum buckling load of composite plates using harmony search algorithm. Compos. Struct. **143**, 287–299 (2016)
3. De Almeida, F.S., Awruch, A.: Design optimization of composite laminated structures using genetic algorithms and finite element analysis. Compos. Struct. **88**, 443–454 (2009)
4. Abachizadeh, M., Shariatpanahi, M., Yousefi-Koma, A., Dizaji, A.F.: Multi-objective optimal design of hybrid laminates using continuous ant colony method. In: Proceedings of the ASME 2010 10th Biennial Conference on Engineering Systems Design and Analysis, Istanbul, Turkey, vol 2, pp. 371–378, 12–14 July 2010
5. Abachizadeh, M., Tahani, M.: An ant colony optimization approach to multi-objective optimal design of symmetric hybrid laminates for maximum fundamental frequency and minimum cost. Struct. Multidiscip. Optim. **37**, 367–376 (2008)
6. Wang, W., Guo, S., Chang, N.: A modified ant colony algorithm for the stacking sequence optimization of a rectangular laminate. Struct. Multidiscip. Optim. **41**, 711–720 (2010)
7. Aymerich, F., Serra, M.: An ant colony optimization algorithm for stacking sequence design of composite laminates. Comput. Model. Eng. Sci. **13**, 49–66 (2006)
8. Aymerich, F., Serra, M.: Optimization of laminate stacking sequence for maximum buckling load using the ant colony optimization (ACO) metaheuristic. Compos. Part A Appl. Sci. Manuf. **39**, 262–272 (2008)
9. Führer, T.: Stiffness degradation of composite skin fields due to strength and buckling onset. Thin Walled Struct. **119**, 522–530 (2017)
10. Nettles, A.T.: Basic Mechanics of Laminated Composite Plates. NASA Reference Publication 1351. National Aeronautics and Space Administration Marshall Space Flight Center MSFC, Alabama, Oct 1994
11. Jones, R.M.: Mechanics of Composite Materials, 2nd edn. Taylor & Francis, Philadelphia, PA (1999)
12. Fragoudakis, R.: Strengths and limitations of traditional theoretical approaches to FRP laminate design against failure. In: Engineering Failure Analysis, IntechOpen, Rijeka (2020)
13. Vasiliev, V.V., Morozov, E.V.: Advanced Mechanics of Composite Materials and Structures, 4th edn. Elsevier Science (2018)
14. Rao, R.V.: Jaya: A simple and new optimization algorithm for solving constrained and unconstrained optimization problems. Int. J. Ind. Eng. Computations **7**, 19–34 (2016)

Sustainable Optimum Design of RC Retaining Walls: The Influence of Structural Material and Surrounding Soil Properties

Zülal Akbay Arama, Aylin Ece Kayabekir, and Gebrail Bekdaş

Abstract This paper presents an approach to design reinforced concrete retaining walls based on the attainment of both sustainable geotechnical design and structural optimization simultaneously with the use of a special algorithm Harmony Search to utilize the influence of both the structural material and the surrounding soil properties. Harmony Search algorithm is applied to design problem with the use of two different objective functions. The first function aims to minimize the total material cost and the second function aims to minimize the CO_2 emission value associated with the supplement of geotechnical safety and structural necessities. The envisaged optimization problem considers several different variants such as the geometric variables (top and bottom thickness of the stem, length of the toe and the heel encasements and the height of the stem), the soil variables (unit weight of soil and shear strength angle), the cost of materials (the unit cost of concrete and steel), the amounts of the CO_2 emission (the unit amount of emission for concrete and steel). The outcomes of the analysis show that Harmony Search algorithm is a convenient meta-heuristic optimization method to achieve optimum design for RC retaining walls considering both geotechnical and structural multi-variants. Besides, the significant effects of the geotechnical and structural parameters are emphasized and demonstrated in detail because of their early position through the construction process.

Keywords Retaining walls · Harmony search algorithm (HS) · Geotechnical stability · Structural material · Sustainable design

---

Z. Akbay Arama (✉) · A. E. Kayabekir · G. Bekdaş
Department of Civil Engineering, Istanbul University-Cerrahpaşa, 34320 Avcılar, Istanbul, Turkey
e-mail: zakbay@istanbul.edu.tr

A. E. Kayabekir
e-mail: ecekayabekir@gmail.com

G. Bekdaş
e-mail: bekdas@istanbul.edu.tr

S. M. Nigdeli et al. (eds.), *Advances in Structural Engineering—Optimization*, Studies in Systems, Decision and Control 326,
https://doi.org/10.1007/978-3-030-61848-3_10

1 Introduction

Although being one of the oldest structural systems, nowadays the design of retaining walls is still investigated in order to save the planet taking into consideration the effects of consumed materials during construction. In this context, recently, the design of retaining walls is still remaining as a main subject for both geotechnical and structural engineering discipline. Therefore, multidisciplinary solutions are tried to be performed to ensure both safety and sustainability with the application of contemporary solution technologies and techniques in the design process. In this design process, generally reinforced concrete design of retaining walls is preferred due to the its applicability easiness, labor simplicity and low cost. However, using higher amounts of concrete and steel leads to increase the emission of carbon dioxide (CO_2) that has harmful effects for the environment and also causes global warming. Hence, it becomes significant to minimize the cost and CO_2 emission amount simultaneously associated with both ensuring the geotechnical and structural requirements. However, the minimization of the cost and the gas emission are not easy because there will be many nonlinear functions that have to be solved and discrete design variables makes it hard to reach the result [1]. During the recent years, the tendency has been arising to use optimization techniques to investigate the design process depending on the environmental effects that are involved in all life-cycle stages in contrary with the used traditional design techniques that are frequently depended on the designers' scientific intuition, experience and vision [2, 3]. These performed optimization analyses can define the structural system based on the design variables, automatically determines the safety equations and validates the structural requirements and subsequently redefinition process is conducted according to the means of an envisaged optimization algorithm which can control the flow with a large number of iterations to obtain the optimum result [4]. In a general manner, two methods named exact methods and approximate methods, can be counted to sort the optimization methods. The exact methods are beneficial if there are a few numbers of design variables for the problem. But if the number of design variables is increased, it will be difficult to conduct the iterations in terms of computing time. The approximate methods involve both the heuristic methods and search algorithms [3]. Considering the design of retaining walls, especially, the metaheuristic search techniques are preferred as being an efficient tool to acquire a cost effective and sustainable design. The algorithms that are used to design the retaining systems can be counted as the Artificial Bee Colony algorithm (ABC) [5], Big Bang-Big Crunch algorithm (BB-BC) [5–7], Teaching-Learning-Based Optimization algorithm (TLBO) [5, 8], Harmony Search algorithm (HS) [9, 10], Firefly (F) algorithm [11, 12], Particle Swarm algorithm (PS) [13, 14], Bat algorithm (B) [15], Simulated Annealing algorithm (SA) [16–18] Genetic algorithm (GA) [19–22], Imperialist Competitive algorithm (ICA) [5], Cuckoo Search algorithm (CS) [5], Charged System Search algorithm (CSS) [5, 23], Ray Optimization algorithm (RO) [5], Tug of War Optimization algorithm (TWO) [5], Water Evaporation Optimization algorithm (WEO) [5], Vibrating Particles System algorithm (VPS) [5, 24], Enhanced Colliding Bodies of Optimization (ECBO) [24], Colliding

Bodies of Optimization (CBO) [24], Cyclical Parthenogenesis algorithm (CPA) [5], Biogeography-Based Optimization algorithm (BBO) [25], Grey Wolf Optimization (GWO) [2, 26], Flower Pollination algorithm (FPA) [27], Backtracking Search algorithm (BSA) [28], Social Spiders optimization (SSO) [29], Search Group algorithm (SGA) [28], Ant Colony Optimization (ACO) algorithm [30], Differential Evolution algorithm (DEA) [31], Shuffled Shepherd Optimization algorithm (SSOA) [5], Gravitational Search algorithm (GSA) [32], Dolphin Echolocation Optimization (DEO) algorithm [33].

However, a small number of studies take into consideration the design of RC retaining walls in terms of geotechnical properties of soil. Uray et al. [34] searches the optimum designs for two types of concrete cantilever retaining walls that was conducted utilizing the artificial bee colony algorithm, Gandomi et al. [35] investigate the use of Differential Evolution (DE), Evolutionary Strategy (ES) and Biogeography Based Optimization algorithm (BBO) to determine the nonlinear constrained optimum design problem of a cantilever retaining wall. Gandomi et al. also studies on the change of the surcharge loading amount, the friction angle and backfill slope angle. Purohit [36] and Das et al. [37] studies on the minimization of cost and attainment of safety factors with the use of an evolutionary multi-objective optimization algorithm, non-dominated sorting genetic algorithm (NSGA-II). Purohit [36] and Das et al. [37] also investigates the effects of single and multi-objective function solutions on the design of RC retaining walls depending on the change of both excavation depth and angle of friction. Konstandakopoulou et al. [38] develops a heuristic optimization approach to obtain the optimal design of RC retaining walls taking into consideration four different kinds of soil. But the optimization of environmental effects is not considered in any of this mentioned design studies. The studies taking into consideration the harmful effects of carbon dioxide emitted from reinforced concrete retaining structures have become widespread in recent years. Yepes et al. [3] designs cantilever retaining walls for road construction with the use of a hybrid multistart optimization strategic method performing two different objective functions including the cost and the embedded carbon dioxide (CO_2) emissions. Aydogdu [29] utilizes from Biogeography Based optimization and Social Spiders optimization methods to solve RC retaining wall design with three kinds of objective functions including cost, CO_2 emission and cost-CO_2 emission respectively. Zastrow et al. [39] uses Harmony Search algorithm to design RC retaining walls to minimize both cost and CO_2 emission. Aydogdu and Akin [1] uses Biogeography Based optimization to evaluate the relationship between the cost and the CO_2 emission. Khajehzadeh [40] develops a new hybrid optimization algorithm that integrates pattern search with Gravitational Search algorithm (GSA) to solve the optimization problem of CO_2 emission of RC retaining walls. Yepes et al. [41] performs analysis with Black Hole algorithm to optimize the cost and CO_2 emissions in earth-retaining walls. Villalba et al. [4] utilizes from simulated annealing to design RC cantilever earth-retaining walls used in road construction depending on the minimum embedded CO_2 emissions.

Inspired by all these studies, the influence of the change of the soil geotechnical properties and the characteristics of structural material change is investigated by the use of Harmony Search algorithm (HS) within the scope of this article. The soil is assumed to be pure frictional and Rankine Earth Pressure Theory is used to determine the lateral reactions of the soil mass. The structural design of the wall is arranged based on ACI 318-05 code [42]. The requirements of geotechnical design such as the adequate safety of sliding, overturning and bearing capacity, is taken into consideration. Two different objective functions are developed to query the minimization of cost and CO_2 emission individually and also discussions are conducted with the use of an integrated relationship function of cost and CO_2 emission. The variants of the analyses are selected as the depth of excavation, the shear strength of the soil strata, the unit weight of the soil strata, the unit costs of both concrete and steel and the unit amount of CO_2 emission of both concrete and steel. Totally 384,000 cases are arranged depending on the design variables change and 114,240,000 data are used to evaluate the design specifications for both eco-friendly and sustainable design of the RC retaining walls. These huge data set and the diversity of the foreseen parameters differentiates this paper from the associate ones and supplies the engineers and researchers an integrated design approach for RC retaining systems.

Within the context of this paper, Sect. 2 gives information about the details of used optimization algorithm and the design steps of RC retaining walls. Besides, the constraints and the constants of the design and the objective functions are explained depending on the logic of the study. Sect. 3 presents the data of the parametric analysis and the details of the cases that are arbitrarily selected, are given. Sect. 4 includes the numerical analysis of the optimization problem and the results of the analyses are briefly described with illustrations. Sect. 5 is the conclusion part and the evaluations including the numerical analyses and the novelty of the paper is presented.

2 Materials and Methods

In this study, the design of reinforced concrete retaining walls is investigated based on multivariate parametric analyses conducted with Harmony Search Algorithm (HS) to optimize both the cost and the emission of CO_2. T-Shaped reinforced concrete retaining walls (RW) are taken into consideration to specify both the constants and variables of the conducted analysis. HS algorithm is a meta-heuristic search algorithm which can be an efficient alternative to overcome the difficulties of complex optimization problems. Recently, HS algorithm has attracted the attention of both researchers and designers because of the advantages of usability. In the context of the application schedule, it is a simple process to implement the problem and the convergency of the optimal solution can be obtained quickly and the estimation of a sufficient solution can be acquired in an acceptable computational time [43]. The algorithm is originally presented by Geem et al. [9] and inspired by the music improvisation process. In the mentioned process, there have to be a number of musicians which are preliminary defined to tune the pitch of the instruments to obtain the best state of the harmony.

Besides, in the natural loop of the life, a harmony can be identified with the generation of lots of sound waves which have various frequencies. Then, the quality of the adjusted harmony is calculated by the aesthetic estimation. The musicians make iterative practices in an attempt to increase the aesthetic estimation and to obtain the best probable harmony. There can be formed a similar process to identify the optimization process of problems. For an envisaged optimization problem, the main aim is to acquire the global optimum of the objective function by the improvisation of a predefined number of decision variables. The decision variables constitute a solution vector. The numerical values of the decision variables are placed in the objective function to calculate the qualification of the solution vector. It is aimed to obtain the global optimum therefore; the solution vector is updated through the iterations. The applicable solution of the HS algorithm is named as the harmony and each of the design variables of the solution is equivalent to a note. A harmony memory (HM) is a store that is derived during the optimization process with a pre-calculated number of harmonies (N). This identification state of harmony memory is defined as the initial step of the implementation of HS algorithm to a natural problem. In this first step, the design constants, the boundary values of design variables, maximum iteration number and main parameters are defined. These parameters are named the harmony memory size (HMS), the harmony memory consideration rate (HMCR) and the pitch adjustment rate (PAR). A new harmony temporized from the formation of a harmony vector with the presence of a random (rnd (0,1)) value that can be changed between 0 and 1. According to the Eq. 1, the upper limit is defined by $X_{i,max}$ and the lower limit is defined by $X_{i,min}$ of each design variable (X_i).

$$X_i = X_{i,min} + rnd(0, 1) \cdot \left(X_{i,max} - X_{i,min}\right) \tag{1}$$

The design equations are calculated with the use of the design constants and design variables and following, the solution of the foreseen objective function is achieved. The results of this calculation are collected in the harmony vector. This solution sequence is reproducible based upon the HMS of the harmony memory. All of the produced harmony vectors are collected in a particular matrix named the initial solution matrix. In the second step, the harmony memory consideration rate (HMCR) is used to generate a new vector through the selection of the suitable way of solution. A random value is formed and if the determined HMCR value is more than the random one, the first way of generation (global optimization) of a new vector is selected, if not, the second way (local optimization) is chosen to use. In the third step, a new harmony vector is created via the beginning of the iteration process. The application rules of the HS algorithm are used to obtain a new harmony vector along with the use of different ways. In order to apply the first way, the upper and lower boundaries are used as the limits to constitute the design variables randomly as done in Eq. 1. In order to apply the second way, it is possible to use a new vector that is formed with the use of a selected vector of the solution matrix ($X_{i,old}$). New values ($X_{i,new}$) are calculated by adding the multiplication of pitch adjusting rate (PAR), the difference of the design variable limits and rnd (0,1) (Eq. 2).

$$X_{i,new} = X_{i,old} + rnd(0,1) \cdot PAR \cdot \left(X_{i,max} - X_{i,min}\right) \tag{2}$$

In the fourth step, the comparison of a new vector with the collected vectors in the solution matrix is made. A new vector is used rather than the present vector if the next vector is better than the present one. If not, the present condition of the solution matrix is saved. The values that are acquired from the solution of the objective function is compared and the minimum value of the solutions is chosen as the best one. In the comparison duration, the constraints of the design are also considered. Additionally, the amounts of the violations are controlled and the solution that comprises the minimum violation is selected to be the better one even if the violations of the design constraints exist for a new solution and present solutions. If one of them is violated, the violated solution is eliminated. In the fifth step, the stopping criterion is controlled. The iterations are turned off if a satisfactory stopping criterion is derived. The maximum iteration number is chosen as the stopping criterion of the analyses in this paper, although there are various ways are existing. The flowchart of the application process of HS algorithm is given in Fig. 1.

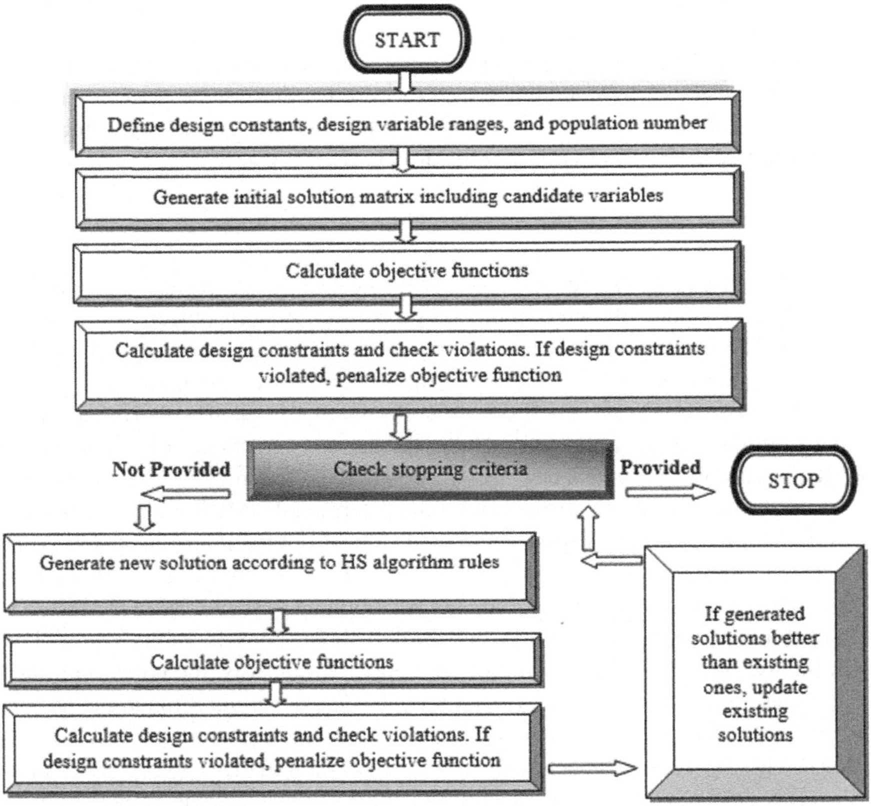

Fig. 1 The application process of HS

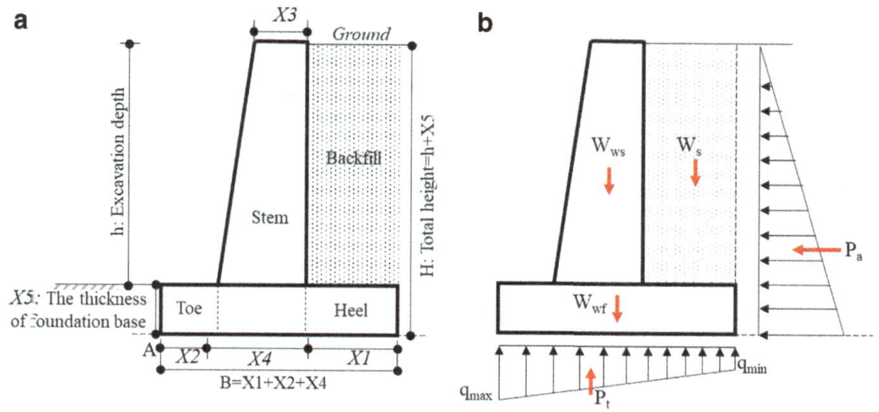

Fig. 2 **a** The cross section of T-shaped retaining wall; **b** The stresses and related forces along the wall

The given application process of HS algorithm is performed for RC retaining walls to query the effects of both structural material and surrounding soil properties. For to reach this purpose, it is a basic requirement to define the design standards of retaining walls. The traditional illustration of a retaining wall and the abbreviations used to define the components of the structure is given in Fig. 2a. Besides the possible features of the forces are shown with the stress distribution caused by the wall system throughout the soil strata is given in Fig. 2b. In Fig. 2, the total height of the wall (H) can be calculated by the sum of the excavation depth (h) and the thickness of the wall foundation (wall base) (X5). The thickness of the wall stem is changed throughout the wall in the vertical axes and the thickness of the wall stem at the top and bottom is abbreviated by X3 and X4 respectively. In addition, the width of the foundation (B) is determined by the sum of toe (X2) and heel (X1) and stem bottom thicknesses (X4) of the foundation. The weight of the wall stem is represented by W_{ws} and the weight of the foundation is abbreviated as W_{wf}, W_s is the soil weight retained on the heel of the wall at the backfill side of the wall, Pa is the active soil force acted to wall, Pt represents the total base pressure and q_{max} and q_{min} is the upper and lower boundaries of ultimate soil base pressure respectively. In a general manner, it is a classical application to design the retaining wall with the satisfaction of the equilibrium of mentioned effective forces on the wall by the use of traditional static limit equilibrium equations to ensure geotechnical safety. In this context, the first step of traditional design consists of the assumption of the dimensions based on the possessed experience of the designer. This initial step can be named as predesign of retaining walls. Correspondingly, in the second step, the stability requirements against sliding and overturning have to be investigated and the adequateness of the safety has to be supplied for probable failure modes. Lateral earth pressure theories [44–46] are used to calculate the effective loads along the wall system. In the context of this paper, Rankine Lateral Earth Pressure Theory (RLEPT) is preferred to be used depending on the envisaged cases. The RLEPT assumes the absence of wall friction, the soil strata pure frictional, the interface of soil and wall vertical, the generated failure surface is straight and the movement of the soil is planar, the resultant total force is affected parallel to the surface inclination of the backfill. Similar to these assumptions, in the wall system evaluated within the scope of this study, the wall system is located in a soil medium that is lying laterally and there is no inclination of soil surface at the backfill side. Besides, pure frictional soils are used for both for being backfill and foundation soil strata. Depending on the RLEPT the lateral active earth pressure is calculated by Eq. 3.

$$K_a = \tan^2\left(45 - \frac{\emptyset}{2}\right) \tag{3}$$

The envisaged wall system is sufficiently flexible so that the retaining wall can rotate enough to allow the active earth-pressure wedge form. In such a situation, the passive stresses can be admissible. In this sense, only the active stresses are taken into account to calculate the forces acted to the wall. The active soil force is calculated with Eq. 4 for pure frictional soils.

$$P_a = \frac{1}{2}\gamma H^2 K_a \tag{4}$$

The terms that are used to calculate active soil forces are γ and K_a represents the unit weight of soil and the active lateral earth pressure coefficient of the soil respectively. The unbalanced lateral earth forces cause the wall to collapse, but the self-weight of the wall resists not to slide, overturn or fail by the inadequateness of the bearing capacity. The lateral active force causes to slide the wall along the foundation. Therefore, it is required to control the sliding safety by the calculation of a safety number with the ratio of the total horizontal resisting forces to the total horizontal sliding forces. In this context, the safety number against sliding (SF_s) can be calculated by Eq. 5. This calculated number has to be bigger than 1.5 value to ensure the static equilibrium against sliding. According to Eq. 5, ΣF_R and ΣF_s represents the total resisting and sliding forces respectively.

$$SF_s = \frac{\Sigma F_R}{\Sigma F_S} > 1.5 \tag{5}$$

In addition, the wall structure tends to overturn about its toe point that is specified as A point in Fig. 1a because of the moment effect. Therefore, it is needed to determine the overturning safety value with the ratio of the total moments generated by the resisting forces to the total moments activated by the sliding forces (Eq. 6). The factor of safety value against overturning (SF_o) has to be bigger than 1.5 for ensuring the static equilibrium. In Eq. 6, ΣM_R and ΣM_s represents the total resisting and sliding moments respectively.

$$SF_o = \frac{\Sigma M_R}{\Sigma M_S} > 1.5 \tag{6}$$

Concurrently, the adequateness of the bearing capacity of the base (foundation) of the wall is a major problem at the design stage of the retaining walls. The ratio of ultimate base pressure ($q_{z,u}$) to the maximum soil pressure ($q_{z,max}$) has to be bigger than 3.0. The factor of safety values is defined by SF_{bc} and determined by Eq. 7.

$$SF_{bc} = \frac{q_{z,u}}{q_{z,max}} > 3.0 \tag{7}$$

The soil base pressure is a computable soil parameter according to the dimensions of the wall system and depending on the soil properties. Besides, the ultimate base pressure has upper and lower boundaries that can be computed based on only the dimensions of the wall. The relationship that is suggested for shallow foundations is given in Eq. 8 to determine the ultimate base pressure. In Eq. 8, ΣV, B, e represents the total vertical forces, the width of the foundation and the eccentricity of the loads respectively. The eccentricity value can be calculated by the multiplication of the half width of the base with the ratio of the net moments to the vertical total loads. This mentioned relationship is also given in Eq. 9 [47–50].

$$q_{max} = \frac{\Sigma V}{B}\left(1 \pm \frac{6e}{B}\right) \tag{8}$$

$$e = \frac{B}{2}\frac{\Sigma M_{R-}\Sigma M_0}{\Sigma V} \tag{9}$$

The satisfaction of the adequate safety degrees of probable geotechnical problems of RC retaining walls led to end the pre-design process. According to the identified appropriate dimensions of the wall, it is required to ensure sufficient shear and moment capacities. Besides, the net bearing pressure of the surrounding soil cannot to be stated a tensile stress and the material of the reinforcing bars has to satisfy the envisaged code necessities [51]. In this context, it is assumed to control the structural requirements depending on the ACI 318-05 code [42] for the analyses conducted within to reach the purpose of this study as before mentioned. Although the requirements of geotechnical and structural design are satisfied with the mentioned information of the wall, they are not enough for designers and engineers to constitute a sustainable design. Recently, the actual studies that are aiming to save both the economy and the environment seems valuable via generating a global protectionist approach for civil engineering applications. From this point of view, the attainment of the optimum costs of structures and formation of less harmful structures to the environment connected by safe sizing has to be also ensured at the design stage. In this study, the geotechnical and structural design necessities are combined with the minimization of the cost and the gas (CO_2) emission phenomena with the performance of the analyses conducted by HS algorithm. In order to adapt HS algorithm to the design process of retaining walls, a distinctive design parameter set is arranged.

The variables of design are classified into two groups. The initial group consists of the parameters in relation to the cross-section of the wall (X1, X2, X3, X4, X5) and the second group consists of the parameters in relation with the reinforced concrete design (X6, X7, X8). The envisaged design parameters are shown in Table 1. In addition, the use of ACI 318-05 [42] code procures to use an equivalent rectangular compressive stress distribution to determine the moment capacity of the wall. According to the Eq. 10, the constraints of the design and m notation represents the number of design constraints.

$$g_{(j)}(x) \leq 0 \quad j = 1, m \tag{10}$$

According to Eq. 10, m is the number of design constraints that are given in Table 2. The inequality of the function is related to the design variable vector which can be identified by $X_T = \{X_1, X_2 \ldots X_n\}$. The critical sections of the stem and foundation is only controlled for design of reinforcement.

Within the context study, the sustainable design of RC retaining walls is investigated with the application of HS algorithm by the use of the mentioned design parameters and constraints. The aim of the paper is to acquire the minimization

Table 1 The design variables of a RC retaining wall

Symbol	Description of the parameter
X1	Length of the wall foundation heel
X2	Length of the wall foundation toe
X3	Thickness of wall stem at the top
X4	Thickness of wall stem at the bottom
X5	Thickness of wall foundation
X6	Area of reinforcing bars of the stem
X7	Area of reinforcing bars of the wall foundation heel
X8	Area of reinforcing bars of the wall foundation toe

Table 2 Design constraints on strength and dimensions

Description	Constraints
Safety for overturning	$g_{1(X)}$: $SF_{o,design} \geq SF_o$
Safety for sliding	$g_{2(X)}$: $SF_{s,design} \geq SF_s$
Safety for bearing capacity	$g_{3(X)}$: $SF_{bc,design} \geq SF_{bc}$
Minimum bearing stress (q_{min})	$g_{4(X)}$: $q_{min} \geq 0$
Flexural strength capacities of critical sections (M_d)	$g_{5-7(X)}$: $M_d \geq M_u$
Shear strength capacities of critical sections (V_d)	$g_{8-10(X)}$: $V_d \geq V_u$
Minimum reinforcement areas of critical sections (A_{smin})	$g_{11-13(X)}$: $A_s \geq A_{smin}$
Maximum reinforcement areas of critical sections (A_{smax})	$g_{14-16(X)}$: $A_s \leq A_{smax}$

relationship of cost and CO_2 emission in the light of the investigations conducted for the attainment of the influence of both structural material and surrounding soil properties.

Two different objective functions are formed to obtain the foreseen design depending on the purpose of the study. The initial objective function that is given with the Eq. 11, investigates the probability of obtaining minimum cost depending on adequate safety.

$$f_{cost}(X) = C_{c,cost}.V_c + C_{s,cost}.W_s \tag{11}$$

According to Eq. 11, $C_{c,cost}$, V_c, $C_{s,cost}$ and W_s is the unit volume cost of the concrete, the volume of the concrete, the unit cost of the reinforcing bars and the unit weight of the reinforcing bars respectively. The second objective function that is given by Eq. 12, investigates the probability of obtaining the minimum CO_2 emission amount depending on adequate safety. According to Eq. 12, C_{c,CO_2}, V_c, C_{s,CO_2} and W_s is the emission caused by the production process of unit volume of concrete, the volume of the concrete, the emission caused by the production process of unit weight of steel and the unit weight of the reinforcing bars respectively.

$$f_{CO_2}(X) = C_{c,CO_2}.V_c + C_{s,CO_2}.W_s \tag{12}$$

3 Parametric Analysis

Parametric analyses are conducted to investigate the influence of structural material and surrounding soil properties change on the sustainable optimum design of RC retaining walls. Approximately 384,000 case analyses are performed and 114,240,000 data are used depending on the arbitrarily selected design parameters that are identified within admissible limits. Matlab 2018a software is used to define the codes of design and the investigations are focused on the changes of cost, CO_2 emission and their interaction with randomly selected various excavation depths, soil and material properties. The backfill and foundation soil is selected to be pure frictional and defined by the term of surrounding soil. The shear strength parameter of pure frictional soil is only the shear strength angle (internal friction angle). Therefore, the shear strength angle of the surrounding soil strata is assumed to be a variant of the analysis and the effect of the change of the shear strength angle is searched through the analysis for the angles beginning from 27° to reach 38°. By the way, the influence of the soil strength on the sustainable optimum design can be investigated. Besides, another effective parameter that is used to calculate the active soil pressure and the weight of retained soil on the heel of the wall, is the unit weight of the surrounding soil. The unit weight of the surrounding soil is taken as a variable and the change of unit weight between 14 and 21 kN/m^3 is considered. The envisaged design variables and the constants of the selected cases are shown in Table 3. The excavation depth

Table 3 The values of design constants and the design variables of the analyses

Symbol	Definition	Value	Unit
f_y	Yield strength of steel	420	MPa
f'_c	Compressive strength of concrete	30	MPa
c_c	Concrete cover	30	mm
E_{steel}	Elasticity modulus of steel	200	GPa
$E_{concrete}$	Elasticity modulus of concrete	23,5	GPa
γ_{steel}	Unit weight of steel	7.85	t/m^3
$\gamma_{concrete}$	Unit weight of concrete	25	kN/m^3
$C_{c,cost}$	Cost of concrete per m^3	50, 75, 100, 125, 150	$
$C_{s,cost}$	Cost of steel per ton	700, 800, 900, 1000, 1100	$
C_{c,CO_2}	CO_2 emission amount of concrete	143.48, 376	kg
C_{s,CO_2}	CO_2 emission amount of steel	352, 3010	kg
X1	Range of width of wall foundation heel	0–10	m
X2	Range of width of wall foundation toe	0–3	m
X3	Range of thickness of wall stem at the top	0.2–3.0	m
X4	Range of thickness of wall stem at the bottom	0.2–3.0	m
X5	Range of wall base thickness	0.2–3.0	m
μ	Concrete-soil friction	tan (2/3) ϕ	–
Φ	Shear strength angle of surrounding soil	27–38	°
γ	Unit weight of surrounding soil	14–21	kN/m^3
h	Depth of excavation	4–10	m
SF_o	Factor of safety for overturning	1.5	–
SF_s	Factor of safety for sliding	1.5	–
SF_{bc}	Factor of safety for bearing capacity	3.0	–

is assumed to change between 4 and 10 m, depending on the application limits of cantilever retaining structures in projects and based on the national and international sources [52, 53].

The unit costs and CO_2 emission amounts of the concrete and the steel is also selected as the variants of the analyses. The unit cost of the concrete for per m^3 is assumed to be $50, $75, $100, $125 and $150 and the unit cost of the steel of the reinforcing bars for per ton is assumed to be $700, $800, $900, $1000 and $1100 respectively. The amounts of these costs are identified depending on the actual sale prices of materials that are existing in different countries. In addition, it is thought that these prices can also represent the upper and lower boundaries of the costs that the constructor can meet. The amounts of the CO_2 emission are also assumed as different values and selected depending on the research studies in the literature [54, 55]. The upper and lower limits of the emission values given in the mentioned studies are selected to represent the probable boundaries of the emission problem. Depending on the selected values of emissions, three group analyses are arranged. In the first group of the analysis, the study of Yeo and Potra [54] is used and the CO_2 emission amount is selected 376 kg for concrete which has 30 MPa strength and the CO_2 emission amount is selected 352 kg for recycled type of steel with 420 MPa strength. In the second group analysis the study of Paya-Zaforteza et al. [55] is used and 3010 kg is selected for the CO_2 emission amount of steel and the CO_2 emission amount 143.48 kg is selected for the concrete (HA-30) that has 30 MPa strength. In the third group analysis, the combination of the emission values is used and it is tried to represent the lower limits of the emissions with the selection of the CO_2 emission amount of steel and concrete 352 kg and 143.48 kg respectively. It has to be remarked that the amounts of CO_2 emission of the first group analyses are used to determine the cost based objective function.

4 Results and Discussion

Within the context of this study, numeric analyses are conducted to obtain the dimensions, total cost and ultimate CO_2 emission amounts of RC retaining walls, depending on the change of both soil properties and structural material properties. In this regard, the analyses are divided into two sections. The first section includes the minimization process of cost and the second section includes the minimization process of CO_2 emission. In addition, several subdivisions are added to the evaluation process depending on the change of the design variables. All the cases are investigated respectively by selecting a reference case study individually to ease the comparisons.

4.1 The Minimization of Total Cost

In order to discuss the effects of surrounding soil and the structural material properties on the design problem of RC retaining walls, all the mentioned design variants are investigated as a sub-title of this section individually. The effect of the excavation depth, the shear strength angle and unit weight of the surrounding soil, the costs of the concrete and steel is determined by the performed optimization analyses aimed to minimize the cost by ensuring both the geotechnical stability and structural requirements. In addition, Eq. 11 is used to conduct the optimization process for cost minimization. The effects of the different CO_2 emission values are not investigated for cost minimization because the objective function of cost minimization does not include any terms about the emission values. Therefore, the amount of CO_2 emissions is not a variable of the analyses conducted in this section.

4.1.1 Effect of the Change of Excavation Depth

The change of excavation depth is investigated subjected to the dimensions, cost and CO_2 emission values of the envisaged RC retaining walls. It is assumed that the excavation depth is increased step by step beginning from 3 m to reach 10 m depth. Some of the parameters of design are selected as constant values to evaluate only the influence of excavation depth change. The unit weight of the soil (γ) and the shear strength angle (Φ) is assumed to be 20 kN/m^3 and 32° respectively. The unit volume cost of the concrete and the unit weight cost of the steel is \$100 and \$900 respectively. The unit volume amount of the CO_2 emission of concrete is selected 376 kg and the unit weight amount of the CO_2 emission of the steel is chosen 352 kg. Figure 3a, b represents the change of the base width and the total height of the wall respectively. According to Fig. 3a, the increase of the base width subjected to the increase of the excavation depth can be seen clearly. The width of the excavation can be calculated by the sum of the foundation toe, heel and bottom stem lengths. In the analysis, it is released to use L-shaped cross section for design to eliminate all the cost increasing conditions. In this point of view, the stability requirements can be ensured with the absence of toe width till 6 m but for bigger excavation depths more than 6 m, the necessity to construct a toe section is arising. If the excavation depth increases 3 times the assumed minimum excavation depth, the total foundation width increases by 350%. In Fig. 3b, the total height of the wall is given by the sum of excavation depth and the thickness of the base. It is assumed in the analysis that the wall is embedded into the soil strata to an equal depth with the thickness of the base. In such a case, if the wall is sufficiently flexible, the wall will rotate enough to allow the active earth-pressure wedge form. In this sense, only active pressures are considered to determine the effective forces of the wall system. Therefore, it is sufficient to calculate only the active lateral soil pressure coefficient for the design [56]. The increase of the depth of the excavation leads to rise the total height of

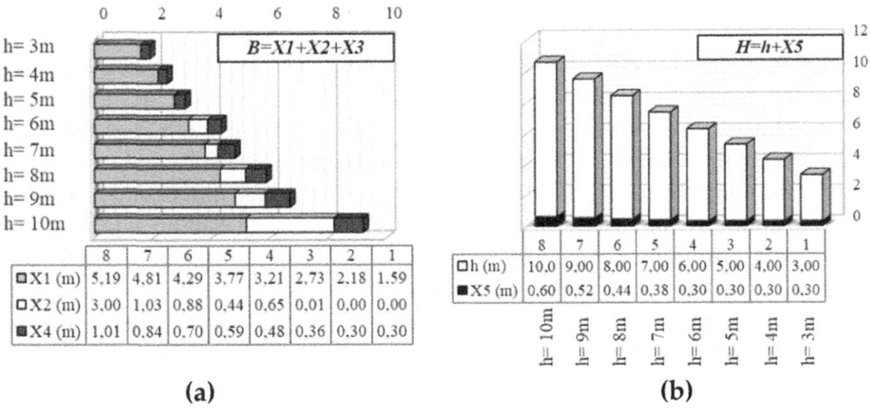

Fig. 3 **a** The change of the foundation width against the excavation depth; **b** The change of total height of the wall against width

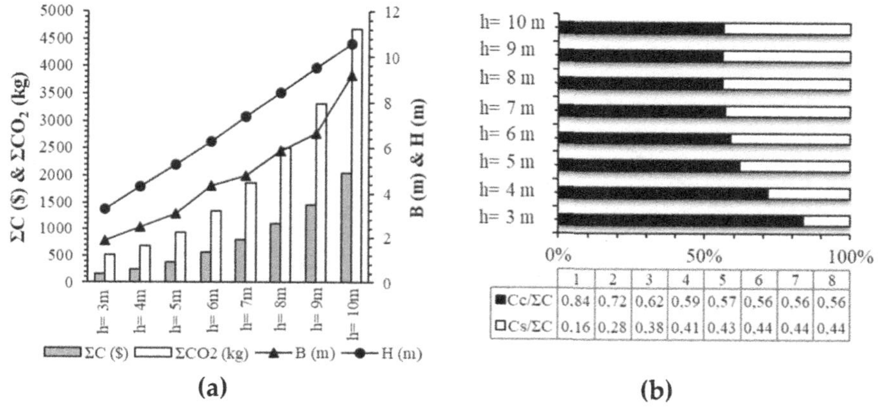

Fig. 4 **a** The change of total cost and CO_2 emission against the excavation depth; **b** The change of the consumed materials against the excavation depth

the wall with a directly proportional relationship. The excavations till the depth 6 m causes a specific situation for the evaluation case of the total height after a limit depth. The thickness of the base remains constant pending 6 m depth.

In Fig. 4a, the change of total cost, CO_2 emission, total height and total foundation width of the wall against the excavation depth is shown. The dimensions of the system are increased depending on the deepen of the excavation depth as before mentioned. This condition leads to rise the total cost and CO_2 emission amount based on the increase of consuming materials in the construction. The relationship of the increment tendencies of the design variables depending on the change of excavation depth is also investigated with the use of the curve fitting option of Microsoft Excel. Figure 5a–d gives the relationship between the total heights, the base width, the total cost and the CO_2 emission amount against the excavation depth. There is a linear relationship between the total height of the wall and the depth of the excavation. Besides, the relationship between the excavation depth and the width of foundation, the total cost and the total CO_2 emission presents a proper compatibility with polynomial curve fitting option.

The R-squared (R^2) is the coefficient of the conducted regression analyses that can be defined as the statistical measure of how close the data are in the fitted regression line, is determined 0.9997, 0.9806, 0.9966, 0.9957 respectively for the envisaged regression models in Fig. 5a–d. The mathematical expressions of the associated cases are also given in the figures and it has been clear to say that the results of the analyses give satisfactory results to predict the dimensions and the cost, the CO_2 emission values by foreseen regression models.

In addition, Fig. 4b shows the cost ratio of consuming materials during construction based on the total costs against the depth change. The ratio of the cost of the concrete materials always remains bigger than the steel ratio depending on the deepen of the excavation depth. But, the steel and the concrete cost rate remain constant after 8 m excavation depth and the values of the ratios comes closer to each other. This

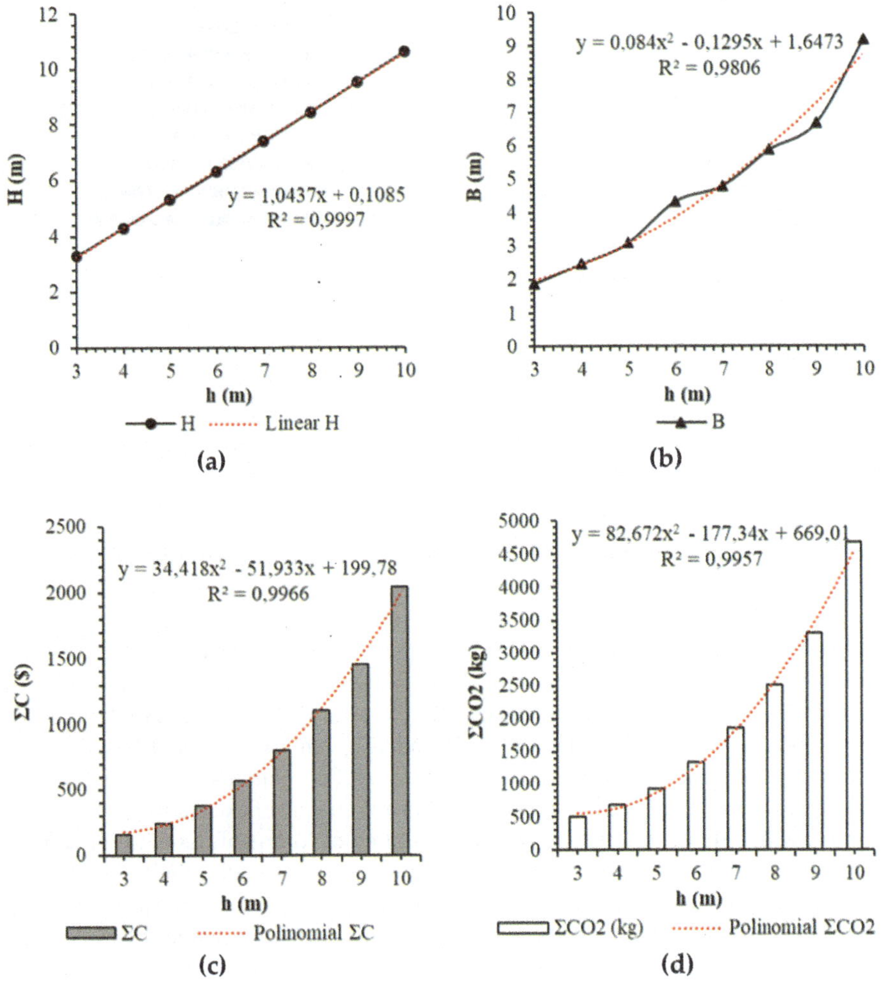

Fig. 5 The change of **a** the total height against the excavation depth; **b** The foundation width against excavation depth; **c** The total cost against the excavation depth; **d** The CO_2 emission against excavation depth

situation is based on the logic of the optimization analyses that applies necessity to ensure structural safety with the construction of minimum dimensions. The required strength of the wall is tried to be provided by strengthen of the materials or by the increase of the usage density of the materials.

4.1.2 Effects of the Change of the Shear Strength Angle of the Surrounding Soil

In the design stage of the retaining walls, the effect of the change of the shear strength angle topic is associated with the determination of the active soil pressures and the base pressure of the envisaged wall system. In the context of this study, the backfill soil and the foundation soil is both named as surrounding soil and they are assumed to be the same and formed by the use of pure cohesionless soils. The distributed loads acted to the wall are activated according to the shear strength angle and the unit weight of the backfill and foundation soil strata for frictional soils. Therefore, the effects of the change of both the shear strength angle and unit weight of the soil have to be investigated. In this section of the analyses, the shear strength of the surrounding soil is assumed to change between 27 and 38° that is representing the granular soils as beginning from very loose to reach very dense [57, 58]. A reference case is arranged to ease the comparisons conducted to find the effectiveness ratio of the change of the shear characteristics of the surrounding soil medium. The excavation depth is assumed to be 4 and 8 m respectively and the unit weight of the soil is 20 kN/m^3, the unit volume cost of the concrete is assumed to be \$100 and the unit weight cost of the steel is assumed to be \$900. The unit volume amount of the CO_2 emission of concrete is selected 376 kg and the unit weight amount of the CO_2 emission of the steel is chosen 352 kg. Figure 6a represents the change of the foundation width against the change of shear friction angle for 4 m excavation depth. According to Fig. 6a, the increase of the shear strength angle leads to narrow down the width of the base. The stability requirements have been ensured by the enlargement of the heel section of the wall till 33–34° shear strength angle existence. After the value of 33–34° shear strength of the surrounding soil the design is formed by the existence of the toe section. This special situation was investigated depending upon both the logic of both the conducted optimization analysis and the design method

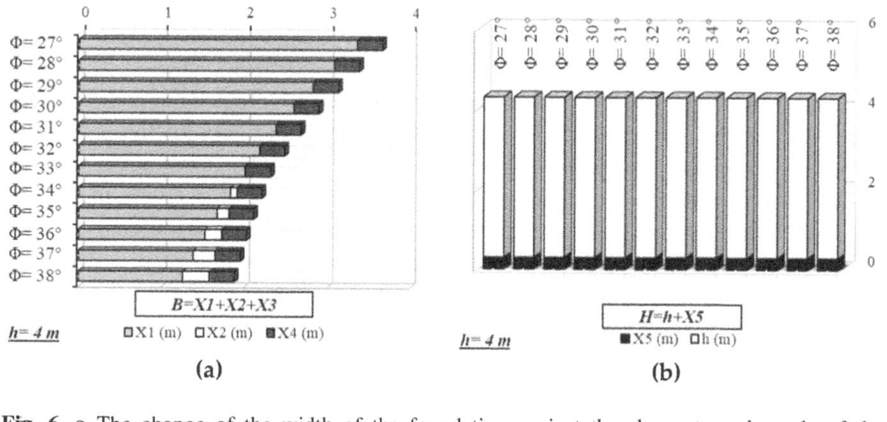

Fig. 6 **a** The change of the width of the foundation against the shear strength angle of the surrounding soil; **b** The change of the total height of the wall against the shear strength angle of the surrounding soil

of the retaining walls. The knowledge that is gained from previous studies of the authors [59] shows that the focused subjects may be the distribution of pressure along the base and the determination way of the necessitated reinforcement bars. The determination process of the reinforcement bars is shaped depending on the components of the structure. The requirement to use a toe as a wall section causes to size a different structural component depending on the maximum bending moment and forces acting on it. In such a case, the distribution of the diameter and length of the reinforcement bars may also lead the cost the high based on the variety of the used sizes of reinforcement. Therefore, the wall section may be formed by the absence of the toe after a limit strength value of the surrounding soil. Besides, the base width of the wall foundation can be changed depending on the necessity of the biggest resistant forces to balance the sliding forces and also the toe of the foundation can be widened based on the activation of the tension forces. In that condition, the necessity of the control of the generated base forces along the wall foundation is born. It is a geotechnical design requirement that the maximum base pressure calculated according to the dimensions of the designed wall components have to be smaller than the determined allowable bearing capacity of the surrounding soil strata and the minimum value of the calculated base pressure has to be bigger than zero not to cause tension. In addition, it is the meaning that the eccentricity of the designed wall section is calculated equal to B/6 or L/6, the minimum base pressure is zero. If, however, the eccentricity is calculated bigger than B/6 or L/6 then the minimum base pressure is smaller than zero. In such a situation, since the tensile strength of the soil is zero so this part of the foundation cannot transmit the loads to the soil strata [60]. The determined values of the minimum base pressures for the envisaged sections of the wall in such a case that the shear strength angle is bigger than 33°, are determined as smaller than zero for the constructions involving only the heel length. But the addition of a toe to the wall section also tolerates the pressures and distributes the loads within positive limits (as a pressure). As a result, this special case may be arose depending on the distribution of the base pressures generated associated with the designed sections for especially smaller excavation depths. In Fig. 6b, the change of the total height of the wall against the change of the soil shear strength angle is given.

According to Fig. 6b, the total height of the wall is not affected depending on the change of the shear strength angle of the soil. The awaited requirements of the wall design based on the change of the shear strength angle of the surrounding soil, are provided by the change of the design of only the base width of the wall. In Fig. 7a, the change in the total cost, the total CO_2 emission amount, the width and the total height of the wall is shown against the change of shear strength of soil. The total height of the wall has become the least affected parameter of the design as before mentioned but this situation is not similar for the other obtained outcomes of the analysis. The amount of CO_2 emission has decreased approximately 27% (from 828 to 604 kg) for the lower and upper limits of friction angle. The total cost of the system has decreased approximately 36% (from \$313 to \$200) for the lower and upper limits of friction angle. Regression analyses have been conducted to achieve the mathematical expressions of both the width of the base, the total cost and the

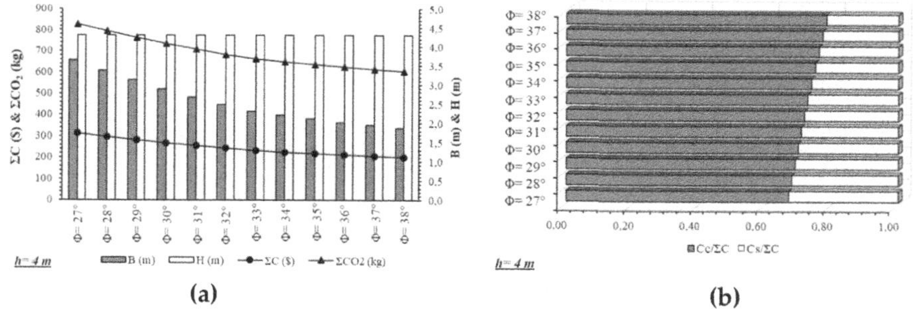

Fig. 7 **a** The change of the total cost and CO_2 emission amount against the shear strength angle of the surrounding soil; **b** The change of the cost rates of the materials against the shear strength angle of the surrounding soil

CO_2 emission amount depending on the change of the shear strength angle of the soil. Figure 8a–c represents the change of total cost, total CO_2 emission amount and the total width of the foundation change against the shear strength angle and the obtained relationships between the related parameters are also given in the figure. The R^2 value is determined 0.9997, 0.9995 and 0.9994 respectively for Fig. 8a–c.

The given relationships in Fig. 8 are obtained for only 4 m excavation depth and the mathematical expression are valid for the envisaged shear strength angles. These shear strength angles also represent the probable limits of friction that can be defined for different densities of sandy soils.

The change of shear strength of the surrounding soil has been investigated also according to the change of excavation depth. The excavation depth is assumed to be 8 m. The change of the width of the foundation and the total height of the wall is given in Fig. 9a, b respectively. The change of the width of the foundation is differentiated from the previous case analysis. The stability of the wall system is tried to be ensured with the existence of both heel and toe for all the assumed frictional change cases. The increase of the friction values also decreases the total length of the width of the foundation but there is a directly proportional increase of the length of the toe component although the bottom thickness of the stem and the heel is decreasing. This condition may be related to the comparison process of the base pressures with the allowable bearing capacity value as before mentioned. The increase of the toe of the wall section may rise depending on the determination of the base pressures not to generate tension along the foundation. In Fig. 9b, the ineffectuality of the change of the friction value on the total height of the wall can be seen.

Figure 10a is achieved to show the change of total cost, total amount of CO_2 amount and the dimensions depending on the change of friction of the soil strata. It can be said that the total amount of CO_2 emission is affected more than the cost of the system based on the change of friction. The amount of CO_2 emission has decreased approximately 39% (from 3311 to 2006 kg) for the lower and upper limits of friction angle. The total cost of the system has decreased approximately 44% (from \$1470 to \$829) for the lower and upper limits of friction angle. It is an interesting result

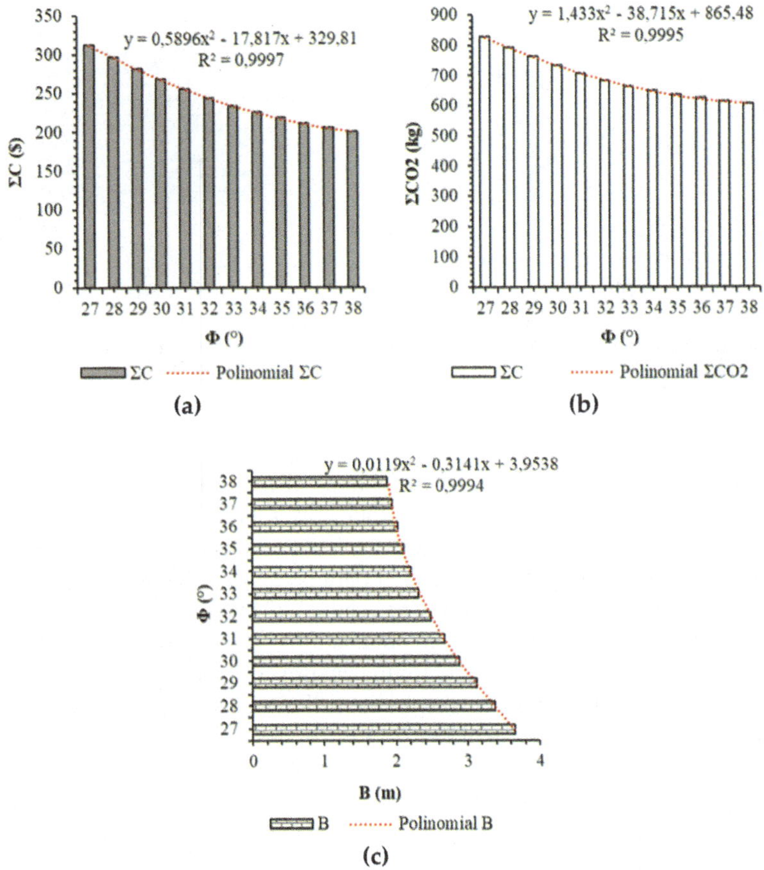

Fig. 8 The relationship between the shear strength angle of surrounding soil and **a** the total cost; **b** The total amount of CO_2 emission; **c** The width of the foundation

that this decreasement ratio is valid for the excavation depths equal and bigger than 6 m excavations. Figure 10b shows the rate of consuming materials to the total cost. The used amount of the concrete will be increased depending on the increase of the strength of soil, but the consumed amount of the steel is decreased. In addition, regression analyses are conducted to achieve the mathematical relationship between the total cost, the amount of the total CO_2 emission, the width of the foundation and the shear strength angle individually. The curve fitting option has been ensured with proper correctness with the use of polynomial functions for all the foreseen variants in Fig. 11.

In Fig. 11, the R^2 values of the attained mathematical expressions are 0.9999, 0.9982 and 0.9998 respectively. These results are satisfactory for the excavation depths equal and bigger than 6 m depth within the envisaged limits of shear strengths of this study.

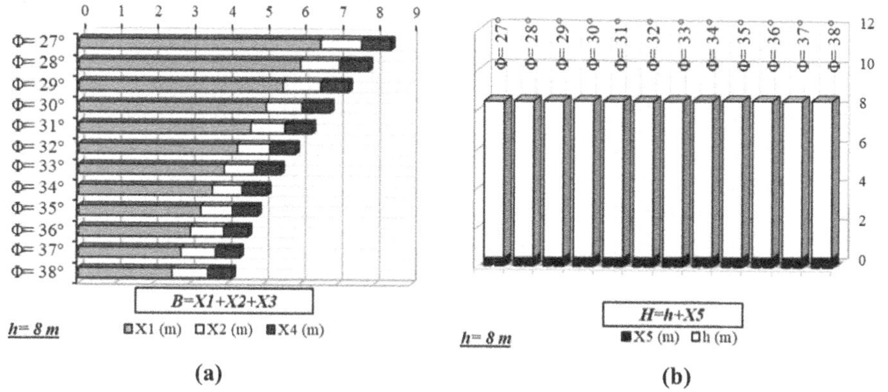

Fig. 9 **a** The change of the width of the foundation against the shear strength angle of the surrounding soil; **b** The change of the total height of the wall against the shear strength angle of the surrounding soil

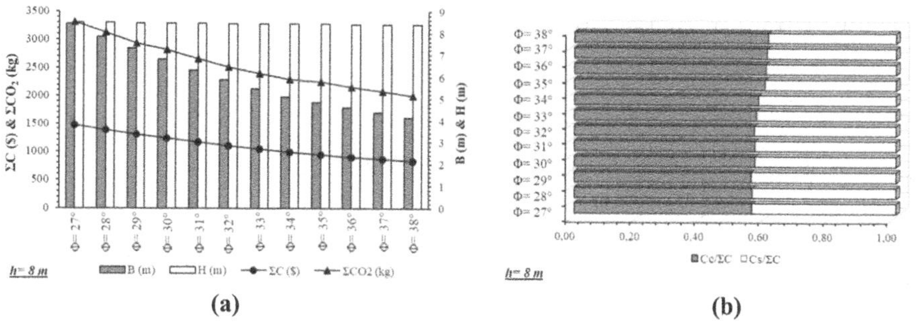

Fig. 10 **a** The change of the total cost and CO_2 emission amount against the shear strength angle of the surrounding soil; **b** The change of the cost rates of the materials against the shear strength angle of the surrounding soil

4.1.3 Effects of the Change of the Unit Weight of the Surrounding Soil

The change of the unit weight of the soil is also investigated within the context of this study. Because, the unit weight of the soil is a direct component of the expressions that are used to calculate the activated forces based on the soil existence. The unit weight of the surrounding soil has been changed beginning from 14 kN/m^3 to reach 21 kN/m^3. In order to ease the comparison of the effect of the soil unit weight, the other variants of the wall design problem have been selected as constant values. The excavation depth is selected 4 m and 8 m respectively and the shear strength angle of the surrounding soil is 32°. The unit volume cost of the concrete is assumed to be \$100 and the unit weight cost of the steel is assumed to be \$900. The unit volume amount of the CO_2 emission of concrete is selected 376 kg and the unit weight amount of the CO_2 emission of the steel is chosen 352 kg. Figure 12a shows the

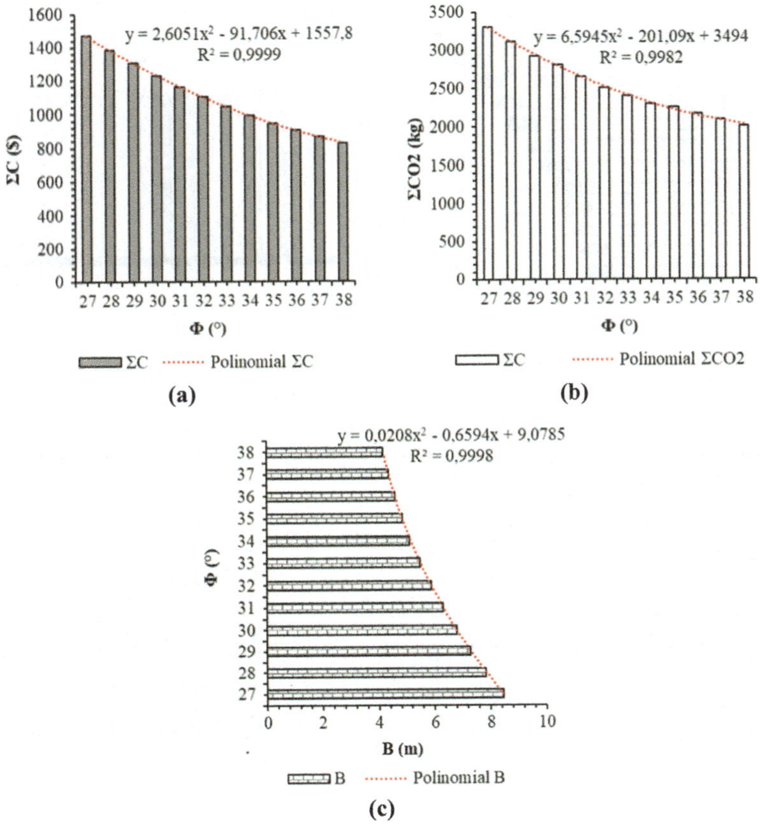

Fig. 11 The relationship between the shear strength angle of surrounding soil and **a** the total cost; **b** The total amount of CO_2 emission; **c** The width of the foundation

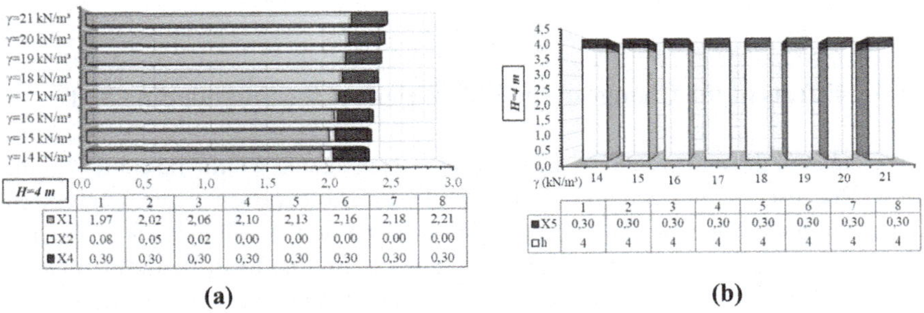

Fig. 12 **a** The change of total width of the foundation against the change of soil unit weight; **b** The change of the total height of the wall against the unit weight of soil

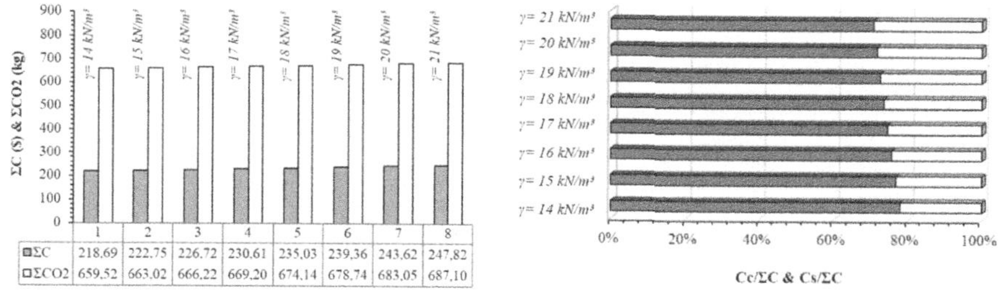

(a) (b)

Fig. 13 **a** The change in the total cost and the amount of total CO_2 emission against the unit weight of the surrounding soil; **b** The change of cost rates of the materials against the unit weight of the surrounding soil

change of the width of the wall foundation against the change of soil unit weight for 4 m excavation depth. The increase of the soil unit weight leads to raise the effected horizontal active force so it is an expected behavior to decrease the width of the foundation against the decrease of the soil unit weight. But, in a contrary way, the increase of the unit weight of the soil also increases the soil weight retained on the heel of the wall at the backfill side of the section (W_s). This is the meaning that the resistant forces are raised due to lengthen of the heel side if the base pressures are not generated as tension. In such a case, the increase of the soil unit weight to 21 kN/m^3 from 14 kN/m^3 causes to increase the foundation width of the wall only 6%. In Fig. 12b the irresponsive situation of the total height of the wall is given depending on the change of the soil unit weight.

Figure 13a represents the situation of the total cost and the amount of the total CO_2 emission against the change of the soil unit weight for 4 m excavation depth. The total cost of the system has increased 13% (from $219 to $248) in such a case that if the unit weight of the soil is raised to the envisaged upper limit from the lower limit. The amount of the total CO_2 emission is similarly increased but the ratio of the increment is lower than the total cost.

The mentioned rates of the alteration are determined 4% (from 659 to 687 kg) for the envisaged boundaries of soil unit weight. Figure 13b represents the cost rates of the consumed materials against the unit weight of the surrounding soil. There is a significant ratio difference between the concrete and steel material. The increase of the soil unit weight leads to decrease the amount of the used concrete material but this change has not generated an occurrence to make the steel material denser than concrete. The increase ratio of the $C_c/\Sigma C$ can be determined 10% between the upper and lower boundaries of soil unit weight.

The analyses are repeated in the case that the excavation depth is assumed to be 8 m. In such a case, the increase of the soil unit weight to 21 kN/m^3 from 14 kN/m^3 causes to increase the foundation width of the wall 15%. The total heights of the wall are also given for different soil unit weights in Fig. 14b. The increase of the

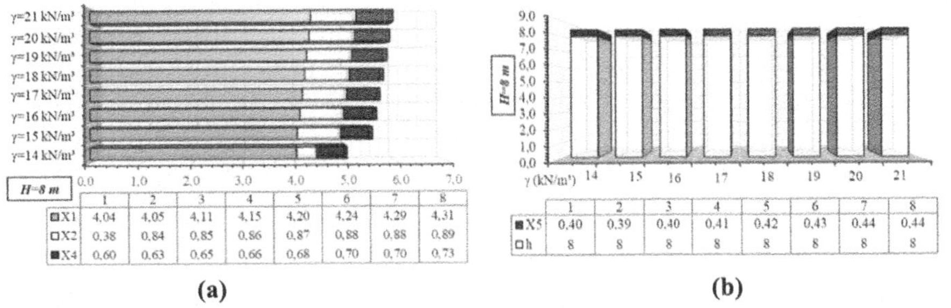

Fig. 14 **a** The change of total width of the foundation against the change of soil unit weight; **b** The change of the total height of the wall against the unit weight of soil

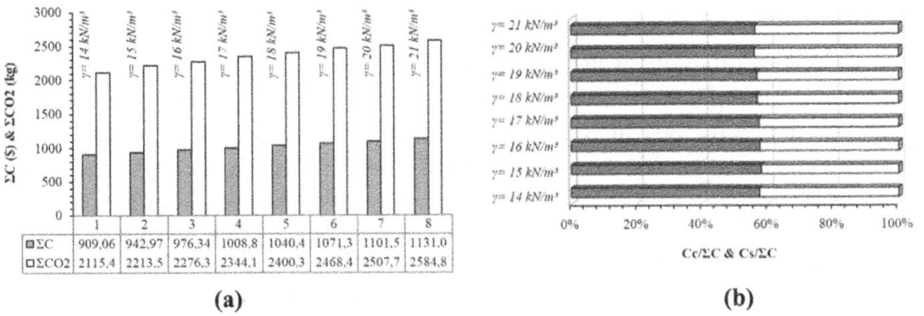

Fig. 15 **a** The change in the total cost and the amount of total CO_2 emission against the unit weight of the surrounding soil; **b** The change of cost rates of the materials against the unit weight of the surrounding soil

unit weight of the soil also increases the thickness of the foundation. The maximum calculated increase of the foundation thickness is approximately 10%.

The change in the total cost and the amount of the CO_2 emission is given in Fig. 15a. The total cost of the system has increased 24% (from $909 to $1131) in such a case that if the unit weight of the soil is raised to the envisaged upper limit from the lower limit. The mentioned rates of the CO_2 emission alteration are determined 22% (from 2115 to 2585 kg) for the envisaged boundaries of soil unit weight.

4.1.4 Effects of the Change of the Unit Volume Cost of the Concrete

The unit volume cost of the concrete is a direct component of Eq. 11 which is used to find the ultimate value of total costs. Therefore, it is important to obtain the difference that can be occurred depending upon the change of concrete cost. The shear strength of the surrounding soil and the unit weight is selected as 32° and 20 kN/m^3 respectively for composing a reference case. The unit volume cost of the concrete is selected $50, $75, $100, $125, $150 and the unit weight cost of the steel is assumed to be $700 and $1100 respectively. These values of the steel costs represent the upper and lower limits of the envisaged costs that are selected considering the literature studies [54, 55]. The evaluations are done according to the change of 4–6–8–10 m excavation depths. The results of the analyses are shown with arranged tables. The outcomes of the analyses that are performed with the use of the C_s value as constant $700, is given in Table 4. The increase of the unit volume cost of the concrete to threefold value causes to increase the total costs of the wall as expected. These increase ratio is determined 123%, 91%, 88%, 88% for 4, 6, 8, 10 m excavation depths respectively. The width of the foundation is decreased to 0.2, 1.3, 6.7, 1.81% and the total height of the wall is reduced to 0.0001%, 0.98%, 0.96%, 1.29% for 4, 6, 8, 10 m excavation depths respectively. The width of the foundation has raised to 0.2, 1.3, 6.7, 1.81% of the width of the base calculated for the case that the cost of the concrete is assumed $50. In addition, the total height of the wall

Table 4 The effects of the change of concrete materials costs for different depths ($C_s = \$700$)

Cc ($)	ΣC ($)	ΣCO_2 (kg)	B (m)	H (m)	Cc/ΣC	Cs/ΣC
h = 4 m						
50	140.93	693.52	2.49	4.30	0.63	0.37
75	184.64	683.05	2.48	4.30	0.71	0.29
100	228.26	683.05	2.48	4.30	0.76	0.24
125	271.87	683.05	2.48	4.30	0.80	0.20
150	315.49	683.05	2.48	4.30	0.83	0.17
h = 6 m						
50	343.92	1473.75	3.95	6.36	0.54	0.46
75	432.19	1355.84	4.35	6.30	0.58	0.42
100	510.77	1254.81	3.91	6.31	0.60	0.40
125	585.57	1213.44	3.90	6.30	0.63	0.37
150	658.00	1191.91	3.90	6.30	0.65	0.35
h = 8 m						
50	671.72	2872.54	6.08	8.48	0.54	0.46
75	839.36	2547.01	5.89	8.44	0.56	0.44
100	989.31	2358.71	5.77	8.42	0.58	0.42
125	1128.24	2257.05	5.68	8.40	0.60	0.40
150	1260.46	2206.03	5.68	8.39	0.62	0.38
h = 10 m						
50	1144.39	4846.32	7.72	10.65	0.53	0.47
75	1434.76	4397.04	7.66	10.59	0.57	0.43
100	1691.31	4061.01	7.63	10.56	0.58	0.42
125	1927.64	3834.59	7.60	10.53	0.59	0.41
150	2149.89	3681.54	7.58	10.51	0.60	0.40

is increased to 0.0001%, 0.98%, 0.96%, 1.29% for 4, 6, 8, 10 m excavation depths respectively. Depending on this small change of the dimensions of the system, the total amount of the CO_2 emission has been decreased to 1.51%, 19.1%, 23.2%, 24% for 4, 6, 8, 10 m excavation depths respectively.

The evaluation of the cost rates determined for 4 m excavation depth shows that the cost rate of the concrete has increased approximately 32% and the cost rate of the steel has decreased 46% between the lower and upper limits of the defined concrete unit cost values. If the individual cost values are calculated for each of the materials, the ultimate change in the concrete cost is determined 195% bigger (from $88.73 to $261.70) and the ultimate change in the steel cost is determined 3% bigger (from $52.20 to $53.79) for a threefold increment of the concrete cost. If the evaluation of the cost rates determined for 6 m excavation depth, the cost rate of the concrete has increased approximately 21% and the cost rate of the steel has decreased 25% between the lower and upper limits of the defined concrete unit cost values.

If the individual cost values are calculated for each of the materials, the ultimate change in the concrete cost is determined 131% bigger (from $185.38 to $429.70) and the ultimate change in the steel cost is determined 44% bigger (from $158.55 to $228.30) for a threefold increment of the concrete cost. This evaluation is applied for also 8 m excavation depth and the cost rate of the concrete has increased approximately 16% and the cost rate of the steel has decreased 18% between the lower and upper limits of the defined concrete unit cost values. If the individual cost values are calculated for each of the materials, the ultimate change in the concrete cost is determined 117% bigger (from $361.22 to $784.60) and the ultimate change in the steel cost is determined 53% bigger (from $310.5 to $475.86) for a threefold increment of the concrete cost. The last one is the evaluations for 10 m excavation depth and the cost rate of the concrete has increased approximately 13% and the cost rate of the steel has decreased 15% between the lower and upper limits of the defined concrete unit cost values. If the individual cost values are calculated for each of the materials, the ultimate change in the concrete cost is determined 113% bigger (from $608.63 to $1297.75) and the ultimate change in the steel cost is determined 59% bigger (from $535.76 to $852.14) for a threefold increment of the concrete cost. All these results show that the increase in the unit cost of concrete doesn't reflect on the total cost at the same increase rate. This outcome can be defined as the advantage of the use of an optimization design process. In addition, deepen of the excavation leads to reduce the relative cost rate difference between concrete and steel. It will be true to say that the amount of the steel is increased during the rise of the concrete unit prices. This situation can be directly seen from the determined individual costs. Table 6 gives the results of the analyses conducted to investigate the change of the unit volume cost of the concrete again but the constant value of the unit weight cost of steel is changed to be $1100. These analyses are performed to find the differentiation tendency of the design between the upper and lower bounds of the material costs. The calculations are conducted for four different excavation depths.

If the excavation depth is assumed to be 4 m, the relative change in the cost of the consumed concrete has increased 174% (from $95.59 to $261.70) and the cost of the consumed steel has increased 15% (from $73.44 to $84.52). In addition, if the unit volume cost of the concrete is raised three times more than to its first cost, the rate of the concrete cost to the total cost has increased approximately 34% and the rate of the cost of the steel to the total cost has decreased 44%. Then the excavation depth is assumed to be 6 m, the relative change in the cost of the consumed concrete is increased 107% (from $224.14 to $466.12) and the cost of the consumed steel has increased 59% (from $197.16 to $314.44). In addition, if the unit volume cost of the concrete is raised three times more than to its first cost, the rate of the concrete cost to the total cost has increased approximately 12% and the rate of the cost of the steel to the total cost has decreased 14%.

The excavation depth is assumed to be 8 m, the relative change in the cost of the consumed concrete has increased 102% (from $431.43 to $872.92) and the cost of the consumed steel has increased 60% (from $400.58 to $640.29). In addition, if the unit volume cost of the concrete is raised three times more than to its first cost, the rate of the concrete cost to the total cost has increased approximately 11% and the rate of the cost of the steel to the total cost has decreased 12%. The excavation depth is assumed to be 10 m, the relative change in the cost of the consumed concrete has increased 107% (from $721.15 to $1493.06) and the cost of the consumed steel has increased 55% (from $701.32 to $1093.85). In addition, if the unit volume cost of the concrete is raised three times more than to its first cost, the rate of the concrete cost to the total cost has increased approximately 13% and the rate of the cost of the steel to the total cost has decreased 14%. The similar change tendency is also obtained based on the rise of the unit weight cost of steel from $700 to $1100. However, another significant discussion has to be done between the consumed material ratios that are determined for the same depths of excavation. The comparison of Tables 4 and 5 shows that the increase of both the concrete and steel unit costs leads the design to consume less concrete. The consumed material ratio of concrete is decreased in such a case that if the steel unit cost is raised to its highest amount. For example, if the depth of excavation has been selected to be 4 m, the $C_c/\Sigma C$ value is calculated 0.63 and 0.83 and the $C_s/\Sigma C$ value is calculated 0.37 and 0.17 depending on the lower and higher limits of the unit volume cost of the concrete respectively. The increase of the unit costs of the materials leads the optimization algorithm to search for approximately equivalent costs for consumed materials individually to obtain the cost effective design.

4.1.5 Effects of the Change of the Unit Weight Cost of the Steel

The unit weight cost of the steel is also a direct component of Eq. 11 that is used to find the ultimate value of total cost. Therefore, it is significant to obtain the difference that can be occurred depending upon the change of steel cost too. The shear strength of the surrounding soil and the unit weight is selected as 32° and 20 kN/m^3 respectively for composing a reference case. The unit weight cost of the steel is selected $700, $800, $900, $1000, $1100 and the unit volume cost of the concrete is assumed to be $50 and $150 respectively. These values of the steel costs represent the upper and lower limits of the envisaged costs that are selected considering the literature studies.

Table 5 The effects of the change of concrete materials costs for different depths (C_s = \$1100)

Cc (\$)	ΣC (\$)	ΣCO_2 (kg)	B (m)	H (m)	Cc/ΣC	Cs/ΣC
h = 4 m						
50	169.03	742.31	2.51	4.30	0.57	0.43
75	215.10	697.85	2.49	4.30	0.62	0.38
100	258.99	683.05	2.48	4.30	0.67	0.33
125	302.61	683.05	2.48	4.30	0.72	0.28
150	346.23	683.05	2.48	4.30	0.76	0.24
h = 6 m						
50	421.30	1748.61	4.69	6.36	0.53	0.47
75	524.98	1540.75	4.51	6.33	0.56	0.44
100	617.54	1406.12	4.40	6.32	0.57	0.43
125	703.47	1288.62	3.90	6.30	0.56	0.44
150	780.56	1269.02	3.91	6.31	0.60	0.40
h = 8 m						
50	832.01	3372.53	6.36	8.52	0.52	0.48
75	1029.53	2917.46	6.11	8.48	0.54	0.46
100	1203.77	2658.14	5.96	8.45	0.55	0.45
125	1363.88	2508.84	5.86	8.43	0.57	0.43
150	1513.21	2393.01	5.79	8.42	0.58	0.42
h = 10 m						
50	1422.47	5647.48	8.01	10.69	0.51	0.49
75	1754.61	4887.62	7.72	10.64	0.53	0.47
100	2053.96	4616.59	7.70	10.64	0.56	0.44
125	2329.98	4309.15	7.66	10.60	0.57	0.43
150	2586.91	4092.64	7.63	10.57	0.58	0.42

The evaluations are done according to the change of 4–6–8–10 m excavation depths. Table 6 gives the results of the analyses conducted for C_c = \$50 (the lower limit of the unit volume cost of the concrete) and includes the data of all the envisaged excavation depths. Besides, Table 7 shows the results of the analyses performed for C_c = \$150 (the upper limit of the unit volume cost of the concrete). The excavation depth is assumed to be 4 m and C_c = \$50, the relative change in the cost of the consumed concrete is increased 108% (from \$88.73 to \$195.59) and the cost of the consumed steel has increased 41% (from \$52.20 to \$140.7). In addition, if the unit volume cost of the steel is raised 57% more than to its first cost, the rate of the concrete cost to the total cost has decreased approximately 10% and the rate of the cost of the steel to the total cost has increased 17%. The excavation depth is assumed to be 4 m and C_c = \$150, the relative change in the cost of the consumed concrete is not changed (\$261.70) and the cost of the consumed steel has increased 57% (from \$53.79 to \$84.52). In addition, if the unit volume cost of the steel is raised 57%

Table 6 The effects of the change of concrete materials costs for different depths ($C_c = \$50$)

Cs ($)	ΣC ($)	ΣCO_2 (kg)	B (m)	H (m)	Cc/ΣC	Cs/ΣC
h = 4 m						
700	140.93	693.52	2.49	4.30	0.63	0.37
800	148.26	706.40	2.49	4.30	0.61	0.39
900	155.37	718.87	2.50	4.30	0.59	0.41
1000	162.28	730.99	2.50	4.30	0.58	0.42
1100	169.03	742.31	2.51	4.30	0.57	0.43
h = 6 m						
700	343.92	1473.75	3.95	6.36	0.54	0.46
800	363.88	1581.50	4.54	6.34	0.55	0.45
900	383.88	1643.25	4.59	6.34	0.55	0.45
1000	402.97	1688.76	4.63	6.35	0.54	0.46
1100	421.30	1748.61	4.69	6.36	0.53	0.47
h = 8 m						
700	671.72	2872.54	6.08	8.48	0.54	0.46
800	714.80	3003.48	6.15	8.49	0.53	0.47
900	755.65	3125.81	6.22	8.50	0.53	0.47
1000	794.62	3241.26	6.28	8.51	0.52	0.48
1100	832.01	3372.53	6.36	8.52	0.52	0.48
h = 10 m						
700	1144.39	4846.32	7.72	10.65	0.53	0.47
800	1219.21	5054.66	7.79	10.66	0.52	0.48
900	1290.16	5273.33	7.87	10.67	0.52	0.48
1000	1358.03	5397.15	7.93	10.68	0.51	0.49
1100	1422.47	5647.48	8.01	10.69	0.51	0.49

more than to its first cost, the rate of the concrete cost to the total cost has decreased approximately 9% and the rate of the cost of the steel to the total cost has increased 43%.

The excavation depth is assumed to be 6 m and $C_c = \$50$, the relative change in the cost of the consumed concrete has increased 20% (from \$185.38 to \$224.14) and the cost of the consumed steel has increased 24% (from \$158.55 to \$197.16). In addition, if the unit volume cost of the steel is raised 57% more than to its first cost, the rate of the concrete cost to the total cost has decreased approximately 1% and the rate of the cost of the steel to the total cost has increased 2%.

The excavation depth is assumed to be 6 m and $C_c = \$150$, the relative change in the cost of the consumed concrete has increased 8% (from \$429.70 to \$466.12) and the cost of the consumed steel has increased 37% (from \$228.30 to \$314.44). In addition, if the unit volume cost of the steel is raised 57% more than to its first cost, the rate of the concrete cost to the total cost has decreased approximately 9% and the

Table 7 The effects of the change of concrete materials costs for different depths (C_c = \$150)

Cs (\$)	ΣC (\$)	ΣCO_2 (kg)	B (m)	H (m)	Cc/ΣC	Cs/ΣC
h = 4 m						
700	315.49	683.05	2.48	4.30	0.83	0.17
800	323.17	683.05	2.48	4.30	0.81	0.19
900	330.86	683.05	2.48	4.30	0.79	0.21
1000	338.54	683.05	2.48	4.30	0.77	0.23
1100	346.23	683.05	2.48	4.30	0.76	0.24
h = 6 m						
700	658.00	1191.91	3.90	6.30	0.65	0.35
800	690.11	1207.36	3.90	6.30	0.63	0.37
900	721.26	1223.31	3.90	6.30	0.62	0.38
1000	751.58	1239.33	3.90	6.30	0.60	0.40
1100	780.56	1269.02	3.91	6.31	0.60	0.40
h = 8 m						
700	1260.46	2206.03	5.68	8.39	0.62	0.38
800	1327.72	2236.36	5.68	8.40	0.60	0.40
900	1392.29	2290.63	5.72	8.40	0.59	0.41
1000	1454.00	2339.41	5.76	8.41	0.58	0.42
1100	1513.21	2393.01	5.79	8.42	0.58	0.42
h = 10 m						
700	2149.89	3681.54	7.58	10.51	0.60	0.40
800	2267.87	3801.09	7.59	10.53	0.60	0.40
900	2379.97	3888.59	7.61	10.55	0.59	0.41
1000	2485.60	3983.05	7.62	10.56	0.58	0.42
1100	2586.91	4092.64	7.63	10.57	0.58	0.42

rate of the cost of the steel to the total cost has increased 16%. The excavation depth is assumed to be 8 m and C_c = \$50, the relative change in the cost of the consumed concrete has increased 20% (from \$361.22 to \$431.43) and the cost of the consumed steel has increased 29% (from \$310.50 to \$400.58). In addition, if the unit volume cost of the steel is raised 57% more than to its first cost, the rate of the concrete cost to the total cost has decreased approximately 3% and the rate of the cost of the steel to the total cost has increased 4%.

The excavation depth is assumed to be 8 m and C_c = \$150, the relative change in the cost of the consumed concrete has increased 11% (from \$784.60 to \$872.92) and the cost of the consumed steel has increased 34% (from \$475.86 to \$640.29). In addition, if the unit volume cost of the steel is raised 57% more than to its first cost, the rate of the concrete cost to the total cost has decreased approximately 7% and the rate of the cost of the steel to the total cost has increased 12%. The excavation depth is assumed to be 10 m and C_c = \$50, the relative change in the cost of the consumed concrete has increased 190% (from \$608.63 to \$721.15) and the cost of the consumed steel has increased 30% (from \$535.76 to \$701.32).

In addition, if the unit volume cost of the steel is raised 57% more than to its first cost, the rate of the concrete cost to the total cost has decreased approximately 5% and the rate of the cost of the steel to the total cost has increased 5%. The excavation depth is assumed to be 10 m and C_c = \$150, the relative change in the cost of the consumed concrete has increased 15% (from \$1397.75 to \$1493.06) and the cost of the consumed steel has increased 28% (from \$852.14 to \$1093.85). In addition, if the unit volume cost of the steel is raised 57% more than to its first cost, the rate of the concrete cost to the total cost has decreased approximately 5% and the rate of the cost of the steel to the total cost has increased 6%. These results of the analyses also indicate that there is a tendency to equal the costs of consuming concrete and steel with the increasing of the excavation depth. The increasing value of the steel costs leads to use concrete more than steel to ensure the requirements of design. But, in addition, the rising of the concrete costs coupled with the steel costs causes to balance both of the costs to an equal value of cost.

4.2 *The Minimization of CO_2 Emission*

In order to discuss the effects of surrounding soil and the structural material properties on the sustainable design problem of RC retaining walls, all the mentioned design variants are investigated as a sub-title of this section individually. The effect of the excavation depth, the shear strength angle and unit weight of the surrounding soil, the unit emission amounts of the concrete and steel is determined with the performed optimization analyses aimed to minimize the CO_2 emission by ensuring both the geotechnical stability and structural requirements. In addition, Eq. 12 is used to conduct the optimization process for the minimization of CO_2 emission. The effects of the different unit cost values of the materials are not investigated for the minimization of CO_2 emission because the objective function of emission minimization does not include any terms about the cost values. Therefore, the unit cost of the materials is not selected as a variable of the analyses conducted in this section. In addition, three different CO_2 emission value couples have been used to define the CO_2 emission amounts of both the concrete and the steel as mentioned in Sect. 3. The analyses conducted for the different couples of CO_2 emission amounts are defined as groups. The first group (Type 1) analysis considers the CO_2 emission amount of concrete and steel, 376 kg (for 30 MPa strength) and 352 kg (for recycled type of steel with 420 MPa strength) respectively. The second group (Type 2) analysis assumes the CO_2 emission amount of concrete and steel; 143.48 kg (for the concrete (HA-30)) and 3010 kg (for 420 MPa strength) respectively. The last group (Type 3) considers the CO_2 emission amount of concrete and steel, 143.48 kg (for the concrete (HA-30)) and 352 kg (for recycled type of steel with 420 MPa strength) respectively. In order to compare the results of the solutions of objective function 1 and 2, the results of the analyses of objective function 1 is also given in this section. The results of the objective function 1 are defined as Type 0.

4.2.1 Effect of the Change of Excavation Depth

The effects of the change of excavation depth on the minimization process of the CO_2 emission are investigated subjected to the dimensions, cost and CO_2 emission values of the envisaged RC retaining walls. The analyses performed between a specific excavation depth range (from 3 to 10 m). Some of the parameters of design are selected as constant values to evaluate only the influence of excavation depth change on the minimization process of total CO_2 emission. The unit weight of the soil (γ) and the shear strength angle (Φ) is assumed to be 20 kN/m^3 and 32° respectively. The unit volume cost of the concrete and the unit weight cost of the steel is \$100 and \$900 respectively. The effects of the change of the types of the analysis depending on the envisaged values of the CO_2 emissions are also investigated. In this context, Fig. 16a represents the change of the total amount of the CO_2 emission value against the type of analysis and the change of excavation depth. The numbers that are beginning from 1 to reach 8 represents the excavation depths beginning from 3 m to reach 10 m respectively. According to Fig. 16a, Type 3 analysis ensures to determine the minimum amount of CO_2 emission value. This situation is due to the fact that the minimum amounts of material emission values are defined to define Type 3 analysis. Besides, the minimum unit emission value of the concrete is defined in the Type 2 analysis and the minimum unit emission value of the steel is defined in Type 1 analysis. The Type 2 analysis gives the smallest amounts of CO_2 emission (except Type 3) till 6 m excavation depth. After this depth, the emission value of the system is increased unexpectedly in comparison with Type 1. Although the total amount of CO_2 emission in Type 1 remains bigger than Type 2 results, the depth 6 m forms a boundary situation for both analyses. Besides Type 1 analysis gives the similar emission values with Type 0 analysis till the depth 5 m. Then, the maximum value

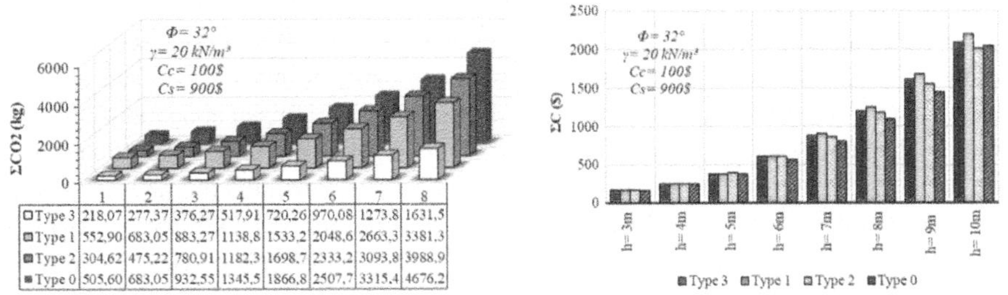

(a) (b)

Fig. 16 **a** The change of the CO_2 emission values against the type of the analysis and the excavation depth; **b** The change of the total cost of the wall system against the type of the analysis and the excavation depth

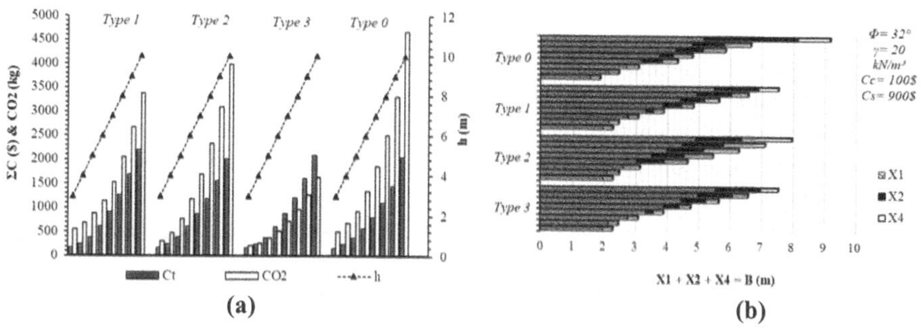

Fig. 17 **a** The integrated presentation of both the total cost-the ultimate CO_2 emission amount-the excavation depth; **b** The change of the width of the foundation against the depth of excavation

of CO_2 emission is obtained by Type 0 analysis. The results show that, in the view to minimize the CO_2 emission amount, it will be beneficial to utilize from the Type 3 analysis which uses the minimum value of material emission values. But the total cost of the system has to be controlled also to achieve both sustainable and cost-effective design methods. In Fig. 16b, the change in the total cost of the wall system is given against the type of the analysis and the change of excavation depth. It is clear that the cost of the system remains constant till 6 m excavation depth. The total cost of the system is decreased for Type 0 in comparison with the other types of the analyses after 6 m excavation depth. This result is predictable because the main aim of Type 0 analysis is to minimize the cost and the objective function didn't take into consideration the minimization process of the CO_2 emission. Besides, Type 2 analyses gives close results with Type 0 analysis. But Type 2 analysis is performed with the calculation of objective function 2 that is focused on the minimization of the CO_2 emission. The closeness of the results of two types of analyses makes think about the similarity of the dimensions suggested for the design. This condition guides the authors to compare the change of dimensions of the system associated with the total cost and total CO_2 emission amount. In this context, Fig. 17a, b represents the change of bot the amount of CO_2 emission and total cost against the depth of the excavation and the width of the foundation against the change of excavation depth.

The advantage of the usage of Type 3 analysis can be seen obviously from Fig. 17a. The relationship of cost and CO_2 emission can be said to be balanced for Type 3 analysis. This figure also proofs the disadvantage of the usage of the design methods that are only focused on the minimization of cost. Moreover, another specific condition has to be pointed. The steel materials that are used to model Type 3 analysis has been differentiated from other kinds by being manufactured with a recycling process. By the way, it is a significant outcome that to show the importance of selecting the proper structural materials to decrease the consumed material and to ensure a sustainable design. In Fig. 17b, the difference of the calculated widths of Type 0 analysis from other types can be seen significantly. Although the aim of Type 0 analysis is to minimization of the cost, this minimization process is not tried to be ensured by making the dimensions of the system smaller. Depending on the foreseen unit costs of materials for this part of the study, the optimization process tends to reduce the amount of steel due to the higher cost rate used at the design stage and ensure to raise the amount of concrete to stay at the cost-effective side. This condition may be the reason of the attainment of the biggest CO_2 emission value depending on the huge volume of the concrete used at the design stage if the unit emission value of the concrete was the biggest.

In Fig. 18a, the change of the total wall height is shown depending on the change of the excavation depth and the type of the analysis. The excavation depths suggested for the wall systems are all the same for different types of the analysis. Therefore, the difference between the wall heights are arisen depending on the change of the thicknesses of the foundation. The change of the thicknesses of the foundation against the analysis types is shown in Fig. 18b.

In Fig. 18b, the thickness of the foundation has remained same till 6 m excavation depth. But the deepening of the excavation leads to increase the thickness of the base of the foundation approximately 100%, 66%, 126%, 66% for Type 0, Type 1, Type 2 and Type 3 analysis respectively. These increments of the percentages are determined for the condition that the rise of the excavation depth from 6 to 10 m. In Fig. 19 the change of the cost rates of the consuming materials during the design against the excavation depth and the types of the analysis is given. The change of the cost rates of the materials presents a dual similarity tendency for Type 0 and Type 2 analysis and Type 1 and Type 3 analysis individually. The change of the inclination of the Type 0 analysis given in Fig. 19, reflects that the increase of the excavation depth leads the design to make approximately equivalent both the costs of the concrete and the steel. The same inclination shape with Type 0 analysis is obtained for Type 2 analysis but the rates of the costs can't come closer. But the cost rate of the steel is tried to be remained at smaller degrees because Type 2 analysis includes the upper amount of the CO_2 emission of steel material. Therefore, the optimization analysis may try to conduct the analysis with the use of the smallest cost rate of the steel within possible safety limits. In Type 1 and Type 3 analysis similar change inclinations are achieved approximately for the same excavation depths. This condition is depending upon the

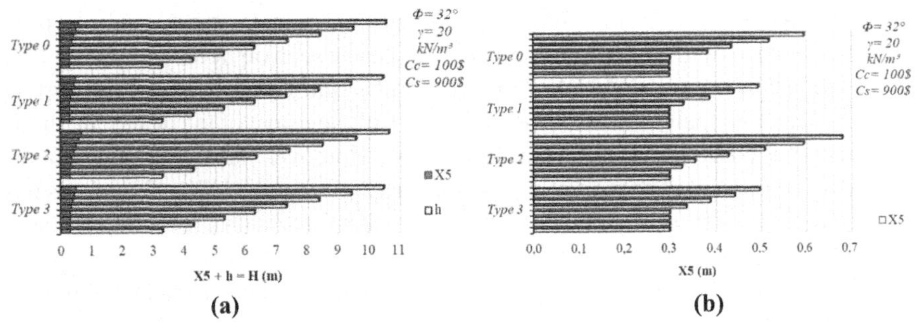

Fig. 18 **a** The change of the total height of the wall against the depth and the type of the analysis; **b** The change of the foundation thickness against the depth and the type of the analysis

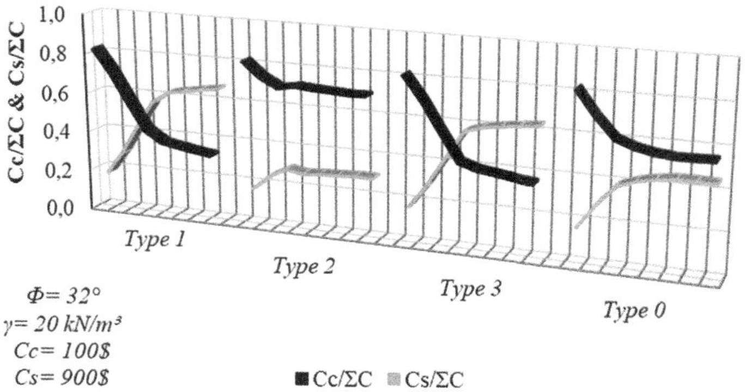

Fig. 19 The relationship between the rates of costs against the types of the analysis and the excavation depth

similarity of the dimensions and amounts of materials used for the design and also can be seen from the explanation of both Figs. 17b and 18b.

Figure 20 shows the change of the costs of materials against the change of both excavation depth and type of the analysis. According to Fig. 20, the costs of all the materials increase with increasing excavation depth as expected. It should be noted that Fig. 20 is valid only in the aforementioned conditions, as a cost pair is used to

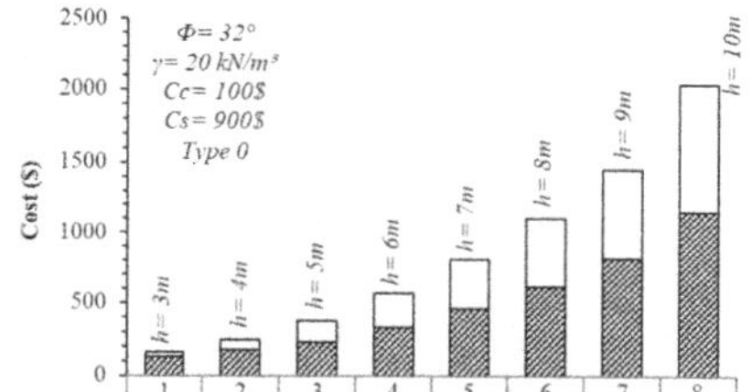

	1	2	3	4	5	6	7	8
□ Type 0 (Cs)	25,82	69,16	142,64	234,16	347,72	485,09	638,40	890,73
▨ Type 0 (Cc)	131,78	174,47	233,18	333,50	460,33	616,50	815,37	1151,0

(a)

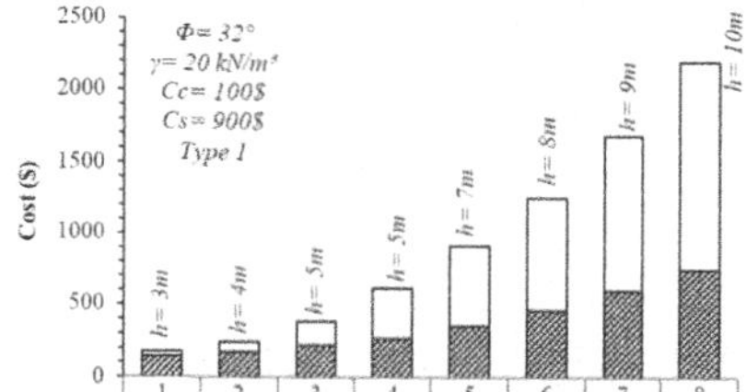

	1	2	3	4	5	6	7	8
□ Type 1 (Cs)	29,31	69,16	162,14	344,59	564,41	795,72	1090,0	1446,5
▨ Type 1 (Cc)	144,00	174,47	218,05	267,03	349,07	462,09	594,96	748,82

(b)

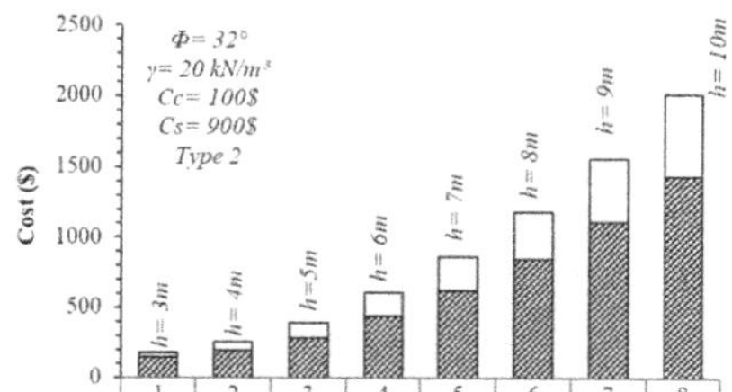

	1	2	3	4	5	6	7	8
□ Type 2 (Cs)	29,31	60,74	112,84	164,38	241,21	334,56	449,67	577,01
▨ Type 2 (Cc)	144,00	189,63	281,25	440,89	621,67	846,34	1108,1	1435,1

(c)

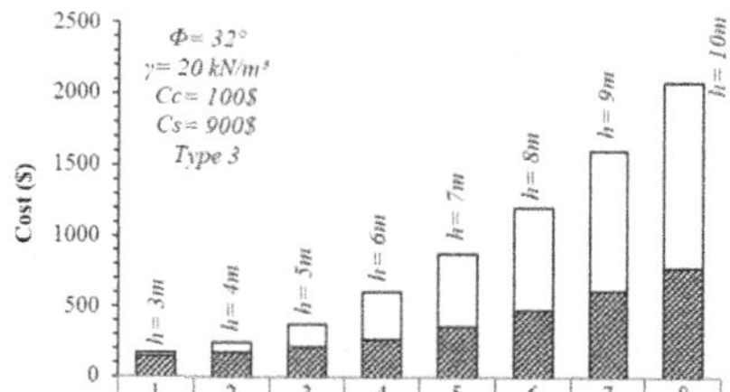

	1	2	3	4	5	6	7	8
□ Type 3 (Cs)	29,31	69,16	162,14	344,59	520,25	727,27	994,03	1305,2
▨ Type 3 (Cc)	144,00	174,47	218,05	267,03	360,18	477,86	616,87	781,35

(d)

Fig. 20 The change of the concrete and steel costs against the excavation depth and type of analysis; **a** Type 0; **b** Type 1; **c** Type 2; **d** Type 3 define concrete and steel materials. For this reason, different rates can be obtained by evaluating other analyzes made within the scope of the study. The maximum value of the cost of the consumed concrete is determined for Type 2 (Fig. 20c). Type 2 analysis has used the highest amount of CO_2 emission for steel therefore, the optimization process tries to minimize the usage of steel material. Type 0 analysis that is shown in Fig. 20a, has calculated the necessitated amount of the materials depending on the costs.

In these analyses, the unit costs of the materials have been selected as the average values of the envisaged all costs for both steel and concrete. But the increment ratio of the concrete and steel materials have not been chosen as constant. The cost of the concrete material is foreseen to be 300% of the minimum cost, and the cost of steel material is foreseen to be about 50% of the minimum cost depending on the limits that are given in the literature. Type 1 and Type 3 analysis present similar tendencies to ensure the same cost of both concrete and steel.

4.2.2 Effects of the Change of the Shear Strength Angle of the Surrounding Soil

The effects of the change of the shear strength angle of the surrounding soil are investigated based on 12 different values of foreseen frictions. A reference case is assumed to ease the comparisons conducted to find the effectiveness ratio of the change of the shear strength of the surrounding soil strata. The excavation depth is assumed to be 4 and 10 m respectively and the unit weight of the soil is 20 kN/m^3, the unit volume cost of the concrete is assumed to be \$100 and the unit weight cost of the steel is assumed to be \$900. All the analyses are conducted for the change of the type of the analysis. Figure 21a shows the results of Type 0 analysis and gives the change of total cost, total amount of CO_2 emission against the shear strength angle. The shear strength angle plays a key role in the determination process of the active pressures that are effecting along the wall. The shear strength angle is the direct component of

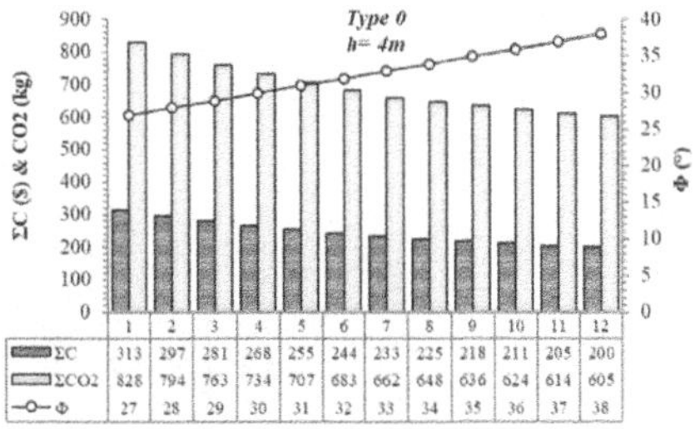

(a)

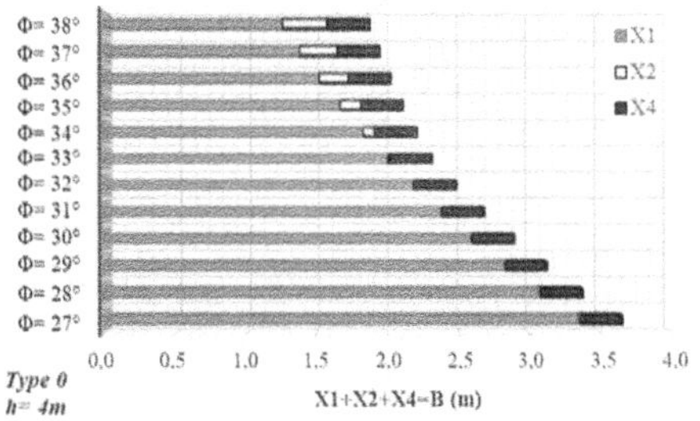

(b)

Fig. 21 **a** The change of total cost and total amount of CO_2 emission against the shear strength angle; **b** The change of the width of the foundation against the shear strength angle (Type 0, 4 m excavation depth)

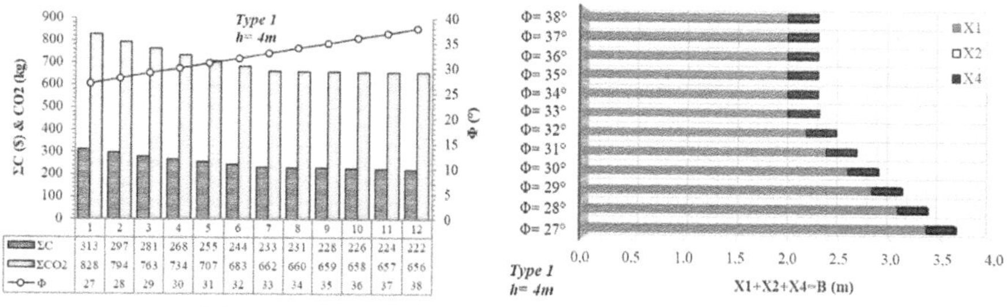

(a) (b)

Fig. 22 **a** The change of total cost and total amount of CO_2 emission against the shear strength angle; **b** The change of the width of the foundation against the shear strength angle (Type 1, 4 m excavation depth)

the mathematical expression of lateral earth pressure coefficient. The increase of the shear strength angle leads to reduce the lateral earth pressure coefficient. Therefore, it is possible to calculate smaller active pressures if the rise of the friction is occurred. This condition affects directly the dimensions of the system. For this reason, it will be a proper choice to evaluate the change of cost and emission values related to the dimensions of the system. In Fig. 21b, the change of the width of the foundation is given for 4 m excavation depth. The total height of the wall is calculated by the sum of the excavation depth and the thickness of the foundation. The excavation depth is assumed to be constant at 4 m to compare the effect of the change of only shear strength angle. The thickness of the foundation was calculated 0.3 m for all types of the analysis. That's why the total height of the wall is not assumed to be an evaluation variable. The change of the shear strength angle between the lower and upper boundaries of the envisaged values leads to reduce the width of the foundation approximately 49%. The total cost decrease happened depending on this increment is determined 36% and also the CO_2 emission has decreased 27%.

Figure 22 represents the results of the same analysis conducted with Type 1 analysis. The change of the shear strength angle between the lower and upper boundaries of the envisaged values leads to reduce the width of the foundation approximately 37%. The total cost decrease happened depending on this increment is determined 29% and also the CO_2 emission has decreased 20%.

Figure 23 represents the results of the same analysis conducted with Type 2 analysis. The change of the shear strength angle between the lower and upper boundaries of the envisaged values leads to reduce the width of the foundation approximately 37%. The total cost decrease happened depending on this increment is determined 29% and also the CO_2 emission has decreased 34%.

Figure 24 represents the results of the same analysis conducted with Type 3 analysis. The change of the shear strength angle between the lower and upper boundaries of the envisaged values leads to reduce the width of the foundation approximately 37%. The total cost decrease happened depending on this increment is determined 29% and the CO_2 emission has decreased 23%.

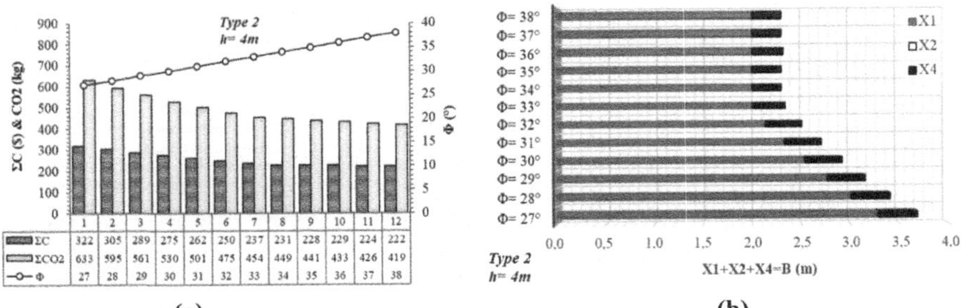

Fig. 23 **a** The change of total cost and total amount of CO_2 emission against the shear strength angle; **b** The change of the width of the foundation against the shear strength angle (Type 2, 4 m excavation depth)

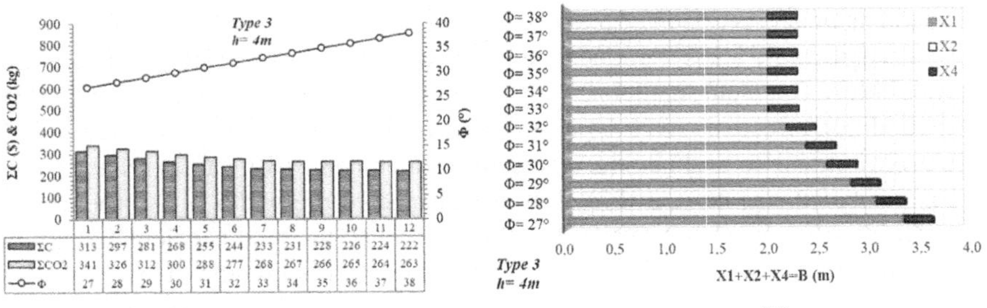

Fig. 24 **a** The change of total cost and total amount of CO_2 emission against the shear strength angle; **b** The change of the width of the foundation against the shear strength angle (Type 3, 4 m excavation depth)

As a result, the decrease of the foundation width has been determined maximum and the also the maximum cost decrease is obtained for Type 0 analysis. The decrease rate of the width of the foundation and the cost is achieved as the same for other types of the analysis but the CO_2 emission values has changed according to the used material types for design. Besides, there happened a sudden dimensional change for the condition that the shear strength angle is equal to 33°. The cost, the CO_2 emission and the dimensions of the wall system remains the same if the shear strength angle of the foundation soil strata is 33° and the excavation depth is 4 m. This condition is only available for Type 1–2–3 analysis and means that the change of soil properties especially the shear strength angle hasn't got an influential effect after a limit value of shear friction. In addition, the change that is happened for Type 0 analysis is determined 20% decrease, for the increase of friction angle from 33° to 38°. The change in the cost of the consumed materials in design is also investigated in Table 8 for some of the selected shear strength angles depending on the types of the analysis. The results obtained in Table 8 demonstrate that, in terms of the 33° shear strength

Table 8 The costs of consuming materials against the change of shear strength angle (4 m excavation depth)

Φ (°)	Type 1		Type 2		Type 3		Type 0	
	Cc ($)	Cs ($)	Cc ($)	Cs ($)	Cc ($)	Cs ($)	Cc ($)	Cs ($)
28	201	95	222	83	201	95	201	95
30	187	81	205	71	187	81	187	81
32	174	69	190	61	174	69	174	69
34	169	62	169	62	169	62	166	59
36	169	57	169	57	169	57	161	51
38	169	53	169	53	169	53	156	44

angle, the design does not change in terms of the utilization rate of concrete and steel material at the greater shear strength angles.

The same analyses are conducted for the assumption of the excavation depth 10 m. In this context, the effects of the change of the shear strength angle of the soil are evaluated with Figs. 25, 26, 27 and 28 and the figures are obtained for Type 0, Type 1, Type 2 and Type 3 analysis respectively.

Depending on the increase of the excavation depth, the behavior tendency seen for 10 m excavation depth is changed in comparison with the analyses conducted for 4 m excavation depth. The relative change that has occurred between the obtained width results of the lower and upper boundaries of shear strength is determined nearly 50% for all the types of the analyses.

On the other hand, the change of the widths of the foundation components, the amounts of the total cost and the total CO_2 emission values presents a directly proportional decrease tendency depending on the increase of the shear strength angle. These results may also exhibit that the sustainable design in the view of both the total cost and the CO_2 emission can be obtained by the use of Type 3 analysis.

In Table 9, the change of the costs of the consumed materials is given depending on the different shear strength angle values. According to Table 9, the decrease of

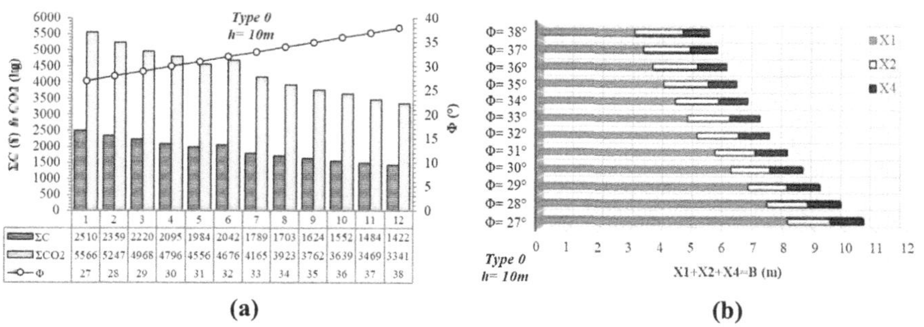

Fig. 25 **a** The change of total cost and total amount of CO_2 emission against the shear strength angle; **b** The change of the width of the foundation against the shear strength angle (Type 0, 10 m excavation depth)

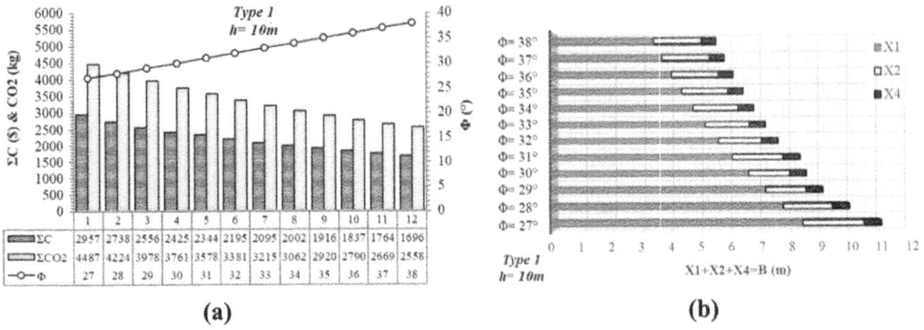

Fig. 26 **a** The change of total cost and total amount of CO_2 emission against the shear strength angle; **b** The change of the width of the foundation against the shear strength angle (Type 1, 10 m excavation depth)

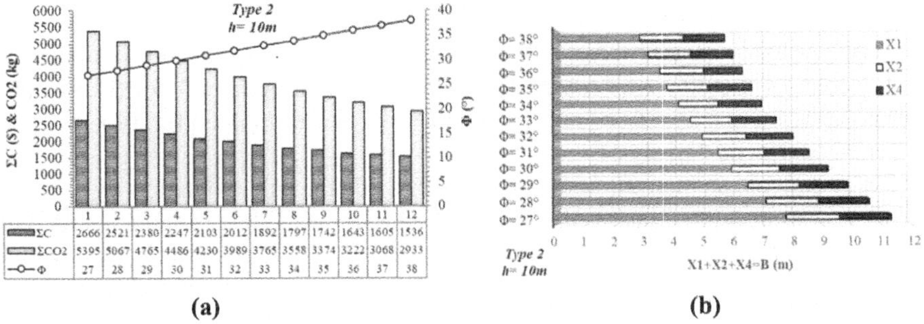

Fig. 27 **a** The change of total cost and total amount of CO_2 emission against the shear strength angle; **b** The change of the width of the foundation against the shear strength angle (Type 2, 10 m excavation depth)

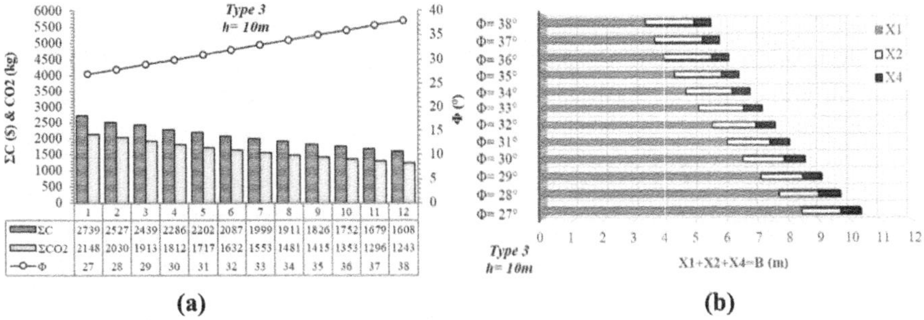

Fig. 28 **a** The change of total cost and total amount of CO_2 emission against the shear strength angle; **b** The change of the width of the foundation against the shear strength angle (Type 3, 10 m excavation depth)

Table 9 The costs of consuming materials against the change of shear strength angle (10 m excavation depth)

Φ (°)	Type 1		Type 2		Type 3		Type 0	
	Cc ($)	Cs ($)	Cc ($)	Cs ($)	Cc ($)	Cs ($)	Cc ($)	Cs ($)
28	936	1802	1761	760	998	1529	1284	1075
30	835	1590	1585	661	879	1407	1180	915
32	749	1446	1435	577	781	1305	1151	891
34	676	1325	1284	513	703	1209	967	737
36	615	1222	1190	453	639	1113	900	652
38	562	1134	1153	382	589	1019	827	596

the friction of the soil leads to consume more material. Besides, the comparison of Tables 8 and 9 shows the change of the selected dominant kind of material is also changed due to the excavation depth.

4.2.3 Effects of the Change of the Unit Weight of the Surrounding Soil

The effects of the change of the soil unit weight are also investigated depending on the minimization process of the CO_2 emission. The unit weight of the surrounding soil has been changed beginning from 14 kN/m^3 to reach 21 kN/m^3. The other variants of the wall design problem have been selected as constant values. The excavation depth is selected 4–6–8–10 m respectively and the shear strength angle of the surrounding soil is 32°. The unit volume cost of the concrete is assumed to be \$100 and the unit weight cost of the steel is assumed to be \$900. In Figs. 29, 20, 21 and 22 gives the change of the total cost and the CO_2 emission against the change of the soil unit weight. In addition, the effects of the different types of the analyses are also

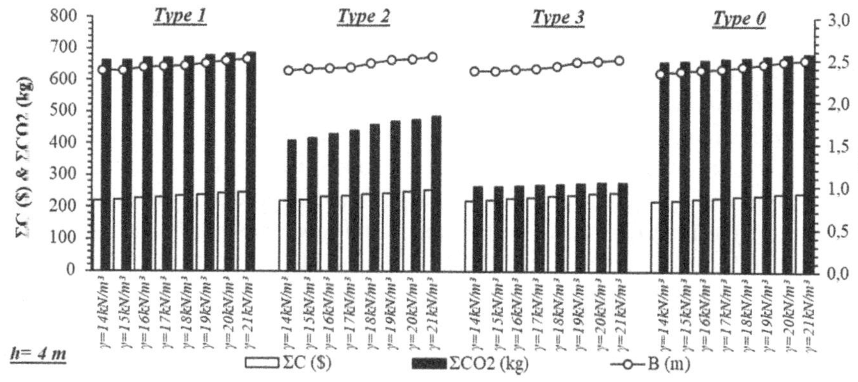

Fig. 29 The change of the total cost, the amount of CO_2 emission and foundation width against the unit weight of surrounding soil for 4 m excavation depth

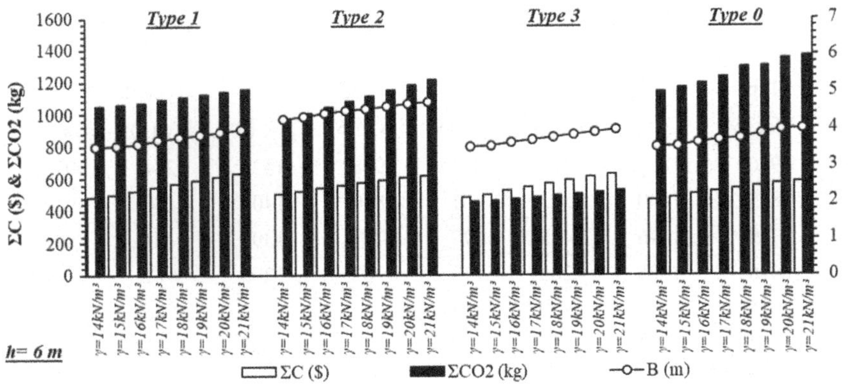

Fig. 30 The change of the total cost, the amount of CO_2 emission and foundation width against the unit weight of surrounding soil for 6 m excavation depth

exhibited. The total cost, the CO_2 emission and the width of the base has increased depending on the increase of the unit weight of the soil. According to Fig. 29, Type 1, Type 2, Type 3 and Type 0 analysis, the relative increment of the total cost between the upper and lower boundaries of envisaged soil unit weights is determined 13%, 16%, 13% and 13% respectively. The relative increment of the amount of the CO_2 emission between the upper and lower boundaries of envisaged soil unit weights is determined 4%, 9%, 6% and 4% respectively. In addition, the relative increment of the width of the wall base between the upper and lower boundaries of envisaged soil unit weights is determined 6%, 7%, 6% and 7% respectively. These results show that the most affected parameter due to the change of soil unit weight is the total cost for 4 m excavation depth.

According to Fig. 30, Type 1, Type 2, Type 3 and Type 0 analysis, the relative increment of the total cost between the upper and lower boundaries of envisaged soil unit weights is determined 31%, 22%, 30% and 23% respectively. The relative increment of the amount of the CO_2 emission between the upper and lower boundaries of envisaged soil unit weights is determined 11%, 26%, 16% and 19% respectively. In addition, the relative increment of the width of the wall base between the upper and lower boundaries of envisaged soil unit weights is determined 13%, 11%, 13% and 14% respectively. These results show that the most affected parameter due to the change of soil unit weight is the width of the foundation for 6 m excavation depth.

According to Fig. 31, Type 1, Type 2, Type 3 and Type 0 analysis, the relative increment of the total cost between the upper and lower boundaries of envisaged soil unit weights is determined 25%, 24%, 25% and 24% respectively. The relative increment of the amount of the CO_2 emission between the upper and lower boundaries of envisaged soil unit weights is determined 21%, 27%, 23% and 22% respectively. In addition, the relative increment of the width of the wall base between the upper and lower boundaries of envisaged soil unit weights is determined 15%, 12%, 15% and 18% respectively. These results show that the most affected parameter due to the change of soil unit weight is the width of the foundation for 8 m excavation depth.

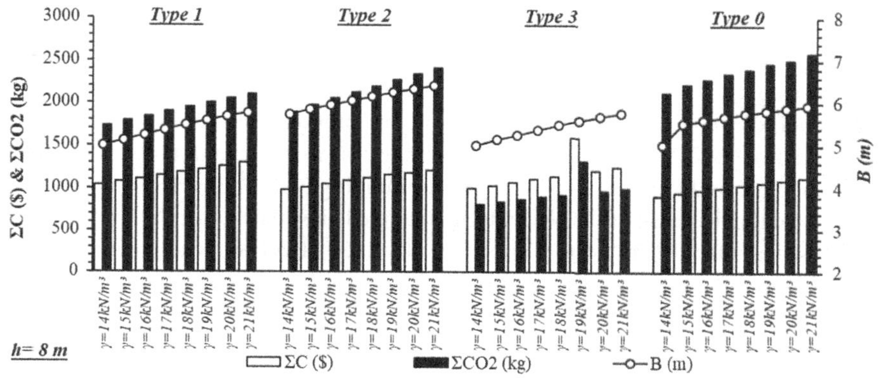

Fig. 31 The change of the total cost, the amount of CO_2 emission and foundation width against the unit weight of surrounding soil for 8 m excavation depth

According to Fig. 32, Type 1, Type 2, Type 3 and Type 0 analysis, the relative increment of the total cost between the upper and lower boundaries of envisaged soil unit weights is determined 28%, 25%, 28% and 27% respectively. The relative increment of the amount of the CO_2 emission between the upper and lower boundaries of envisaged soil unit weights is determined 23%, 28%, 25% and 26% respectively. In addition, the relative increment of the width of the wall base between the upper and lower boundaries of envisaged soil unit weights is determined 15%, 12%, 15% and 14% respectively. These results show that the most affected parameter due to the change of soil unit weight is the width of the foundation for 10 m excavation depth.

As a result, it can be seen that the unit weight of the surrounding soil is also another geotechnical significant factor on the design of RC retaining walls. But the influence ratio of the unit weight of the soil is also associated with the depth of excavation. The increase of the excavation depth leads to increase the influence ratio of the soil

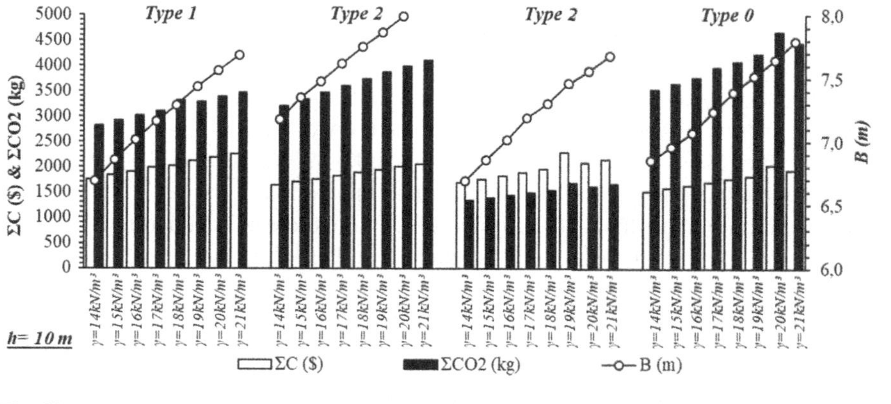

Fig. 32 The change of the total cost, the amount of CO_2 emission and foundation width against the unit weight of surrounding soil for 10 m excavation depth

unit weight. Besides, the change of the shear strength of the surrounding soil has to be evaluated in conjunction with the soil unit weight to measure the changeability of the design. The decrease of the shear strength angle causes to rise the lateral earth coefficient value and therefore, the active pressures are reduced indirectly. This situation induces a contrary case occurrence between the amounts of the shear strength angle and the unit weight of the surrounding soil. The increase of the unit weight and the decrease of the shear strength angle leads to unstable the wall system. For these reasons, the investigation of the integrated effect of these parameters are formed one of the developable side of this paper.

5 Conclusions

An extensive study was presented in this paper to evaluate the optimization of RC retaining wall design problem of sustainability with respect to the CO_2 emissions in relation with cost effectiveness. The analyses are conducted with the Harmony Search Algorithm whose applicability to the retaining wall design problems was supported by many previous studies. The retaining walls are assumed to be embedded in a pure frictional soil formation and Rankine Earth Pressure Theory is used to calculate the lateral earth pressures caused by the soil mass. The focus point of the study is formed by the evaluation of the changes of the structural material and surrounding soil properties. In the context of the paper;

- The depth of excavation is assumed to be 3–4–5–6–7–8–9–10 m to represent the usability limits of retaining walls according to the national and international sources.
- The shear strength angle of the surrounding soil is envisaged 27–28–29–30–31–32–33–34–35–36–37–38° in order to represent frictional soil formations strength characteristics.
- The unit weight of the surrounding soil is assumed to be 14–15–16–17–18–19–20–21 kN/m^3 depending on the well-known geotechnical sources to represent the probable limits of soil unit weights.
- The cost of the concrete is selected to be \$50–\$75–\$100–\$125–\$150 depending on the market research conducted all around the world.
- The cost of the steel is selected to be \$700–\$800–\$900–\$1000–\$1100 depending on the market research conducted all around the world.
- The unit amount of the CO_2 emission of C30 concrete is assumed to be 143.48 and 376 kg based on the recently conducted studies in the literature.
- The unit amount of the CO_2 emission of S420 steel is chosen to be 352 and 3010 kg based on the recently conducted studies in the literature.

The effects of all these parameters in the design problem of RC concrete retaining walls are investigated considering the cost and the CO_2 emission. The investigations are conducted with the use of two different objective functions which the first one is aiming to find the minimum cost and the second one is aiming to find the minimum

CO_2 emission. The outcomes of these solutions can be evaluated as below mentioned depending on the selected specific reference cases throughout this study.

- The change of excavation depth can be selected as the most influencer parameter of RC retaining wall design problem. The increase of the excavation depth causes to increase the total height of the wall and the length of the wall foundation with a directly proportional relationship. This condition also leads to rise the total cost and CO_2 emission amount based on the increase of consuming materials in the construction.
- Regression analyses had also performed to obtain then relationship between the variants and the excavation depth. A linear relationship between the total height of the wall and the depth of the excavation is also obtained, besides the relationship between the excavation depth and the width of the foundation, the total cost and the total CO_2 emission presents a proper compatibility with the polynomial option of conducted curve fitting.
- The change of the shear strength angle was evaluated by the assumption of constant depth of excavations which were representing the upper and lower boundaries of depth. The increase of the shear strength angle had caused to narrow down the width of the base. The cost and the CO_2 emission values have both decreased, but the rate of these decreases has reached bigger by the increase of the excavation depth. Therefore, the effects of the change of the shear strength angle cannot be investigated by the absence of the effect of the depth of excavation. The height of the wall was not affected by the change of soil shear strength angle till 6 m excavation depth. But the deepening of the excavation depth bigger than 6 m has led the wall to design the thickness of the base bigger. For this reason, the total height of the wall was increased depending on the increase of both the excavation depth and the shear strength of the soil.
- The effects of the change of the soil unit weight were investigated by the assumption of constant depth of excavations which were representing the upper and lower boundaries of depth. The increase of the soil unit weight leads to rise the width of the foundation, the total cost and the amount of CO_2 emission, but the total height of the wall is not changed for relatively smaller excavation depths. The increase of the depth of the excavation also raises the increase rate of the variants. This situation has seemed to be similar to the previously defined case about the change of the shear strength angle of the soil.
- The change of the unit volume cost of the concrete increased the total cost of the wall system. In addition, increasing the depth of the excavation further increases the effect of the increase in the unit cost of concrete. Besides, the increase of the unit volume cost of the concrete has led the design to change, depending on to ensure safety requirements with the use of more steel material based on the cost minimization scheme.
- The increasing value of the steel costs has led to use concrete more than steel to ensure the requirements of the design. But, in addition, the rising of the concrete costs coupled with the steel costs has cause to balance both of the costs to an equal value of cost.

- As a result of the conducted analyses, for the evaluation of unit cost changes of the structural materials have shown that the design of the system is modelled depending on the change rates of the costs. The further increase of the concrete costs has led the system to be narrowed if the steel costs are relatively less, to increase the amount of steel density for the attainment of structural requirements. In contrary situation, if the rising of the steel costs has been the matter of the case, the usage of the concrete material has been increased and the sections of the wall are enlarged.
- The second objective function was used to discuss the effects of the change of the emission values of structural materials to minimize the total amount of CO_2 emission. The results of the analyses showed that the attainment of minimum CO_2 emission was significantly depended on the characteristics of the selected structural materials. It must be noted that the design is not only based on the selection of appropriate techniques used to find the optimal solution, but also depending on the selection of the materials that is fitting to purpose.
- It can be an important result to prefer special materials in construction with low CO_2 emission characteristics to get an eco-friendly design, but the cost-effective design must be procurable to ensure production efficiency.
- The usage of optimization techniques including both the effects of cost and CO_2 emission in relation, seems to be more useful to find the most appropriate solution results for design considerations.
- Concordantly, the usage of HS seems to be preferable according to the presentation of user-friendly easiness to define multi-objective functions.
- In future works, the robustness of different metaheuristic algorithms can be investigated for sustainable eco-friendly optimum design of RC retaining walls. Additionally, integrated multi-variant combinations of the cost and CO_2 emission can be investigated.

References

1. Aydogdu, I., Akin, A.: Biogeography based CO_2 and cost optimization of RC cantilever retaining walls. Part VII. **17**(7) (2015). https://doi.org/10.5505/pajes.2016.25991 (Paris France, 20–21 July)
2. Kalemci, E.N., İkizler, S.B., Dede, T., Angın, Z.: Design of reinforced concrete cantilever retaining wall using grey wolf optimization algorithm. Structures 245–253 (2020). https://doi.org/10.1016/j.istruc.2019.09.013
3. Yepes, P.V., Gonzalez, V.F., Alcalá, G.J., Villalba, I.P.: CO_2-Optimization design of reinforced concrete retaining walls based on a VNS-Threshold acceptance strategy. J. Comput. Civ. Eng. **26**(3), 378–386 (2012). https://doi.org/10.1061/(ASCE)CP.1943-5487.0000140
4. Villalba, P., Alcalá, J., Yepes, V., González-Vidosa, F.: CO_2 optimization of reinforced concrete cantilever retaining walls. In: 2nd International Conference on Engineering Optimization, Lisbon, Portugal, 6–9 Sept 2010
5. Kaveh, A., Hamedani, K.B., Bakhshpoori, T.: Optimal design of reinforced concrete cantilever retaining walls utilizing eleven meta-heuristic algorithms: a comparative study. Periodica Polytech. Civ. Eng. **64**(1), 156–168 (2020). https://doi.org/10.3311/PPci.152

6. Dorigo, M., Maniezzo, V., Colorni, A.: The ant system: optimization by a colony of cooperating agents. IEEE Trans. Syst. Man Cybern. B. **26**, 29–41 (1996). https://doi.org/10.1109/3477.484436
7. Camp, C.V., Akin, A.: Design of retaining walls using big bang-big crunch optimization. J. Struct. Eng. ASCE. **138**(3), 438–448 (2012). https://doi.org/10.1061/(ASCE)ST.1943-541X.0000461
8. Temur, R., Bekdaş, G.: Teaching learning-based optimization for design of cantilever retaining walls. Struct. Eng. Mech. **57**(4), 763–783 (2016). https://doi.org/10.12989/sem.2016.57.4.763
9. Geem, Z.W., Kim, J.H., Loganathan, G.V.A.: New heuristic optimization algorithm: harmony search. Simulation **76**, 60–68 (2001). https://doi.org/10.1177/003754970107600201
10. Kaveh, A., Abadi, A.S.M.: Harmony search based algorithms for the optimum cost design of reinforced concrete cantilever retaining walls. Int. J. Civ. Eng. **9**(1), 1–8 (2011). Corpus ID: 54802626
11. Yang, X.S.: Firefly algorithms for multimodal optimization. In: Stochastic Algorithms: Foundations and Applications, pp. 169–178. Springer, Berlin, Heidelberg (2009). https://doi.org/10.1007/978-3-642-04944-6_14
12. Sheikholeslami, R., Gholipour Khalili, B., Zahrai, S.M.: Optimum cost design of reinforced concrete retaining walls using hybrid firefly algorithm. Int. J. Eng. Technol. **6**(6), 465–470 (2014). https://doi.org/10.7763/IJET.2014.V6.742
13. Kennedy, J., Eberhart, R.C.: Particle swarm optimization. In: Proceedings of IEEE International Conference on Neural Networks, No. IV, Perth Australia, Nov 27–Dec 1, pp. 1942–1948 (1995). https://doi.org/10.1109/icnn.1995.488968
14. Ahmadi-Nedushan, B., Varaee, H.: Optimal design of reinforced concrete retaining walls using a swarm intelligence technique. In: The First International Conference on Soft Computing Technology in Civil, Structural and Environmental Engineering, Stirlingshire, Scotland (2009). https://doi.org/10.4203/ccp.92.26
15. Yang, X.S.A.: New metaheuristic bat-inspired algorithm. In: Nature Inspired Cooperative Strategies for Optimization (NISCO 2010). Studies in Computational Intelligence, pp. 65–74. Springer, Berlin (2010). https://doi.org/10.1007/978-3-642-12538-6_6
16. Ceranic, B., Fryer, C., Baines, R.W.: An application of simulated annealing to the optimum design of reinforced concrete retaining structures. Comput. Struct. **79**(17), 1569–1581 (2001). https://doi.org/10.1016/S0045-7949(01)00037-2
17. Yepes, V., Alcala, J., Perea, C., Gonzalez-Vidosa, F.: A parametric study of optimum earth-retaining walls by simulated annealing. Eng. Struct. **30**(3), 821–830 (2008). https://doi.org/10.1016/j.engstruct.2007.05.023
18. Pei, Y., Xia, Y.: Design of cantilever retaining walls using heuristic optimization algorithms. Procedia Earth Planet. Sci. **5**, 32–36 (2012). https://doi.org/10.1016/j.proeps.2012.01.006
19. Sasidhar, T., Neeraja, D., Samba Murthy Sudhindra, V.: Application of genetic algorithm technique for optimizing design of reinforced concrete retaining wall. Int. J. Civ. Eng. Tech. **8**(5), 999–1007 (2017). https://doi.org/10.20528/cjsmec.2016.11.031
20. Goldberg, D.E.: Genetic Algorithms in Search, Optimization and Machine Learning. Addison Wesley, Boston (1989)
21. Holland, J.H.: Adaptation in Natural and Artificial Systems. University of Michigan Press, Ann Arbor (1975)
22. Kaveh, A., Kalateh-Ahani, M., Fahimi-Farzam, M.: Constructability optimal design of reinforced concrete retaining walls using a multi-objective genetic algorithm. Struct. Eng. Mech. **47**(2), 227–245 (2013). https://doi.org/10.12989/sem.2013.47.2.227
23. Talatahari, S., Sheikholeslami, R.: Optimum design of gravity and reinforced retaining walls using enhanced charged system search algorithm. KSCE J. Civ. Eng. **18**(5), 1464–1469 (2014). https://doi.org/10.1007/s12205-014-0406-5
24. Kaveh, A., Jafarpour Laiden, D.: Optimal design of reinforced concrete cantilever retaining walls using CBO ECBO and VPS algorithms. Asian J. Civil Eng. **18**(4), 657–671 (2017). Corpus ID: 136118632

25. Aydogdu, İ.: Cost optimization of reinforced concrete cantilever retaining walls under seismic loading using a biogeography-based optimization algorithm with Levy flights. Eng. Optim. **49**(3), 381–400 (2017). https://doi.org/10.1080/0305215X.2016.1191837
26. Bekdas, G., Temur, R.: Grey wolf optimizer for optimum design of reinforced concrete cantilever retaining walls. In: International Conference of Numerical Analysis and Applied Mathematics. AIP Conference Proceedings. American Institute of Physics, Melville (2018). https://doi.org/10.1063/1.5043893
27. Mergos, P.E., Mantoglou, F.: Optimum design of reinforced concrete retaining wall with the flower pollination algorithm. Struct. Multi. Optim. **61**, 575–585 (2019). https://doi.org/10.1007/s00158-019-02380-x
28. Holdorf Lopez, R.: Optimum project of cantilever retaining wall using search group algorithm and backtracking search algorithm. Rev. Int. Methods Numer. Calc. Disenoing (2017). https://doi.org/10.1016/j.engstruct.2015.08.034
29. Aydogdu, İ.: Comparison of metaheuristics on multi-objective (cost&CO_2) optimization of RC cantilever retaining walls. Pamukkale Univ. J. Eng. Struct. **23**(3), 221–231 (2016)
30. Ghazavi, M., Bonab, B.S.: Optimization of reinforced concrete retaining walls using ant colony method. In: Vogt, N., Schuppener, B., Straub, D., Brau, G. (eds.) Geotechnical Safety and Risk. ISGSR 2011, Bundesanstalt fur Wasserbau, Karlsruhe, pp. 297–305 (2011). https://doi.org/10.1007/s12205-015-1163-9
31. Kumar, V.N., Suribabu, C.R.: Optimal design of cantilever retaining wall using differential evolution algorithm. Int. J. Optim. Civ. Eng. **7**(3), 433–49 (2017). Corpus ID: 56437534
32. Khajezadeh, M., Taha, M.R., Eslami, M.: Efficient gravitational search algorithm for optimum design of retaining walls. Struct. Eng. Mech. **45**(1), 111–127 (2013). https://doi.org/10.12989/sem.2013.45.1.111
33. Kaveh, A., Farhoudi, N.: Dolphin echolocation optimization for design of cantilever retaining walls. Asian J. Civ. Eng. **17**(2), 193–211 (2016). https://doi.org/10.1007/s00366-020-00999-9
34. Uray, E., Çarbaş, S., Erkan, İ.H., Olgun, M.: Investigation of optimal designs for concrete cantilever retaining walls in different soils. Challange J. Concr. Res. Lett. **11**(2), 39–45 (2020)
35. Gandomi, A.H., Kashani, A.R., Mousavi, M., Jalalvandi, M.: Slope stability analysis using evolutionary optimization techniques. Int. J. Numer. Anal. Meth. Geomech. **41**, 251–264 (2017). https://doi.org/10.1002/nag.2554
36. Purohit, S.: Multi-objective optimization of reinforced cement concrete retaining wall. Bachelor thesis, Department of Civil Engineering National Institute of Technology Rourkela, Orissa, India, May 2014. https://doi.org/10.1007/s40098-015-0178-y
37. Das, M.R., Purohit, S., Das, S.K.: Multi-objective optimization of reinforced cement concrete retaining wall. Indian Geotech. J. **46**(4), 354–368 (2016). https://doi.org/10.1007/s40098-015-0178-y
38. Konstandakopoulou, F., Tsimirika, M., Pnevmatikos, N., Hatzigeorgiou, G.D.: Optimization of reinforced concrete retaining walls designed according to european provisions. Infrastructures **5**(46), 1–16 (2020). https://doi.org/10.3390/infrastructures5060046
39. Zastrow, P., Molina-Moreno, F., García-Segura, T., Martí, J.V., Yepes, V.: Life cycle assessment of cost-optimized buttress earth-retaining walls: a parametric study. J. Cleaner Prediction **140**(3), 1037–1048 (2017). https://doi.org/10.1016/j.jclepro.2016.10.085
40. Khajehzadeh, M.A.: New hybrid algorithm for CO_2 emissions optimization of retaining walls. Int. J. Adv. Mech. Civ. Eng. **3**(1), 7–11 (2016). https://doi.org/10.1080/10286608.2013.853746
41. Yepes, V., Martí, J.V., García, J.: Black hole algorithm for sustainable design of counterfort retaining walls. Sustainability **12**, 2767 (2020). https://doi.org/10.3390/su12072767
42. Building Code Requirements for Structural Concrete and Commentary, ACI Committee, ACI 318 (2014)
43. Askarzadeh, A., Rashedi, E.: Harmony search algorithm: basic concepts and engineering applications (2017). https://doi.org/10.4018/978-1-5225-5643-5.ch001
44. Rankine, W.J.M.: On the stability of loose earth. Philos. Trans. Roy Soc Lond **147**, 9–27 (1857). https://doi.org/10.1098/rstl.1857.0003

45. Coulomb, C.A.: Essai sur une application des regles des maximis et minimis a quelques problemes de statique relatifs a l'artitecture. Mem. Acad. Royal Pres. Div. Sav. Paris, 7 (1776)
46. Boussinesq: Sur la détermination de l'épaisseur minima que doit avoir un mur vertical d'une hauteur et d'une densité données, pour contenir un massif terreux sans cohésion, dont la surface supérieure est horizontale. Ann. des Ponts et Chaussées, t. III, 625 (1882)
47. Terzaghi, K.V.: General wedge theory of earth pressure. Trans. ASCE **106**, 68–97 (1941). Corpus ID: 127332748
48. Jaky, J.: The coefficient of earth pressure at rest. J. Soc. Hung. Arch. Engrs. Bp. Hung. 355–358 (1944). https://doi.org/10.1061/asce1090-02412005131:111429
49. Dembicki, E., Chi, T.: System analysis in calculation of cantilever retaining walls. Int. J. Numer. Anal. Meth. Geomech. **13**(6), 599–610 (1989). https://doi.org/10.1002/nag.1610130603
50. Powrie, W.: Limit equilibrium analysis of embedded retaining walls. Geotechnique **46**(4), 709–723 (1996). https://doi.org/10.1680/geot.1996.46.4.709
51. Bowles, J.E.: Foundation Analysis and Design. McGraw-Hill, New York (1988)
52. FHWA-IF-99-015: Geotechnical Engineering Circular No. 4 Ground Anchors and Anchored Systems. Federal Highway Administration, Washington, DC, USA (1999)
53. British Standards Institution: Eurocode 7: Part 1, General Rules. British Standards Institution, London, UK (1995)
54. Yeo, D.H., Potra, F.A.: Sustainable design of reinforced concrete structures through CO_2 emission optimization. J. Struct. Eng. **141**(3) (2015). https://doi.org/10.1061/(asce)st.1943-541x.0000888
55. Paya-Zaforteza, I., Yepes, V., Hospitaler, A., Gonzalez-Vidosa, F.: CO_2-optimization of reinforced concrete frames by simulated annealing. Eng. Struct. **31**(7), 1501–1508 (2009)
56. Azizi, F.: Applied Analysis in Geotechnics, E & FN Spon. Taylor and Francis Group (2000)
57. Bowles, J.E.: Physical and Geotechnical Properties of Soils, 2nd edn. Civil Engineering Series. McGraw-Hill International Editors (1984)
58. Das, B.M.: Principals of Geotechnical Engineering, 25th edn. Cengage Learning (2020)
59. Bekdas, G., Akbay Arama, Z., Kayabekir, A.E., Geem, Z.W.: Optimal design of cantilever soldier pile retaining walls embedded in frictional soils with harmony search algorithm. Appl. Sci. (Switzerland) **10**(3232) (2020)
60. Budhu, M.: Foundations and Earth Retaining Structures. John Wiley & Sons Inc., Hoboken (2008)

Statistical Evaluation of Metaheuristic Algorithm: An Optimum Reinforced Concrete T-beam Problem

Aylin Ece Kayabekir and Müge Nigdeli

Abstract Structural engineering is an area which have problems that are nonlinear in optimization. Due to that reason, metaheuristic methods are a major application source. In addition to finding the best suitable solution, different algorithms are needed to be verified and compared via statistical evaluation. In this chapter, optimum design of reinforced concrete (RC) beams that have design constraints formulated according to stress–strain capacity of members in the design regulations are investigated. Several metaheuristic algorithms such as harmony search (HS), teaching–learning-based optimization (TLBO), flower pollination algorithm (FPA) and Jaya algorithm (JA) are compared via statistical methods such as Friedman ranking, one-way ANOVA, post hoc Bonferroni test and independent samples t-test.

Keywords Structural optimization · Metaheuristic algorithms · Friedman ranking · One-way ANOVA · Post hoc bonferroni test · T-test

1 Introduction

In the design of reinforced concrete (RC) members, minimum cost and structural safety are the two most important factors. In the design process, iterative methods are needed in order to provide these two factors together and to reach the optimum results in the fastest way.

Metaheuristic algorithms are one of the most effective iterative methods in finding the optimum result. Metaheuristic algorithms, inspired by a life metaphor, enable the

A. E. Kayabekir
Department of Civil Engineering, Istanbul University-Cerrahpaşa, 34320 Avcılar/Istanbul, Turkey
e-mail: ecekayabekir@gmail.com

M. Nigdeli (✉)
Institute of Environmental Sciences, Boğaziçi University, Hisar Campus, 34342 Bebek/Istanbul, Turkey
e-mail: muge.ozdemir@boun.edu.tr

S. M. Nigdeli et al. (eds.), *Advances in Structural Engineering—Optimization*, Studies in Systems, Decision and Control 326,
https://doi.org/10.1007/978-3-030-61848-3_11

generation of design variables by controlling the design constraints in the optimization process. In this process, optimum design variables that provide the best objective function value are obtained by eliminating weak solutions in terms of objective function. The different stress–strain behaviors of the material (steel and concrete) constituting RC members are the design constraints taken into account in the optimum design. The use of materials with different unit cost and mechanical behavior requires finding the most suitable material combination. Up to now, several scientific studies have been conducted by using metaheuristic algorithms for the optimum design of RC members. Some of these will be summarized in this section.

RC beams were designed optimally with the best known evolutionary based algorithm called Genetic Algorithm (GA) [1–3]. For the optimum design of RC continuous beam, hybrid metaheuristic algorithm combinating genetic algorithm (GA) and simulating annealing (SA) was used by Leps and Sejnoha [4]. Harmony search (HS) algorithm was also employed for the optimization of continuous beams [5] and T-beams [6, 7]. By considering different flexural moment cases, doubly reinforced beams were optimized by Bat algorithm [8]. In another study, the efficiency of the TLBO algorithm on optimum beam design has been tested [9]. The optimum T-beam design was also done with JA, TLBO, modified HS and Flower Pollination Algorithm (FPA) according to Eurocode 2 constraints [10–12].

In addition to these studies in which minimum cost is defined as the objective function in optimum design, CO2 emission can also be the objective function of optimization [13, 14]. Differently form RC beams, metaheuristic-based methods were used for the other RC members such as columns [15, 16], frames [17, 18], retaining walls [19–22] and slabs [23].

In the optimization process, metaheuristic algorithms search optimum values by generating values randomly. Since the process is iterative, it takes time and requires the selection of the most appropriate algorithm for the solution of the problem. In other words, it is critical to determine the most suitable algorithm in terms of performance among the algorithms used in the optimization process. In this study, a statistical study is presented to compare the performance of four different algorithms used for the T-beam problem optimized by Kayabekir et al. [11]. In this comparison, the data of 30 independent run results by Kayabekir et al. for HS [24], TLBO [25], FPA [26] and JA [27] algorithms were used. Statistical evaluation was carried out according to the Friedman ranking test, one-way ANOVA, post hoc Bonferroni test and independent samples t-test.

2 The Optimization Process

The optimization process of the T-beam shown in Fig. 1 can be summarized with five steps as follows.

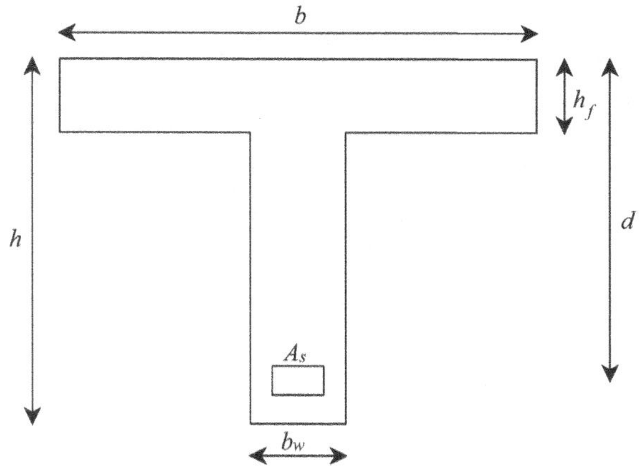

Fig. 1 T-shaped cross section [11]

Step 1. Design constants (Table 1), design variable ranges (Table 2), stopping criteria of the optimization problem and other constant parameters related with algorithms are defined.

Step 2. The design variables (X_i) are randomly (with rand) generated within the minimum ($X_{i,\min}$) and maximum ($X_{i,\max}$) limit values defined in step 1 (Eq. 1).

$$X_i = X_{i,\min} + rand\left(X_{i,\max} - X_{i,\min}\right) \quad (1)$$

Then, generated variables stored in a matrix called initial solution matrix. This matrix given as symbolically

Table 1 Design constants

Symbol	Definition
f_{ck}	Characteristic compressive strength for concrete
f_{cd}	Allowable compressive strength for concrete
f_{yd}	Characteristic yield strength of reinforcement
ρ_{max}	The maximum reinforcement ratio
ρ_{min}	The minimum reinforcement ratio
L	The length of beam
E_s	Young's elastic modules for steel
M_{Ed}	The ultimate bending moment capacity
V_{Ed}	The ultimate bending moment capacity
C_s	The unit total cost of reinforcing steel
C_c	The unit total cost of concrete

Table 2 Design variables

Symbol	Definition	Range
B	Effective with of compressive flange (mm)	$b_w \le b \le \min[0.2L + b_w, 8h_f]$
b_w	Web width [mm]	$0.20d \le b_w \le 0.40d$
H	Height [mm]	$L/16 \le h \le 2.0$
h_f	Flange depth [mm]	$0.15 \le h_f \le d$
D	Effective depth [mm]	$d = 0.9h$
d_s	Cover of reinforcements [mm]	$d_s = 0.1h$
A_s	Area of reinforcing steel [mm^2]	$0 \le A_s \le 0.1$

$$CL = \begin{bmatrix} X_{1,1} & X_{1,2} & \dots & X_{1,vn} \\ X_{2,1} & X_{2,2} & \dots & X_{2,vn} \\ . & . & \dots & . \\ . & . & \dots & . \\ X_{pn-1,1} & X_{pn-1,2} & \dots & X_{pn-1,vn} \\ X_{pn,1} & X_{pn,2} & \dots & X_{pn,vn} \end{bmatrix} \tag{2}$$

where each column shows a candidate solution vector including different design variables ($X_1 - X_{pn}$). The number of candidate solution vectors is determined by the vector number (vn) defined in the first step.

Step 3. The objective function of each solution is calculated, and design constraints are checked. Objective functions of solutions that do not provide design constraints are assigned a penalized value.

The objective function of the optimization is total material cost (C) of the beam and is written as in the Eq. (3).

$$C = b_w d + (b - b_w)h_f + (C_s/C_c)A_s \tag{3}$$

In the design, Eurocode 2 [12] design constraints given in Table 3 were considered. According to Eurocode 2, w, p, M_{Ed1} and V_{Ed1} are calculated with formulations given in Eqs. (4–9).

Table 3 Design constraints

Constraint Number	Constraints
1	$\omega(1 - 0.5\omega) \le 0.392$
2	$0.0035(0.8 - \omega)/\omega \ge f_{yd}/E_s$
3	$\rho_{\min} \le \rho \le \rho_{\max}$
4	$M_{Ed} \le M_{Ed1}$
5	$V_{Ed} \le V_{Rd\max}$

$$\omega = (f_{yd}/f_{cd})(A_s/b_w d) - (b - b_w)h_f/(b_w d) \tag{4}$$

$$\rho = A_s/(b_w d) \tag{5}$$

$$M_{Ed1} = f_{cd}(b - b_w)h_f\left(d - 0.50h_f\right) + f_{cd}b_w d^2\omega(1 - 0.5\omega) \tag{6}$$

$$V_{Rd\,\max} = \nu_1 f_{cd} b_w z/(\tan(45) + \cot(45)) \tag{7}$$

$$\nu_1 = 0.6(1 - f_{ck}/250) \tag{8}$$

$$z = 0.9d \tag{9}$$

Step 4. In this step, new solution matrix including new design variables are generated according to algorithm rules. For that reason, this step is specific for each algorithm. For this step, details can be found in Refs. [24–27].
Step 5. In the last step, new solutions are compared with existing one. If new solution gives better solutions in terms of objective function, existing solutions are updated.

Last two steps are continued until providing stopping criteria. The stopping criterion of this optimization problem is maximum iteration number. All optimization process can be also seen in Fig. 2. The optimization code used in this study was developed by using MATLAB [28].

3 The Numerical Example

The values of deign constants used in the numerical example are presented in Table 4.

The minimum, average and standard deviation values of 30 independent analysis results for four algorithms, average iterations values that obtained optimum results are given in Table 5.

Although the algorithms found approximately similar results in terms of objective function, the JA and TLBO algorithms obtained slightly better results. When the mean value and standard deviation results were evaluated, it was seen that the TLBO algorithm found slightly better results.

However, only the results given in the table may not be sufficient to reach an effective result in terms of evaluating the performance of the algorithms. In addition, it will be useful to do a detailed statistical evaluation and the algorithm results should be also interpreted according this evaluation. For this purpose, in the next section, the results obtained from the algorithms are evaluated with a statistical approach.

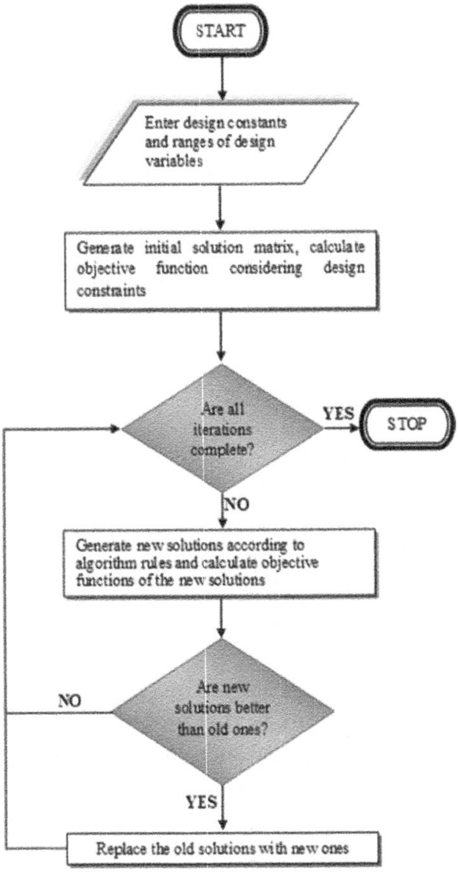

Fig. 2 The flowchart of the optimization process [11]

Table 4 Design constants of numerical example [3]

Symbol	Values
f_{ck}	20 MPa
f_{cd}	11.33 MPa
f_{yd}	348 MPa
ρ_{max}	0.04
ρ_{min}	0.0013
L	20 m
E_s	200,000 MPa
M_{Ed}	4.991 N m
V_{Ed}	1.039 N
C_s/C_c	36

Table 5 Optimum design variables [11]

Design variables	FPA	TLBO	HS	JA
b (m)	1.13879373	1.137632212	1.144541099	1.137605021
b_w (m)	0.30432074	0.304358948	0.304095428	0.304361011
h (m)	1.69067078	1.690883044	1.689419046	1.690894506
h_f (m)	1.5216037	1.52179474	1.520477141	1.521805056
d (m)	0.15	0.15	0.15	0.15
A_s (m^2)	0.01141132	0.011413083	0.011405725	0.01141303
ω	0.48660829	0.486991788	0.485020591	0.486991176
C/C_c	0.99903391	0.999033809	0.999043105	0.999033809
F_{min}	0.99903391	0.999033809	0.999043105	0.999033809
F_{mean}	1.00182853	0.999033816	1.000081551	1.000120713
F_{std}	0.00546358	5.27E-09	0.005435734	0.005917898
Mean iteration	22,869.9	41,595.36667	53,446.31333	83,928.8

4 Statistical Evaluation

In this section, the results of Friedman ranking, and parametric tests such as one-way ANOVA, post hoc Bonferroni test, independent samples t-test of objective function and iteration numbers of each algorithm are investigated in order to quantify the performances of the algorithms during optimization processes.

SPSS [29], Statistical Package for the Social Sciences, was used for the statistical treatment of the data.

4.1 Friedman Ranking as Non-parametric Test

Friedman ranking is a non-parametric statistical test which is developed by Milton Friedman in 1937 [30, 31]. It is used for performing multiple statistical comparisons between more than two algorithms [32]. Application of Friedman test is more common in machine learning that it is not needed normality of sample means and elimination of outliers because it is a nonparametric test [33]. The test procedure is applied in two steps: first ranking the data in each row of a two-way table, second testing different columns of the table of ranks come from the same universe by obtaining the mean rank of each column [30]. The means came from the same universe or not can be tested by the equation of

Table 6 Mean, standard deviation, and friedman ranking of objective functions of algorithms

Objective functions/algorithms	Mean	Standard deviation	Friedman ranking
FPA	1.00182	0.00546	4.30
TLBO	0.09903	0.00000	1.65
HS	1.00008	0.00544	4.33
JA	1.00012	0.00592	1.52
Modified HS	1.00102	0.00753	3.20

Table 7 Mean, standard deviation, and friedman ranking of iteration numbers of algorithms

Iteration numbers/algorithms	Mean	Standard deviation	Friedman ranking
FPA	22,869.90	30,146.53	1.53
TLBO	41,595.37	6860.86	2.40
HS	534,463.13	257,517.76	4.90
JA	83,928.80	16,453.52	3.87
Modified HS	44,660.93	19,365.87	2.30

$$X_r^2 = \frac{12n}{np(p+1)} \sum_{j=1}^{p} \left(\sum_{i=1}^{n} r_{ij} \right)^2 - 3n(p+1) \tag{10}$$

where n the number of rows or the number of ranks averaged, p is the mean value, r_{ij} is the rank entered in the i-th row and j-th column [30].

In this paper, Friedman ranking is used in this research to rank the algorithms based on the results data. So, Friedman test is conducted at the 95% comparison level. Mean, standard deviation, and Friedman ranking values of the objective function and iteration numbers of algorithms are represented in Tables 6 and 7, respectively. Based on the Friedman test results, JA was ranked as best in terms of objective function values, and followed by TLBO, modified HS, FPA, and HS, respectively. Also, FPA was ranked as best by looking iteration numbers of each algorithms with Friedman test, and followed by modified HS, TLBO, JA, and HS, respectively.

4.2 Parametric Tests: One-Way ANOVA, Post hoc Bonferroni Test, and Independent Samples T-test

ANOVA, or Analysis of Variance, is a statistical test which assumes all groups have same population means null hypothesis is rejected when the population means do not equal each other, and alternative hypothesis is accepted. Similarly, null hypothesis is accepted when population means are identical for all groups and variability between means would be used to estimate population variances. The purpose of ANOVA is to compare variance estimated between group means and variance estimated variability within groups by dividing them each other and obtaining and F value:

$$F = \frac{\sum_{j=1}^{k} n_j (M_j - M_G)^2}{k-1} \Big/ \sum_{j=1}^{k} \frac{\sum_{i=1}^{n_j} (x_{ij} - M_j)^2}{n_j - 1} \quad (11)$$

where k is the number of groups (three tree populations), n_j is the number of scores in group j, M_j is the mean of group j, M_G is the grand mean of all scores pooled together, and x_{ij} is the ith score for group j [34].

Before parametric tests, rank transformation approach as a distribution free procedure is applied in order to deal with the non-normal data [35]. So, one-way ANOVA, Post hoc Bonferroni test, and independent samples t-test are carried out on the rank transformed data. After ranking, objective functions and iteration numbers of algorithms significantly differ due to p value of less than 0.001 at the 0.05 alpha level, as a result of one-way ANOVA.

Since one-way ANOVA found a significant result, the null hypothesis can be rejected which indicates all means are equal. Null hypothesis can be demonstrated like that:

$$H_0 : \mu_1 = \mu_2 = \mu_3 \quad (12)$$

It means one of the conditions or more of them written below could be true as an alternative hypothesis:

$$H_1 : \mu_1 \neq \mu_2 = \mu_3 \quad (13)$$

$$H_1 : \mu_1 = \mu_2 \neq \mu_3 \quad (14)$$

$$H_1 : \mu_1 \neq \mu_3 = \mu_2 \quad (15)$$

$$H_1 : \mu_1 \neq \mu_2 \neq \mu_3 \quad (16)$$

In order to find which possible alternative hypothesis should be chosen, post hoc tests could be applied in order to compare pairs of means. Post hoc test helps to compute pairwise comparisons [34]. Bonferroni comparison is applied as a post hoc test to correct experiment-wise or family-wise error rates during multiple comparisons [36]. In order to compensate for type I error, new threshold the adjusted *p*-value is calculated by dividing the original alpha level by the number of tests being performed for a single test to be classed as significant [36].

The post hoc test results are represented in Table 8. According post hoc Bonferroni test results, objective functions of the algorithms are significantly different from each other, except between FPA and HS, and between TLBO and JA due to higher *p*-values

Table 8 Post hoc bonferroni test results

Compared algoritms/*p*-values	Objective functions	Iteration numbers
FPA-TLBO	<0.001	0.001
FPA-HS	1.000	<0.001
FPA-JA	<0.001	<0.001
FPA-Modified HS	<0.001	<0.001
TLBO-HS	<0.001	<0.001
TLBO-JA	1.000	<0.001
TLBO-Modified HS	<0.001	1.000
HS-JA	<0.001	<0.001
HS-Modified HS	<0.001	<0.001
JA-Modified HS	<0.001	<0.001

than the significance threshold of 0.05. Also, the iteration numbers of the algorithms are significantly different from each other, except between TLBO and modified HS ($p = 0.515$) at the significance level of 0.05.

The results of independent samples t-test that are represented in Table 9 justify the post hoc Bonferroni test results that objective functions of the algorithms are significantly different from each other, except between FPA and HS, and between TLBO and JA due to higher *p*-values at the level of 0.05. Also, the iteration numbers of the algorithms are significantly different from each other, except between TLBO and modified HS with the *p*-values of 0.515 at the significance level of 0.05.

Table 9 Independent samples t-test results

Compared algorithms/*p*-values	Objective functions	Iteration numbers
FPA-TLBO	<0.001	0.001
FPA-HS	0.751	<0.001
FPA-JA	<0.001	<0.001
FPA-Modified HS	<0.001	0.001
TLBO-HS	<0.001	<0.001
TLBO-JA	0.588	<0.001
TLBO-Modified HS	<0.001	0.515
HS-JA	<0.001	<0.001
HS-Modified HS	<0.001	<0.001
JA-Modified HS	<0.001	<0.001

5 Conclusions

The statistical evaluation of optimization process was done by conducting 30 cycles for the employed algorithms via minimum objective function (F_{min}) and mean number of iterations. In conclusion, as a result of Friedman ranking, objective function of JA is the best comparing with other algorithms, although the objective function of TLBO is the best according to their mean values. The iteration numbers of FPA are the most superior one according to its mean and Friedman ranking results. According to parametric test results, there is not statistically significant difference between FPA-HS, and TLBO-JA, by looking their objective functions. Also, TLBO and modified HS do not differ significantly due to the iteration numbers.

References

1. Coello, C.C., Hernandez, F.S., Farrera, F.A.: Optimal design of reinforced concrete beams using genetic algorithms. Expert Syst. Appl. **12**, 101–108 (1997)
2. Govindaraj, V., Ramasamy, J.V.: Optimum detailed design of reinforced concrete continuous beams using genetic algorithms. Comput. Struct. **84**, 34–48 (2005). https://doi.org/10.1016/j.compstruc.2005.09.001
3. Fedghouche, F., Tiliouine, B.: Minimum cost design of reinforced concrete T-beams at ultimate loads using eurocode2. Eng. Struct. **42**, 43–50 (2012). https://doi.org/10.1016/j.engstruct.2012.04.008
4. Leps, M., Sejnoha, M.: New approach to optimization of reinforced concrete beams. Comput. Struct. **81**, 1957–1966 (2003). https://doi.org/10.1016/S0045-7949(03)00215-3
5. Akin, A., Saka, M.P.: Optimum Detailed Design of Reinforced Concrete Continuous Beams using the Harmony Search Algorithm. In: The Tenth International Conference on Computational Structures Technology, Paper 131 (2010)
6. Bekdaş, G., Nigdeli, S.M.: Cost optimization of t-shaped reinforced concrete beams under flexural effect according to ACI 318. In: 3rd European Conference of Civil Engineering (2012)
7. Bekdaş, G., Nigdeli, S.M.: Optimization of T-shaped RC flexural members for different compressive strengths of concrete. Int. J. Mech. **7**, 109–119 (2013)
8. Bekdaş, G., Nigdeli, S.M., Yang, X.: Metaheuristic optimization for the design of reinforced concrete beams under flexure moments
9. Bekdaş, G., Nigdeli, S.M.: Optimum design of reinforced concrete beams using teaching-learning-based optimization. In: 3rd International Conference on Optimization Techniques in Engineering (OTENG'15), pp. 7–9 (2015)
10. Kayabekir, A.E., Bekdaş, G., Nigdeli, S.M.: Optimum design of T-beams using Jaya algorithm. In 3rd International Conference on Engineering Technology and Innovation (ICETI), Belgrad, Serbia (2019)
11. Kayabekir, A.E., Bekdaş, G., Nigdeli, S.M.: Evaluation of metaheuristic algorithm on optimum design of T-beams. In: 6th International Conference on Harmony Search, Soft Computing and Applications (ICHSA 2020), Istanbul, Turkey, 16–17 July 2020
12. EN (Veranst.): EN 1992-1-1 Eurocode 2: design of concrete structures. CEN, Brussels (2005)
13. Koumousis, V.K., Arsenis, S.J.: Genetic algorithms in optimal detailed design of reinforced concrete members. Comput. Aided. Civ. Inf. **13**, 43–52 (1998)
14. Yepes, V., Martí, J.V., García-Segura, T.: Cost and CO2 emission optimization of precast–prestressed concrete U-beam road bridges by a hybrid glowworm swarm algorithm. Autom Constr. **49**, 123–134 (2015)

15. Rafiq, M.Y., Southcombe, C.: Genetic algorithms in optimal design and detailing of reinforced concrete biaxial columns supported by a declarative approach for capacity checking. Comput. Struct. **69**, 443–457 (1998)
16. Gil-Martin, L.M., Hernandez-Montes, E., Aschheim, M.: Optimal reinforcement of RC columns for biaxial bending. Mater. Struct. **43**, 1245–1256 (2010)
17. Camp, C.V., Pezeshk, S., Hansson, H.: Flexural design of reinforced concrete frames using a genetic algorithm. J. Struct. Eng. ASCE **129,** 105–11 (2003)
18. Govindaraj, V., Ramasamy, J.V.: Optimum detailed design of reinforced concrete frames using genetic algorithms. Eng. Optimiz. **39**(4), 471–494 (2007)
19. Ceranic, B., Fryer, C., Baines, R.W.: An application of simulated annealing to the optimum design of reinforced concrete retaining structures. Comput. Struct. **79**, 1569–1581 (2001)
20. Camp, C.V., Akin A.: Design of retaining walls using big bang-big crunch optimization. J. Struct. Eng. ASCE. **38**(3), 438–448 (2012)
21. Kaveh, A., Abadi, A.S.M.: Harmony search based algorithms for the optimum cost design of reinforced concrete cantilever retaining walls. Int. J. Civil Eng. **9**(1), 1–8 (2011)
22. Talatahari, S., Sheikholeslami, R., Shadfaran, M, Pourbaba, M.: Optimum design of gravity retaining walls using charged system search algorithm. Math. Probl. Eng. **2012** (2012). (Article ID 301628)
23. Sahab, M.G., Ashour, A.F., Toropov, V.V.: Cost optimisation of reinforced concrete flat slab buildings. Eng. Struct. **27**, 313–322 (2005). https://doi.org/10.1016/j.engstruct.2004.10.002
24. Geem, Z.W., Kim, J.H., Loganathan, G.V.: A new heuristic optimization algorithm: harmony search. Simulation **76**, 60–68 (2001)
25. Rao, R.V., Savsani, V.J., Vakharia, D.P.: Teaching–learning-based optimization: a novel method for constrained mechanical design optimization problems. Comput. Aided Des. **43**(3), 303–315 (2011)
26. Yang, X.S.: Flower pollination algorithm for global optimization. In International Conference on Unconventional Computing and Natural Computation, pp. 240–249. Springer, Berlin (2012)
27. Rao, R.: Jaya: a simple and new optimization algorithm for solving constrained and unconstrained optimization problems. Inte. J. Ind. Eng. Comput. **7**(1), 19–34 (2016)
28. The MathWorks: Matlab R2018a, Natick, MA (2018)
29. IBM Corp.: IBM SPSS Statistics for Windows, Version 25.0. IBM Corp., Armonk, NY (2017)
30. Friedman, M.: The use of ranks to avoid the assumption of normality implicit in the analysis of variance. J. Am. Stat. Assoc **32**, 674–701 (1937)
31. Friedman, M.: A comparison of alternative tests of significance for the problem of m rankings. Ann. Math. Stat. **11**, 86–92 (1940)
32. García, S., Fernández, A., Luengo, J., Herrera, F.: Advanced nonparametric tests for multiple comparisons in the design of experiments in computational intelligence and data mining: Experimental analysis of power. Inf. Sci. **180**(10), 2044–2064 (2010)
33. Benavoli, A., Corani, G., Mangili, F.: Should we really use post-hoc tests based on mean-ranks? J. Mach. Learn. Res. **17**(1), 152–161 (2016/2017)
34. Herzog, M.H., Francis, G., Clarke, A.: Understanding Statistics and Experimental Design: How to Not Lie with Statistics, p. 142. Springer (2019)
35. Conover, W.J., Iman, R.L.: Rank transformations as a bridge between parametric and nonparametric statistics. Am. Statist. **35**(3), 124–129 (1981)
36. Armstrong, R.A.: When to use the Bonferroni correction. Ophthalmic Physiol. Opt. **34**(5), 502–508 (2014)

The manufacturer's authorised representative in the EU is Springer Nature Customer Service Centre GmbH, Europaplatz 3, 69115 Heidelberg, Germany. If you have any concerns regarding our products, please contact ProductSafety@springernature.com

Printed and bound by CPI Group (UK) Ltd, Croydon, CR0 4YY
30/09/2026
02225289-0007